Handbook on Mechanical Properties of Rocks

Other volumes in the
Series on Rock and Soil Mechanics

W. Dreyer:
The Science of Rock Mechanics
Part I Strength Properties of Rocks
1972

T. H. Hanna:
Foundation Instrumentation
1973

C. E. Gregory:
Explosives for North American Engineers
1973

M. & A. Reimbert:
Retaining Walls Vol. I
– Anchorages and Sheet Piling –
1974

Vutukuri, Lama, Saluja:
Handbook on Mechanical Properties
of Rocks Vol. I
1974

M. & A. Reimbert:
Retaining Walls Vol. II
– Study of Passive Resistance in Foundation Structures –
1976

H. R. Hardy, Jr. & F. W. Leighton:
First Conference on Acoustic Emission/Microseismic Activity
in Geologic Structures and Materials
1977

L. L. Karafiath & E. A. Nowatzki:
Soil Mechanics for Off-Road Vehicle Engineering
1978

Baguelin, Jézéquel, Shields:
The Pressuremeter and Foundation
Engineering
1978

R. D. Lama & V. S. Vutukuri:
Handbook on Mechanical Properties
of Rocks Vol. II
1978

Editor-in-Chief
Professor Dr. H. Wöhlbier

Series on Rock and Soil Mechanics
Vol. 3 (1978) No. 2

HANDBOOK ON MECHANICAL PROPERTIES OF ROCKS

– Testing Techniques and Results –

Volume III

by

R. D. Lama

CSIRO
Division of Applied Geomechanics
Australia

V. S. Vutukuri

Department of Mining Engineering
Broken Hill Division
University of New South Wales
Australia

First Printing
1978

TRANS TECH PUBLICATIONS

Distributed by
TRANS TECH S. A.
CH-4711 Aedermannsdorf, Switzerland

International Standard Book Number

ISBN O-87849-022-1

Printed in Germany
by Druckerei E. Jungfer, Herzberg

FOREWORD

Rock mechanics is an essentially applied science, directed towards the improved understanding and prediction of the behaviour of rock in a wide range of practical fields in mining, engineering, geology and seismology.

In major civil engineering works such as tunnels and underground power plants and dams, and in mining underground or in large open cuts, it is necessary to base designs for these works on reliable predictions of the subsequent behaviour of rocks during construction and in later operation.

Predictions are based on site observations of rock conditions, test results, analysis and calculation of rock behaviour, and also comparison with similar works constructed elsewhere in the world in related rock types.

Accumulated experience has long shown that prototype rock behaviour may be more closely predicted from tests on rocks in-situ rather than from laboratory tests—a wide range of in-situ tests have accordingly been developed and refined, but it has been found that the various techniques give somewhat differing results.

Nevertheless, it is only by the improved understanding of these in-situ tests that we can more confidently predict the behaviour of the rocks in prototype structures.

Rocks also exhibit time-dependent behaviour, such as can be seen in the slow deterioration and closure of coal mine workings, and the slow diminishing adjustment of a rockfill dam to water load. The understanding and prediction of time dependent behaviour is also necessary for engineering works in those rocks showing prospensity for continued deformation with time.

Rock tests, particularly in-situ tests, are rather expensive, and it is desirable that we develop our understanding of rock behaviour to such an extent that, as far as possible, our predictions can be based on a few carefully selected and less expensive tests.

With this objective in mind, there has long been a need for a critical review of all the various methods of testing the mechanical properties of rock, both in the laboratory and in-situ.

Dr. Lama and Mr. Vutukuri have taken on this task and succeeded admirably. Their volumes in this series are all exceedingly well done, and constitute a most thorough and comprehensive review of rock testing and rock mechanical properties.

The subject of rock mechanics has developed in an essentially observational pattern. It is now an appropriate time to move more strongly towards classification of rocks and rock masses, the classification of rock mechanical properties and the rationalisation of testing procedures.

This valuable book by Dr. Lama and Mr. Vutukuri provides an ideal base for such continuing work.

L.A. Endersbee
Dean, Faculty of Engineering
and Professor of Civil Engineering
Monash University
Melbourne, Victoria, Australia

January 1978

CONTENTS

CHAPTER 8

In Situ Testing of Rock

8.1. Introduction

In general, the mechanical behaviour of rock mass cannot be determined purely from laboratory tests. Large scale in situ tests form an extremely important part in design considerations of all major projects.

Major developments in in situ testing have been with their applications in civil engineering projects. The need to build higher and larger dams, the necessity of constructing dams at places having deficient foundation and surrounding rock characteristics, the driving of larger hydraulic pressure tunnels, large diameter multiple transportation tunnels and the belief that a great deficiency exists in the knowledge of the behaviour of rock masses, their heterogeneity and discontinuity have contributed to the development of the techniques.

Two parameters of great importance are the deformability of the rock mass and its shear characteristics. As such these are the two aspects which have been studied in great detail.

In mining engineering application, the strength of large pillars is an important mine design parameter and most of the work in the field of strength of rock, in situ, has been done from the point of view of design of mine pillars.

This chapter deals with the various methods used in the determination of deformability of rock masses. The commonly used methods such as plate bearing test, pressure tunnel test along with their modifications are discussed in detail. Large scale in situ shear tests are described. The results of a limited number of triaxial in situ tests on rocks are given. The mechanical behaviour of rocks in uniaxial compression obtained from in situ tests is discussed in detail along with limited in situ tension and torsion tests. The chapter should be read in conjunction with chapter 10 where more details about the shear behaviour of joints and jointed rock are given.

8.2. Types of Large Scale In Situ Tests

The need for in situ testing arises when large-scale effects are anticipated which cannot be ascertained from small scale laboratory tests. It is not always essential to conduct a large scale test. The sensitivity of a structure to any particular parameter plays a dominant role in decision making. For example, in concrete dams, the ratio between moduli of deformation of rock mass (E_m) and concrete (E_c) becomes important only when $E_m/E_c < 1/4$. For $E_m/E_c \geq 1/4$, the modulus of deformation of foundations plays a very slight role. For $E_m/E_c < 1/16$, the behaviour of the dam is entirely governed by the modulus of deformation of the foundation. This is particularly important when the foundation rock is nonhomogeneous. For homogeneous rock, the structure is not much influenced even by values as low as $E_m/E_c < 1/16$ (ROCHA, 1974).

The sensitivity of the structure is very important in deciding upon the number of tests and the test locations. The greater the number of tests, the higher the reliability of the results will be. Because of the marked heterogeneity of the rock mass, and the need to test large volumes, it becomes essential to first establish three dimensional zoning of the rock mass based upon geological information available from initial borings. Dilatometers or other borehole devices can be used in the initial stage of zoning before any plate bearing or other large scale tests are planned. It is possible that the zoning from dilatometer test results may not coincide with the geological division of the rock mass. But zoning from *RQD* values (Chapter 11, Volume IV) may agree fairly well with dilatometer zoning.

The problem of arriving at a representative value in any test is fairly complex and there is no quick answer. Each case has got to be studied separately. The ratio of laboratory to in situ values of modulus of deformation may vary in wide limits; 10:1 or even more. In general, the ratio is lower for high modulus rocks.

The number of tests depends upon the sensitivity of the structure, the relative homogeneity of the various zones and within the zone and the modulus of deformation of the rock. When the anticipated modulus of deformation exceeds 10000 MPa (1450000 lbf/in^2), in situ tests can be dispensed with or their number reduced only to confirm the anticipated values (ROCHA, 1974). When modulus of deformation is lower, 2 to 6 tests may be performed in each zone..

Large scale in situ tests can be divided into 3 main categories.

(1) Deformability tests

(2) Shear tests

(3) Strength tests

Deformability tests are conducted to obtain deformation parameters of the foundation and surrounding rocks on which a heavy structure is to be placed (e.g. dam foundation) and give an important input parameter for design.

Shear tests are mainly conducted on existing planes of discontinuities such as bedding planes, joints, foliations etc. and have the purpose of obtaining information about the stability of the system when subjected to loads brought about by the structure.

Strength parameters such as compressive strength are important in case where rock mass is used as a supporting structure, for example pillars left in situ to support the surrounding rocks when mineral around them has been extracted.

Depending upon the type of information required, the test must be designed to provide the information which can be accepted and is reliable. Any in situ test must fulfil a number of conditions. Among others, the following seem to be most important.

(1) The test conditions must be as close as possible to the theory available for the interpretation of the results obtained from the test.

(2) The test must affect a rock-mass-volume large enough so that it represents the behaviour of the representative volume of the rock mass or a zone of the rock mass that is geologically and structurally differentiable. Consequently the method of testing and the dimensions of the surface to be loaded are conditioned by the type of rock mass.

(3) The cost of testing must be low. This involves both the preparation of the test site and the conduct of tests. Tests, when conducted during the periods when other excavation works are going on, should have least interference. It should have maximum speed in its preparation and conduct but it should not compromise the main features of the test (such as time in drained shear test).

(4) The loading system and other necessary equipment should be simple and easy to operate. The deformation measuring equipment must be able to operate under severe in situ conditions. Certain equipment may have to conform to special statutory regulations (e.g. flame safety law for underground coal mines).

(5) The measurement of deformation must be referred to fixed points i.e. these must not be affected by the test.

(6) The loading equipment must not be too heavy. These must allow easy transportation and installation (weight of each unit limited to 80 kgf (175 lbf) maximum).

(7) The test should reproduce as faithfully as possible the state of stress that shall be occurring during and after completion of the structure.

It is advisable to consider creep tests while planning deformability tests. It is also advantageous to obtain data about rock anisotropy by conducting tests in different directions.

The types of large-scale in situ tests required for different types of structures as suggested by the International Society for Rock Mechanics are given in Table 1.

8.3. Selection of Test Site

Selection of test site is extremely important. Prior to selecting a site, it is essential that all available surface and sub-surface geological data is compiled and analysed. In general, tests are conducted in exploratory headings, adits or pits. It is essential that in tests loads are applied to the rock surface in the same direction as loads from the proposed structure. If possible, loads may also be applied parallel and at right angle to the geologic structure. The test site should be so located that it is a representative of the conditions to be found in the significant portion of the rock mass. In this case, significance of a portion of the rock mass is determined by its relative influence on the deformability of the rock mass and its effect on the structure stability. A small volume of highly deformable material may be as important as a large mass of relatively competent rock mass. In other cases e.g. shear testing, certain weaker joints or thin layers lying in between competent rock mass may determine overwhelmingly the overall stability of the structure. A certain joint set may be more important than the other, not merely because it is more developed, but just because it is more unfavourably placed.

In test site evaluation, the following points should be carefully considered.

(1) Spatial orientation and intensity of loads to be transmitted to the rock mass by the proposed structure.

(2) Types of rock materials and their relative volume.

(3) Spatial orientation of rock structures i.e. bedding planes, foliation, joints, etc. and their relationship to the load applied by the structure.

(4) Joint continuity and joint density.

TABLE 1

Rock Mechanics In Situ Tests

(after Int. Soc. Rock. Mech., 1975)

Legend:

Test importance
n – necessary
a – advisable
oi – of interest
( ) – alternative

Stage of the work
F – Feasibility
DD – Detailed design
DC – During construction
AC – After completion

TYPE OF WORK / IN SITU MECHANICAL TESTS	FOUNDATIONS		NATURAL AND ARTIFICIAL ROCK SLOPES			UNDERGROUND WORK		ROCK EXCAVATION	HARBOURS & OTHER SUBMARINE WORKS
	gravity dams	arch dams	large structures	involving reservoirs	involving other works	large underground works	tunnels, shafts, underground mining	open air mining, quarries, large surface excavations	
1. Deformability Tests									
1.1 Static method	nDD	aF: nDD; nAC	aDD			nDC			oiF
1.1.1 Plate bearing (hydraulic jack; flat jack; cable jacking)		(n)DD; (n)AC				(n)DD	nDC		
1.1.2 Pressure tunnel (water loading; radial jacks)		(n)DD; (n)AC				(n)DD			
1.1.3 Pressure borehole (dilatometer)	aF	(n)DD; (n)AC	oiF			aF; (n)DD	oiDC		
1.2 Dynamic method	nDD	nF; nDD; nAC	oiF			aF; nDC	nDC		aF
1.2.1 Measurement of longitudinal waves velocity (geophones)	nDD	nF; (n)AC				aF	nDC		
1.2.2 Measurement of the velocity of longitudinal and transversal waves; Love's waves, Raleigh's (vibrograph)		nDD; (n)AC				nDD			

TABLE 1 (continued)

TYPE OF WORK / IN SITU MECHANICAL TESTS	FOUNDATIONS		NATURAL AND ARTIFICIAL ROCK SLOPES			UNDERGROUND WORK		ROCK EXCAVATION	HARBOURS & OTHER SUBMARINE WORKS
	gravity dams	arch dams	large structures	involving reservoirs	involving other works	large underground works	tunnels, shafts, underground mining	open air mining, quarries, large surface excavations	
1.2.3 Measurement of direct longitudinal waves velocity in borehole (sonic-coring)		aDD; (n)AC				aDC			
1.2.4 Detailed stratigraphic surveys							nDC		
2. Natural Rock Mass Stress Tests									
2.1 Rock surfaces tests						aF;nDD			
2.1.1 Measurement of deformation after over-coring or bond removal (by strain rosette)						(n)F			
2.1.2 Measurement of pressure to balance natural stresses (by flat jack)						a(F)			
2.2 Test inside borehole	oiDD	oiDD	oiDD	oiF		nDD			
2.2.1 Measurement of core deformation after over-coring						nDD			
2.2.2 Measurement of borehole wall deformation after over-coring						nDD			

TABLE 1 (continued)

TYPE OF WORK / IN SITU MECHANICAL TESTS	FOUNDATIONS		NATURAL AND ARTIFICIAL ROCK SLOPES			UNDERGROUND WORK		ROCK EXCAVATION	HARBOURS & OTHER SUBMARINE WORKS
	gravity dams	arch dams	large structures	involving reservoirs	involving other works	large underground works	tunnels, shafts, underground mining	open air mining, quarries, large surface excavations	
3. Strength Tests									
3.1 Compression		aDD							oiDD
3.1.1 Triaxial tests		oiDD				aDD			
3.2 Shear	nDD	nDD	nDD	nDD	nDD	nDD	aDD	aDD	
3.2.1 Rock block test along discontinuity surface	nDD	nDD	nDD	nDD	nDD	nDD	aDD	aDD	
3.2.2 Concrete block test along interface	aDD	aDD	aDD					oiF	
4. Permeability									
4.1 Inside borehole (Lugeon)	nF; nAC	nF; nAC	aF	nF		nF			
4.2 In a joint pumping test	nF; nAC	nF; nAC							
4.3 Piezometric levels & ground-water flow				nAC		nF	nF	oiF	nF
5. Rock Anchor Tests				nF; aDC; oiAC	nF; aDC; oiAC	aDD	aDD	aDC	oiDC
6. Rock Movement Monitoring									
6.1 Long base extensometer	nAC	nAC		nAC	nAC	nAC			
6.2 Inverted pendulum	aAC	aAC	aAC	oiAC					
6.3 Slope indicator				nAC				aAC	
6.4 Blast & ground motion monitoring	oiDD	oiDD		nDC	nDC	nF; nDC	nF; nDC		oiDC
6.5 Rock noise monitoring				aAC				oiAC	oiAC

(5) Presence of major or minor faults in the area and their orientation with respect to the loads imposed by the structure.

(6) State of alteration and moisture content.

The geologic structure at each test site should be thoroughly evaluated. Invariably boreholes drilled for instrumentation purposes should be cored with care to ensure maximum recovery of oriented core. The cores from these boreholes should be logged to determine rock type, strike and dip of foliations, bedding planes and joints, weathering and alteration, etc. A geologic cross-section through the test site should be prepared. Boreholes, if possible, should be inspected using borehole cameras or boroscope devices. The cores should be properly stored for future laboratory tests etc.

In selecting the position of test site, maximum use should be made of the galleries, trenches, etc. excavated for geological investigations.

The division of the whole area into zones of same or similar mechanical state is important. In determining the mechanical state of rock mass, the internal state of stress, alteration of the rock mass, presence of microfissures and joint density, degree of decompression (expansion) and moisture content must be considered. Geophysical investigations can be of extreme value in this case.

8.4. Uniaxial Compressive Strength of Rock In Situ

The uniaxial compressive strength of rocks in situ is important in the design of pillars in mines. In situ test for its determination is an extension of the laboratory test conducted on smaller specimens. Because of the high costs involved, uniaxial compressive strength tests in situ have been conducted only to a limited extent. Most of the studies have been done on coal (GREENWALD, HOWARTH and HARTMANN, 1939, 1941; LAMA, 1966a and b, 1970; BIENIAWSKI, 1968, 1969; COOK, 1967), iron ore (JAHNS, 1966; GIMM et al, 1966; RICHTER, 1968) and granite (GEORGI et al, 1970; PRATT et al, 1970, 1972).

Extensive work has been done in the laboratory for compressive strength of rocks on specimens varying from a few centimeters (an inch) to several centimeters (a foot) or so (Chapter 2, Vol. I) but it is still not possible to predict the failure behaviour of rocks in situ. The main reason behind this is that the strength of rocks depends upon the visible and invisible planes of weakness present, the severest of which determines the strength. Because of the disorderly nature of these planes of weakness, their effect is difficult to interpret by extrapolation of results obtained from the testing of small specimens in the laboratory. These results could possibly be relied upon if it is assumed that the strength of rocks is governed only by the number of planes of weakness present in them.

The failure of rocks in situ takes place along cleavage and other planes of weakness present along which the cohesion is at the minimum. If an accurate value of strength is to be obtained, the specimen must contain at least one such plane. Minimum specimen dimensions should be governed not only by the overall density of the planes of weakness but also by the density of each group of these planes of weakness.

Besides the presence of planes of weakness, rocks contain non-homogeneities (e.g. petrographic constituents) which cause differential stress distribution. The presence of these non-homogeneities results in variation in strength values of specimens, even when planes of weakness are absent. With the increase in the size of the specimen, the non-homogeneities, however, play a lesser role due to their large number which tend to converge the results.

The Austrian School (DENKHAUS, 1962; MÜLLER, 1963) suggested that cross-sectional area of the specimen must contain at least 100–200 planes of weakness.

HOEK (1966) suggested that for materials such as coal, the specimen size should be approximately 50–100 times the spacing between discontinuities, if the compressive strength of the specimen is to be within 10% of the in situ value. Further discussion on this subject is given in Chapter 10, Volume IV.

8.4.1. Specimen Preparation

The method to be adapted for preparation of specimens depends upon the type of rock, shape of the specimens, and the conditions of test (i.e. whether the specimen is to be isolated from all the four sides or is to be allowed to remain in contact with one or two or more sides). Different types of cutting machines have been used for the preparation of the specimens.

GREENWALD, HOWARTH and HARTMANN (1941) prepared square coal pillars by overcutting and sidecutting of coal seams. JAHNS (1966) and PRATT et al (1972) prepared specimens by drilling holes skin to skin and excavating all around. The surfaces of the specimens were finished manually with the help of chisels. LAMA (1966b) used a universal coal cutter for preparing coal specimens. BIENIAWSKI (1968) used a modified forestry chain saw for medium-size coal specimens, 15 to 45 cm (6 to 18 in), and a universal coal cutting machine for large-size specimens, 0.6 to 2.0 m (2.0 to 6.6 ft). The use of chain cutters (e.g. universal cutting machine) for softer rocks and disc cutter (LNEC, Portugal, 1970) for medium to hard rocks seems to be the prevailing practice. For very hard rocks, skin to skin drilling is useful, but such rocks may not require in situ testing unless they are highly fractured.

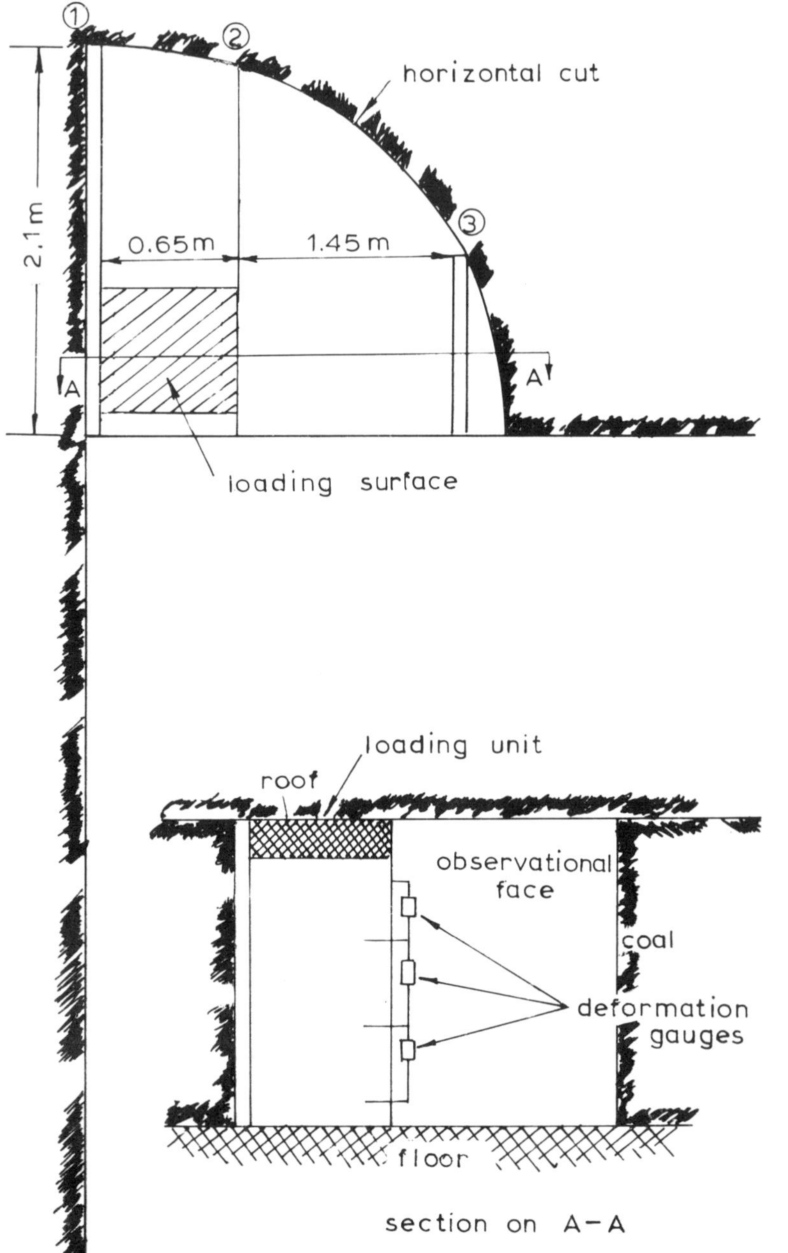

Fig. 8-1. Preparation of specimens for loading under uniaxial conditions; 1, 2, 3 – order of cuts (after LAMA, 1966b).

While preparing specimens, the blasting should be kept to a minimum and limited to at least 2–3 m (6.5–10.0 ft) (in softer rocks) or 1–1.5 m (3.25–5.0 ft) (in harder rocks) away from the actual test site. Vertical cuts are given first followed by horizontal cuts to separate the specimen from the roof. Fig. 8-1 shows the method of preparation of specimens in an area in the wall of a tunnel in a coal seam.

For preparing specimens using skin to skin drilling, a steel template should be used to accurately locate the hole centres and to avoid deviation of holes (Fig. 8-2). Diamond drills mounted on movable arm and capable of drilling vertical, horizontal or inclined holes are best suited for the purpose.

Fig. 8-2. Skin-to-skin drilling in progress (after PRATT, 1976).

It is important that the loading surface is sufficiently uniform. When rock is soft and is loaded by individual cylinders, the surface should be protected from penetration. Different capping materials have been used. A flexible capping material gives more uniform stress distribution, while rigid surface like concrete slabs may simulate conditions of adhesion between roof or floor rocks. LAMA (1966b) used 6 mm (0.24 in) cork sheets placed between two thin steel sheets to ensure no penetration and uniform stress distribution.

BIENIAWSKI (1968) capped the specimens with a cement and sand mix of approximately 7.6 cm (3 in) thickness to minimise the effects of surface irregularities. A lateral restraint was also provided by either steel shuttering fastened around the upper part of the specimen or reinforced concrete capping to simulate the effects of roof on the coal pillar. To ensure uniform load distribution and eliminate corner failures, jacks were not placed directly on the concrete capped surface of the specimen but on a rigid steel channel arrangement. On top of the jacks, steel plates were placed, the remaining space between the plates and the seam being filled with wooden packings.

COOK, HODGSON and HOJEM (1971) used a thin layer (about 1.5 cm (0.6 in) thick) of fine river sand. It is covered with a sheet of 'masonite' 3 mm (0.12 in) thick.

8.4.2. Loading and Displacement Measuring System

Equipment used for loading of specimens in situ usually consists of hydraulic jacks which are placed between the roof or floor and the specimen end surface or within the specimen. Powerful hydraulic props used in underground coal mines may also be used if the space available between the specimen and the roof is sufficient, i.e. for testing of smaller specimens in thick beds (JAHNS, 1966). The two different systems used by LAMA (1966b) are described below.

Low pressure hydraulic assembly unit: The unit consists of 16 separate hydraulic cylinders 21 cm (8.27 in) high, 15 cm (5.91 in) diameter with a piston area of 120 cm^2 (18.6 in^2) and a maximum displacement of 5 cm (1.97 in). Each cylinder is connected through a high pressure copper tube to a common distribution block fed by a hand-operated hydraulic pump. The pressure obtained is equal to the arithmetic mean of the force exerted by each cylinder, the characteristics of which are separately determined in the laboratory. These hydraulic cylinders are designed for maximum pressure of 250 bars (3625 lbf/in^2) and any number of them can be used at a time permitting a variation of the specimen size to be tested. For the most dense combination, the maximum size of the testing area using 16 cylinders is limited to 3632 cm^2 (563 in^2). The total force exerted by the set of 16 cylinders at 250 bars (3625 lbf/in^2) is 480 tonnef (472.5 tonf), giving a stress of about 13.1 MPa (1895 lbf/in^2) (Fig. 8-3).

Fig. 8-3. Low pressure hydraulic loading assembly unit (after LAMA, 1966 b).

High pressure hydraulic assembly unit: This unit consists of 4 individual multi-cylinder sets 27 cm × 33 cm × 12 cm (10.63 in × 13.00 in × 4.72 in) (height) in the shape of a parallelogram. The vertical walls of the sets have been curved following the contours of the cylinders so as to achieve a higher concentration and reduction in weight.

Each set consists of 9 cylinders and 18 pistons of 8 cm (3.15 in) diameter (each cylinder having two pistons which displace on opposite directions) with a maximum working displacement of 9.5 cm (3.74 in). Each set weighs about 77 kgf (169.8 lbf).

These hydraulic sets have been designed for a working pressure of 600 bars (8700 lbf/in^2) and are connected in parallel to a common distribution block which, in turn, is connected to a high pressure pump. All the sets may be used at a time or separately giving a wide choice of testing base area. Maximum area at maximum pressure density is limited to 3360 cm^2 (520.8 in^2). The assembly gives a total force of 1140 tonnef (1122 tonf) and a stress of 35.7 MPa (5176 lbf/in^2) at a pressure of 600 bars (8700 lbf/in^2) (Fig. 8-4).

BIENIAWSKI (1968) also tested coal pillars in situ. He used larger hydraulic jacks (each having a load area of 929 cm^2 (1 ft^2).

COOK, HODGSON and HOJEM (1971) used a 100 MN (10000 tonf), stiff jacking system for testing pillars with cross-sections of upto 2 m × 2 m (6.6 ft × 6.6 ft).

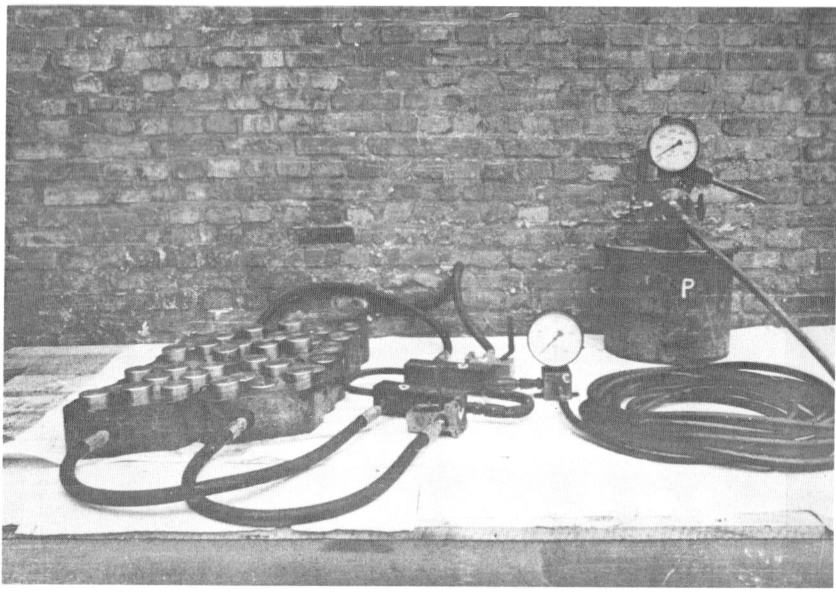

Fig. 8-4. High pressure hydraulic loading assembly unit (after LAMA, 1966b).

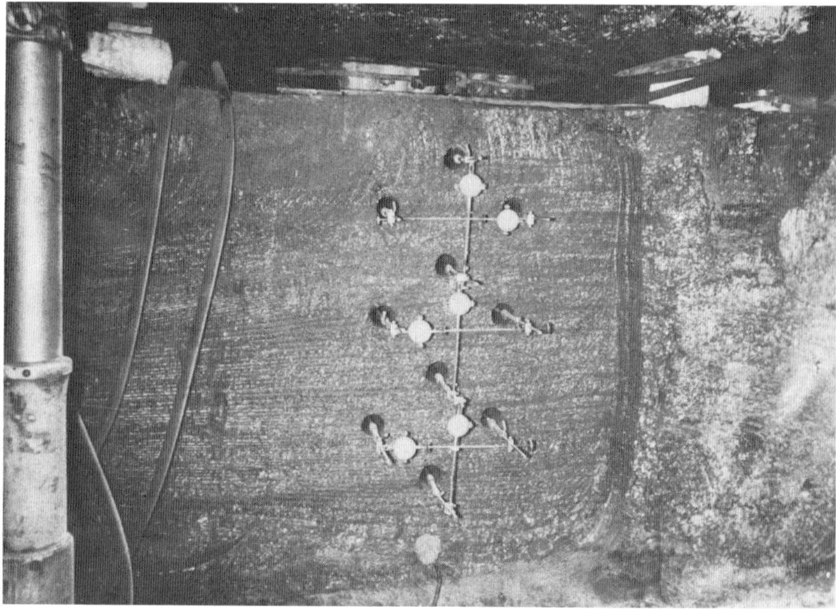

Fig. 8-5. Mounting of gauges and establishing of bases for deformation measurement in in situ tests (after LAMA, 1970).

This system consists of 25 jacks. They are separately and individually supplied with equal amounts of hydraulic fluid. These jacks when placed in the centre of the specimen height deform the specimen uniformly and symmetrically about its mid-plane.

Displacements can be measured using dial gauges mounted on the specimen using appropriate holders. The method adopted by LAMA (1966b) consists of holders mounted in rods equipped with ball and socket joints allowing three dimensional flexibility of arrangement (Fig. 8-5). One end of the rod having ball and socket joints has female threads to which is screwed a threaded reper.

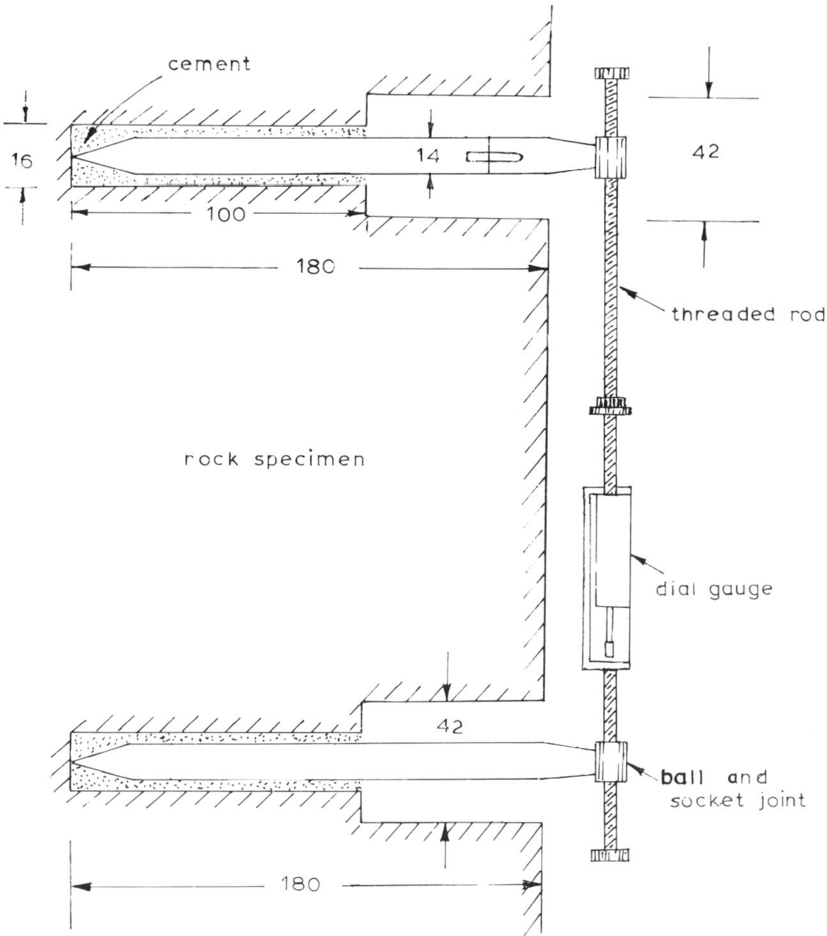

Fig. 8-6. Fixing of measurement points to rock specimen (dimensions in mm) (after LAMA, 1966b).

The top of the dial gauge lever presses against a plate of a threaded rod mounted onto another ball and socket rod (joined to another reper) by means of a bronze block. The distance between the two repers forms the measuring base.

The threaded rod is assembled in two parts; the threaded arm and plate arm, the plate arm being replaceable depending upon the length of the base chosen. The repers are of 14 mm (0.55 in) diameter and 180 mm (7.1 in) long and are cemented in 16/42 mm (0.63/1.65 in) diameter concentric holes (Fig. 8-6).

8.4.3. Results of In Situ Compressive Strength Tests

GREENWALD, HOWARTH and HARTMANN (1941) carried out tests on square specimens in Pittsburgh coal bed. The specimens were prepared by over-cutting and side-cutting and tested at right angles to the bedding planes by hydraulic assembly inserted between the top of the specimen and roof of the seam. The dimensions of the specimens were – height 61 to 162 cm (2.0 to 5.3 ft); width 30 to 162 cm (1.0 to 5.3 ft) (width-to-height ratio was 0.41 to 1.68). They expressed their results in the form

$$\sigma_c = 2800 \sqrt{w}/(h)^{5/6} \text{ lbf/in}^2 \tag{8.1}$$

where σ_c = compressive strength
w = width, in.
h = height, in.

BICZ (1962) carried out a large number of in situ tests in various coal fields of the U.S.S.R. and found that the strength of coal seams varies from 2.45 MPa (355.5 lbf/in^2) to 17.65 MPa (2560 lbf/in^2).

JAHNS (1966) tested ironstone specimens in situ and found that strength drops from 98 MPa (14220 lbf/in^2) or even higher on smaller specimens to 63.7 MPa (9243 lbf/in^2) on specimens 40 cm × 40 cm × 40 cm (15.75 in × 15.75 in × 15.75 in).

Results obtained by LAMA (1966b) show that in situ strength is always lower than the values obtained in the laboratory. He found that while the strength of specimens of size 65 cm × 65 cm × 163 cm (25.6 in × 25.6 in × 64.2 in) (height) of seam 506 (Upper Silesian coal field) was 6.8 MPa (981 lbf/in^2) the compressive strength determined in the laboratory from specimens prepared from the material taken from the same place where underground tests were conducted and of size 10 cm × 10 cm × 10 cm (3.94 in × 3.94 in × 3.94 in), was 24.5 MPa (3555 lbf/in^2). Similarly, the in situ strength value in seam 504 was 11.3 MPa (1635 lbf/in^2), whereas the corresponding laboratory value was 17.4 MPa (2517 lbf/in^2). The predicted values from extrapolation of laboratory tests and those determined from large scale in situ tests are given in Table 2.

TABLE 2

Comparison of Laboratory and In Situ Results on Coal

(after LAMA, 1970)

Seam	Compressive Strength, kgf/cm^2	Deformation Modulus, kgf/cm^2	Condition of Test
506 (in situ)	68.6	21 390	uniaxial 65 cm × 65 cm × 160 cm
504 (in situ)	113.9	35 320	,, ,,
506 (laboratory)	135	32 000	10 cm × 10 cm base; w/h = 2.46
504 (laboratory)	95	7 500	,, ,,

The results of tests on South African coals conducted by BIENIAWSKI (1967, 1968) and VAN HEERDEN (1975) are given in Figs. 8-7 and 8-8 respectively. Tests were conducted on specimens up to 2 m (6.6 ft) cubes. The strength decreases with increasing specimen size but the curve flattens out, eventually approaching an asymptotic value at the specimen size of about 1.5 m (5 ft) (Fig. 8-7). On the basis of these tests, BIENIAWSKI (1968) gave the following relationship:

$$\sigma_c = 1000 \frac{w^{0.16}}{h^{0.55}} \tag{8.2}$$

where σ_c = compressive strength, lbf/in^2
w = width, ft and
h = height, ft

When constant strength is reached,

$$\sigma_c = 400 + 200 \frac{w}{h} \tag{8.3}$$

This Eq. 8.3 is valid for w/h greater than unity and for specimen sizes equal to or greater than 1.5 m (5 ft). For w/h ratios as well as specimen width dimensions of less than those specified above, Eq. 8.2 will apply.

Tests by RICHTER (1968) on specimens of iron ore containing shale and sandstone showed that strength of specimens of 1.25–1.5 m (4.1–4.9 ft) (height) × 1.25–1.4 m (4.1–4.6 ft) × 1.5–2.15 m (4.9–7.1 ft) (length) was 18 times lower than laboratory specimens of 48 mm (1.9 in) diameter (20 MPa vs 360 MPa) (2900 lbf/in^2 vs 52 200 lbf/in^2) (200 kgf/in^2 vs 3600 kgf/in^2).

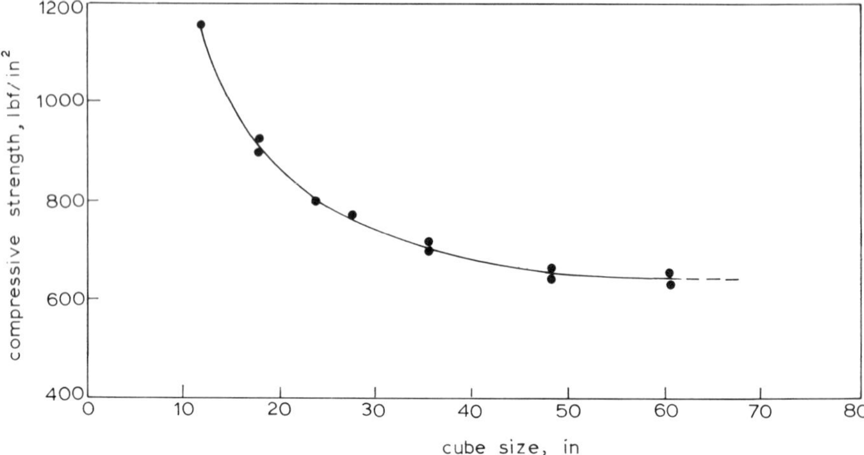

Fig. 8-7. Relationship between compressive strength and cube size of large coal specimens (after BIENIAWSKI, 1967).

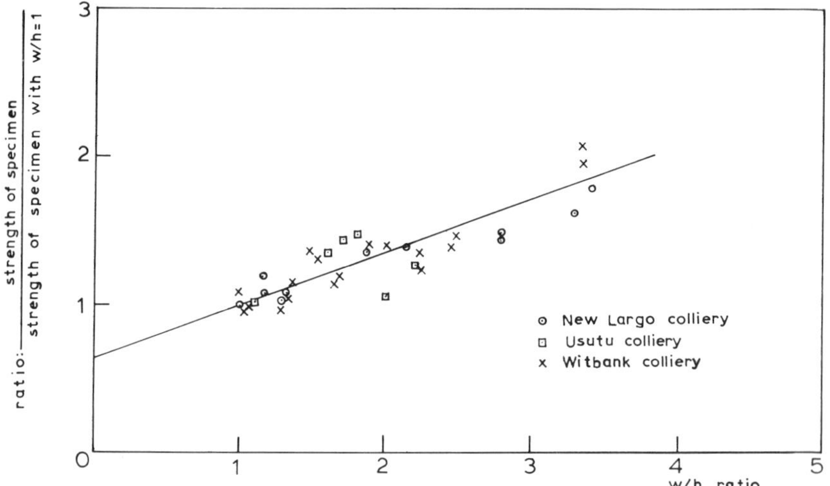

Fig. 8-8. Representation of compressive strength data in dimensionless form (after VAN HEERDEN, 1975).

Tests conducted by CHAOUI et al (1970) on cylindrical marl specimens 0.7 m (27.6 in) diameter and 1.0 m (39.4 in) height gave compressive strength of 3 MPa (435 lbf/in^2) (30 kgf/cm^2) compared to laboratory strength of 10.5 MPa (1523 lbf/in^2) (105 kgf/cm^2).

All these tests and others show the decrease in strength with increase in size of specimen dimension. The influence of specimen size on the compressive strength of coal based on field tests obtained by various investigators is given in Table 3.

TABLE 3

Summary of Strength Formulae Proposed From Large Scale In Situ Tests in Compression

Investigators	Rock tested	Specimen cross-section	Width (m)	Height (m)	Width to height ratio	No. of tests	Formula	Remarks
H. P. GREENWALD H. C. HOWARTH & I. HARTMANN (1939)	Coal (Pittsburgh)	Square	0.81 to 1.61	0.78 to 1.61	0.50 to 1.03	5	$\sigma_c = 700\sqrt{w/h}$ (lbf/in^2) or $\sigma_c = 4 \cdot 8\sqrt{w/h}$ (MPa)	Seven specimens were tested but two were disregarded as they failed due to clay flow in the floor. The equation was originally given in lbf/in^2 with a constant 695 later approximated to 700
H. P. GREENWALD H. C. HOWARTH & I. HARTMANN (1941)	Coal (Pittsburgh)	Square	0.30 to 1.07	0.74 to 0.77	0.41 to 1.68	7	$\sigma_c = 2800(\sqrt{w/\sqrt[6]{h^5}})$ (lbf/in^2) or $\sigma_c = 19.3\sqrt{w/\sqrt[6]{h^5}}$ (MPa)	A relation of $\sigma = 900\sqrt{w/h}$ in lbf/in^2 was also found for $w/h < 1$
Z. T. BIENIAWSKI (1968)	Coal (Witbank)	Square	0.61 to 1.22	0.61 to 1.22	0.5 to 2.0	19	$\sigma_c = 7.6\, w^{0.16}/h^{0.55}$ (MPa) with w & h in m	Valid for: width-to-height ratios < 1 & specimen width < 1.5 m.
Z. T. BIENIAWSKI (1969)	Coal (Witbank)	Square	1.5 to 2.0	0.6 to 2.0	1.0 to 3.1	16	$\sigma_c = 2.5 + 2w/h$ (MPa)	Valid for: width-to-height ratios > 1 & specimen or pillar width > 1.5 m.
M. D. G. SALAMON (1967)	Coal						$\sigma_c = 7.2w^{0.46}/h^{0.66}$	Based on survey of failed pillars
H. WAGNER (1974)	Coal (Usutu)	Square & rectangle	0.6 to 2.0	0.86 to 2.0	0.6 to 2.2	12	$\sigma_c = 11\sqrt{w/h}$ (MPa)	Equation: $\sigma_c = 7 + 4w/h$ (MPa) also fits the data well
W. L. VAN HEERDEN (1974)	Coal (New Largo)	Square	1.4	0.41 to 1.24	1.14 to 3.39	10	$\sigma_c = 10 + 4.2w/h$ (MPa)	

The drop in compressive strength (Fig. 8-9) is smaller for specimens of larger h/d ratio than for cubic specimens, the stress distribution playing an important role. The capping material used between the loading units and specimen surface influences both the results and the mode of failure. Using a thin cork sheet, LAMA (1966b) found that in the absence of well developed cleavage planes in coal, failure occurs with planes parallel to the direction of loading (Fig. 8-10) without formation of any cones as often observed in laboratory tests with high friction at the ends. Tests by BIENIAWSKI (1968) showed a typical such failure (Fig. 8-11). When well developed cleavage planes are present in specimens, these may be the controlling features and failure may occur along these planes (Fig. 8-12).

If the influence of friction at the ends is eliminated altogether or its effect greatly reduced, the drop in strength with increase in height of the specimens may not be significant. Fig. 8-13 shows the results obtained in laboratory using brush platens and ordinary platens. It is therefore important that the end conditions, i.e. contact between pillars and roof and floor rocks must be investigated before using any of the formula which include the influence of height on the compressive strength of specimens.

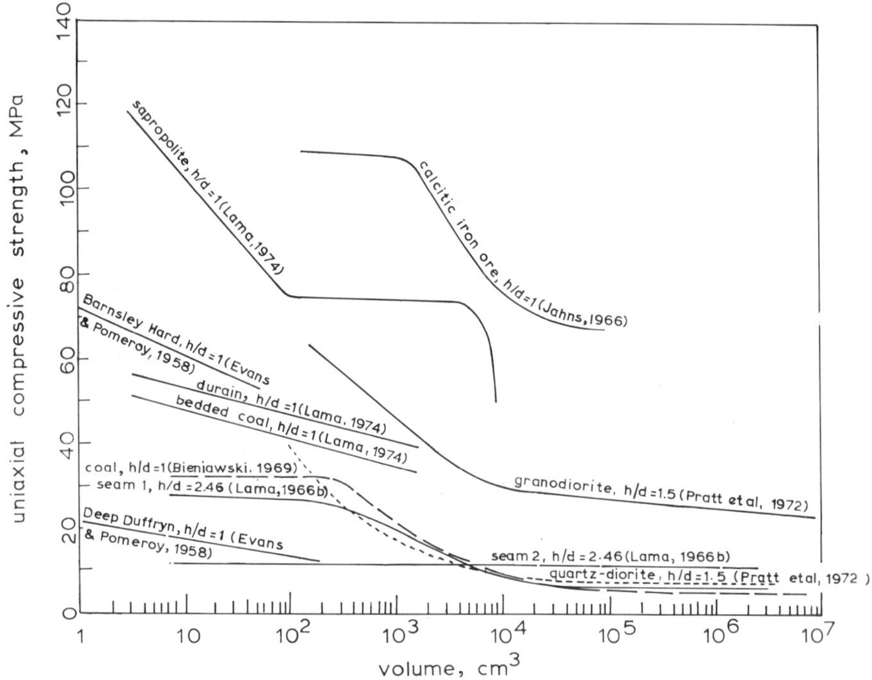

Fig. 8-9. Volume effect on compressive strength in rock specimens (after LAMA and GONANO, 1976).

Fig. 8-10. Failure of a specimen in situ, in the absence of cleavage planes (after LAMA, 1966 b).

Fig. 8-11. A 5 ft cube coal specimen after failure (after BIENIAWSKI, 1968).

Fig. 8-12. Failure of a specimen in situ as a result of dislodging along a cleavage plane (after LAMA, 1966b).

Attempts have been made to determine the complete stress-strain curve for rock specimens in situ (VAN HEERDEN, 1975; BIENIAWSKI and VAN HEERDEN, 1975; COOK et al, 1971). Using a comparatively softer system, the stress-strain curve for specimen of width/height = 3.28 (height ≐ 4.27 mm (16.8 in)) is given in Fig. 8-14. Fig. 8-15 shows the post-failure modulus for different width/height ratios. The slope of post-peak curve drops with increase in width/height ratios.

From the point of view of design of pillars, it is sufficient to test the strength of unit cubes. Tests have shown that the strength of a pillar of any width/height ratio is related to the compressive strength of a cube giving the following relationship:

$$\frac{\sigma_c}{\sigma_{c_1}} = 0.64 + 0.36\,(w/h) \tag{8.4}$$

where σ_c = compressive strength of the pillar, MPa
σ_{c_1} = compressive strength of cube, MPa
w = width of pillar, m and
h = height of pillar, m.

SALAMON formula (Table 3) has been used extensively with success in South African coal mines with safety factor of 1.5–1.7 (average 1.6) when the pillars are left as a permanent support. When pillars are to be extracted at a later stage (depillared) the safety factor of 2.0 should be used (SALAMON, 1976).

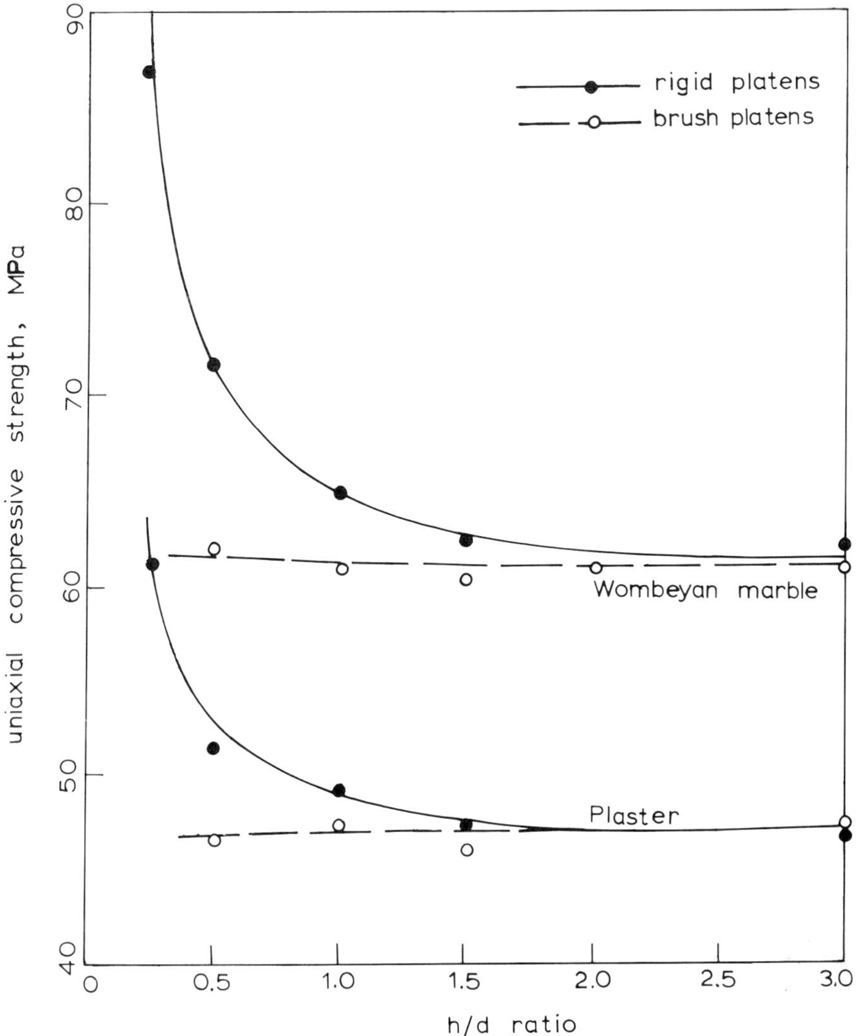

Fig. 8-13. Influence of end conditions (different platens) and height/diameter (h/d) ratios on compressive strength (after LAMA and GONANO, 1976).

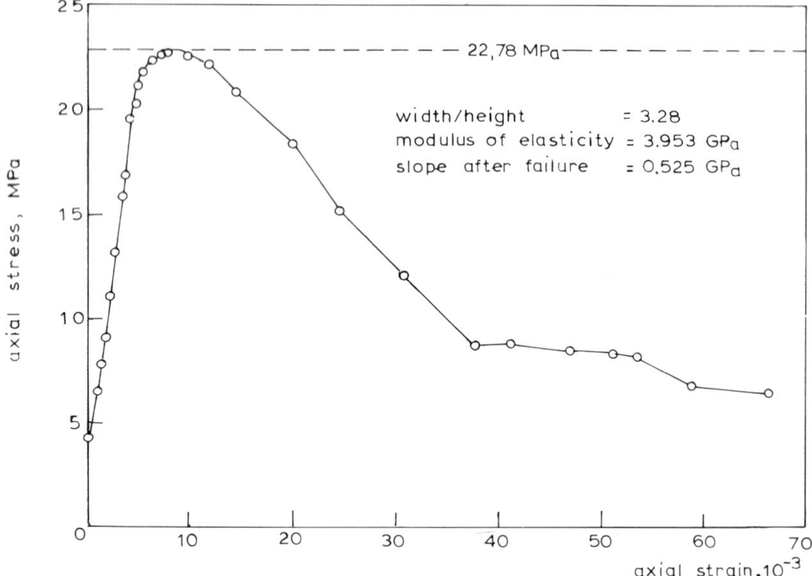

Fig. 8-14. Complete stress-strain curve obtained for coal specimen in situ (after VAN HEERDEN, 1975).

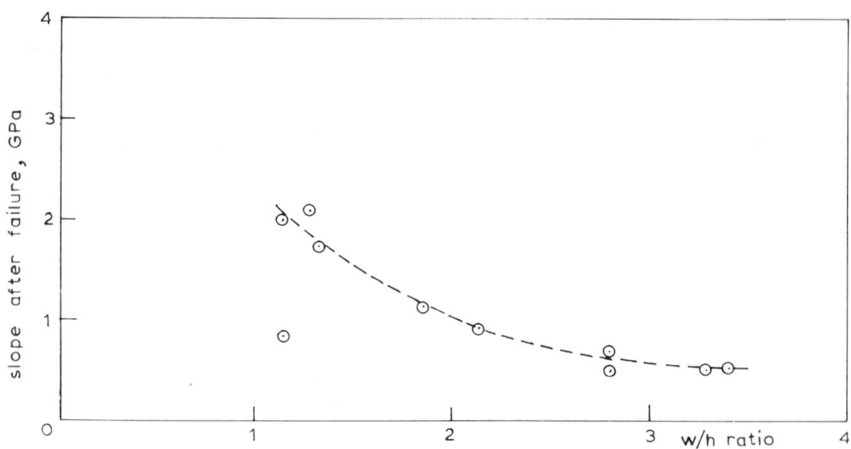

Fig. 8-15. Influence of *w/h* ratio on the slope of the stress-strain curve after failure (after VAN HEERDEN, 1975).

8.5. In Situ Tests for Deformability of Rock

The deformability of rock in situ has been determined using different methods. The tests can be divided into the following three types:

1. Plate bearing test
2. Pressure tunnel test
3. Borehole tests.

These methods along with their variations and interpretation of results obtained from these tests are described below:

8.5.1. Plate Bearing Test

This is one of the simplest and most commonly used methods for determination of the deformability of rock mass in situ. The method consists of applying normal load to a specially prepared flat surface through a rigid or flexible finite plate and the deformation is measured at the centre of the plate or at its edge or at any other convenient point within the rock mass. The deformation modulus can then be calculated using the relationships developed depending upon the shape of the loading plate and the nature of the rock (assuming isotropic or anisotropic behaviour).

8.5.1.1. Theoretical basis

The basic principle underlying a plate bearing test is that if a semi-infinite half space is loaded, the resulting settlement of the plate is a function of the effective modulus of deformation of the material composing the semi-infinite half space called as the subgrade. The theoretical basis is the well known BOUSSINESQ solution (TIMOSHENKO and GOODIER, 1951) where the displacement of a surface of a semi-infinite elastic solid under the action of a point (Fig. 8-16) normal load can be given by

$$\rho_z = \frac{P(1+\nu)}{2\pi ER}\left[2(1-\nu)+\frac{z^2}{R^2}\right] \tag{8.5}$$

$$\rho_r = \frac{P(1+\nu)}{2\pi ER}\left[\frac{rz}{R^2}-\frac{(1-2\nu)r}{R+z}\right] \tag{8.6}$$

where P = normal concentrated load

ν = POISSON's ratio of sub-grade and

E = modulus of deformation of sub-grade.

When $z=0$,

$$\rho_z = \frac{P(1-\nu^2)}{\pi E R} \tag{8.7}$$

$$\rho_r = -\frac{P(1+\nu)(1-2\nu)}{2\pi E R} \tag{8.8}$$

The negative sign means displacement away from the point of application of load.

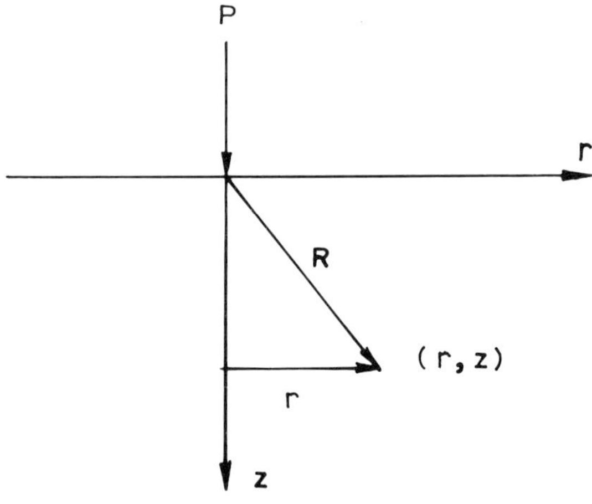

Fig. 8-16. Point load on surface of semi-infinite elastic solid

When load is distributed uniformly over a greater surface (Fig. 8-17), the Eqs. 8.5 and 8.6 can be integrated over prescribed area of the surface. The expressions are cumbersome and simplified version is given by:

For a circular area

$$\rho_z = \frac{p(1+\nu)}{E} a \left[\frac{z}{a} A + (1-\nu) H \right] \tag{8.9}$$

$$\rho_r = \frac{p(1+\nu)}{E} (-r) \left[(1+2\nu) E_1 - D \right] \tag{8.10}$$

where A, D, E_1 and H are dependent upon the dimensions of the loaded surface and the coordinates of the point. Their values are given in Tables 4 to 7. Vertical deflection ρ_z beneath a uniform circular load is given in Fig. 8-18.

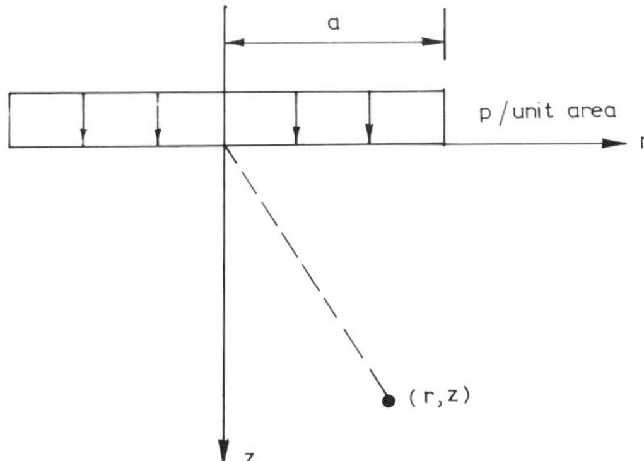

Fig. 8-17. Distributed load on surface of semi-infinite elastic solid – Notation for circular area

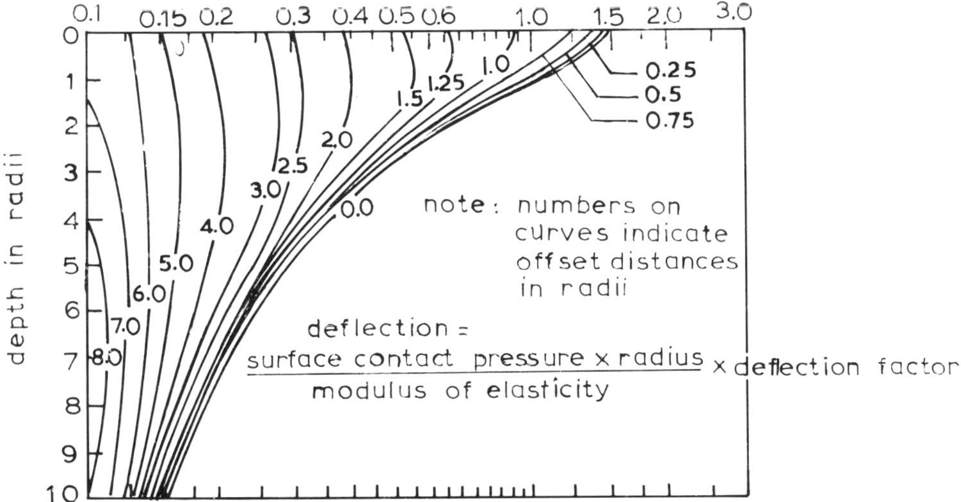

Fig. 8-18. Vertical deflection ρ_z beneath uniform circular load (after FOSTER and AHLVIN, 1954).

For $z=0$, $r/a=0$, $H=2.0$ (i.e. deflection at the centre of the plate)

$$\rho_z = \frac{2ap(1-\nu^2)}{E} \tag{8.11}$$

For $z=0$, $r/a=1$, $H=1.27$ (i.e. deflection at the edge of the plate)

$$\rho_z = \frac{1.27\,ap\,(1-\nu^2)}{E} \tag{8.12}$$

TABLE 4

Function "A"

(after AHLVIN and ULERY, 1962)

z/a \ r/a	0	0.2	0.4	0.6	0.8	1	1.2	1.5	2	3	4	5	6	7	8	10	12	14
0	1.0	1.0	1.0	1.0	1.0	.5	0	0	0	0	0	0	0	0	0	0	0	
0.1	.90050	.89748	.88679	.86126	.78797	.43015	.09645	.02787	.00856	.00211	.00084	.00042						
0.2	.80388	.79824	.77884	.73483	.63014	.38269	.15433	.05251	.01680	.00419	.00167	.00083	.00048	.00030	.00020			
0.3	.71265	.70518	.68316	.62690	.52081	.34375	.17964	.07199	.02440	.00622	.00250							
0.4	.62861	.62015	.59241	.53767	.44329	.31048	.18709	.08593	.03118									
0.5	.55279	.54403	.51622	.46448	.38390	.28156	.18556	.09499	.03701	.01013	.00407	.00209	.00118	.00071	.00053	.00025	.00014	.00009
0.6	.48550	.47691	.45078	.40427	.33676	.25588	.17952	.10010										
0.7	.42654	.41874	.39491	.35428	.29833	.21727	.17124	.10228	.04558									
0.8	.37551	.36832	.34729	.31243	.26581	.21297	.16206	.10236										
0.9	.33104	.32492	.30669	.27707	.23832	.19488	.15253	.10094										
1	.29289	.28763	.27005	.24697	.21468	.17868	.14329	.09849	.05185	.01742	.00761	.00393	.00226	.00143	.00097	.00050	.00029	.00018
1.2	.23178	.22795	.21662	.19890	.17626	.15101	.12570	.09192	.05260	.01935	.00871	.00459	.00269	.00171	.00115			
1.5	.16795	.16552	.15877	.14804	.13436	.11892	.10296	.08048	.05116	.02142	.01013	.00548	.00325	.00210	.00141	.00073	.00043	.00027
2	.10557	.10453	.10140	.09647	.09011	.08269	.07471	.06275	.04496	.02221	.01160	.00659	.00399	.00264	.00180	.00094	.00056	.00036
2.5	.07152	.07098	.06947	.06698	.06373	.05974	.05555	.04880	.03787	.02143	.01221	.00732	.00463	.00308	.00214	.00115	.00068	.00043
3	.05132	.05101	.05022	.04886	.04707	.04487	.04241	.03839	.03150	.01980	.01220	.00770	.00505	.00346	.00242	.00132	.00079	.00051
4	.02986	.02976	.02907	.02802	.02832	.02749	.02651	.02490	.02193	.01592	.01109	.00768	.00536	.00384	.00282	.00160	.00099	.00065
5	.01942	.01938				.01835			.01573	.01249	.00949	.00708	.00527	.00394	.00298	.00179	.00113	.00075
6	.01361					.01307			.01168	.00983	.00795	.00628	.00492	.00384	.00299	.00188	.00124	.00084
7	.01005					.00976			.00894	.00784	.00661	.00548	.00445	.00360	.00291	.00193	.00130	.00091
8	.00772					.00755			.00703	.00635	.00554	.00472	.00398	.00332	.00276	.00189	.00134	.00094
9	.00612					.00600			.00566	.00520	.00466	.00409	.00353	.00301	.00256	.00184	.00133	.00096
10								.00477	.00465	.00438	.00397	.00352	.00326	.00273	.00241			

TABLE 5
Function "D"
(after AHLVIN and ULERY, 1962)

z/a \ r/a	0	0.2	0.4	0.6	0.8	1	1.2	1.5	2	3	4	5	6	7	8	10	12	14
0	0	0	0	0	0	0	0	0	0	0	0	0						
0.1	.04926	.04998	.05235	.05716	.06687	.07635	.04108	.01803	.00691	.00193	.00080	.00041						
0.2	.09429	.09552	.09900	.10546	.11431	.10932	.07139	.03444	.01359	.00384	.00159	.00081	.00047	.00029	.00020			
0.3	.13181	.13305	.14051	.14062	.14267	.12745	.09078	.04817	.01982	.00927	.00238							
0.4	.16008	.16070	.16229	.16288	.15756	.13696	.10248	.05887	.02545									
0.5	.17889	.17917	.17826	.17481	.16403	.14074	.10894	.06670	.03039	.00921	.00390	.00200	.00116	.00073	.00049	.00025	.00015	.00009
0.6	.18915	.18867	.18573	.17887	.16489	.14137	.11186	.07212										
0.7	.19244	.19132	.18679	.17782	.16229	.13926	.11237	.07551	.03801									
0.8	.19046	.18927	.18348	.17306	.15714	.13548	.11115	.07728										
0.9	.18481	.18349	.17709	.16635	.15063	.13067	.10866	.07788										
1	.17678	.17503	.16886	.15824	.14344	.12513	.10540	.07753	.04456	.01611	.00725	.00382	.00224	.00142	.00096	.00050	.00029	.00018
1.2	.15742	.15618	.15014	.14073	.12823	.11340	.09757	.07484	.04575	.01796	.00835	.00446	.00264	.00169	.00114			
1.5	.12801	.12754	.12237	.11549	.10657	.09608	.08491	.06833	.04539	.01983	.00970	.00532	.00320	.00205	.00140	.00073	.00043	.00027
2	.08944	.09080	.08668	.08273	.07814	.07187	.06566	.05589	.04103	.02098	.01117	.00643	.00398	.00260	.00179	.00095	.00056	.00036
2.5	.06403	.06565	.06284	.06068	.05777	.05525	.05069	.04486	.03532	.02045	.01183	.00717	.00457	.00306	.00213	.00115	.00068	.00044
3	.04744	.04834	.04760	.04548	.04391	.04195	.03963	.03606	.02983	.01904	.01187	.00755	.00497	.00341	.00242	.00133	.00080	.00052
4	.02854	.02928	.02996	.02798	.02724	.02661	.02568	.02408	.02110	.01552	.01087	.00757	.00533	.00382	.00280	.00160	.00100	.00065
5	.01886	.01950				.01816			.01535	.01230	.00939	.00700	.00523	.00392	.00299	.00180	.00114	.00077
6	.01333					.01351			.01149	.00976	.00788	.00625	.00488	.00381	.00301	.00190	.00124	.00086
7	.00990					.00966			.00899	.00787	.00662	.00542	.00445	.00360	.00292	.00192	.00130	.00092
8	.00763					.00759			.00727	.00641	.00554	.00477	.00402	.00332	.00275	.00192	.00131	.00096
9	.00607					.00746			.00601	.00535	.00470	.00415	.00358	.00303	.00260	.00187	.00133	.00099
10								.00542	.00506	.00450	.00398	.00364	.00319	.00278	.00239			

TABLE 6

Function "E_1"

(after AHLVIN and ULERY, 1962)

z/a \ r/a	0	0.2	0.4	0.6	0.8	1	1.2	1.5	2	3	4	5	6	7	8	10	12	14
0	.5	.5	.5	.5	.5	.5	.34722	.22222	.12500	.05556	.03125	.02000	.01389	.01020	.00781	.00500	.00347	.00255
0.1	.45025	.44949	.44698	.44173	.43008	.39198	.30445	.20399	.11806	.05362	.03045	.01959						
0.2	.40194	.40043	.39591	.38660	.36798	.32802	.26598	.18633	.11121	.05170	.02965	.01919	.01342	.00991	.00762			
0.3	.35633	.35428	.33809	.33674	.31578	.28003	.23311	.16967	.10450	.04979	.02886							
0.4	.31431	.31214	.30541	.29298	.27243	.24200	.20526	.15428	.09801									
0.5	.27639	.27407	.26732	.25511	.23639	.21119	.18168	.14028	.09180	.04608	.02727	.01800	.01272	.00946	.00734	.00475	.00332	.00246
0.6	.24275	.24247	.23411	.22289	.20634	.18520	.16155	.12739										
0.7	.21527	.21112	.20535	.19525	.18093	.16356	.14421	.11620	.08027									
0.8	.13765	.18550	.13049	.17190	.15977	.14523	.12928	.10602										
0.9	.16552	.16337	.15921	.15179	.14168	.12954	.11634	.09686										
1	.14645	.14483	.14610	.13472	.12618	.11611	.10510	.08865	.06552	.03736	.02352	.01602	.01157	.00874	.00683	.00450	.00318	.00237
1.2	.11589	.11435	.11201	.10741	.10140	.09431	.08657	.07476	.05728	.03425	.02208	.01527	.01113	.00847	.00664			
1.5	.08398	.08356	.08159	.07885	.07517	.07088	.06611	.05871	.04703	.03003	.02008	.01419	.01049	.00806	.00636	.00425	.00304	.00228
2	.05279	.05105	.05146	.05034	.04850	.04675	.04442	.04078	.03454	.02410	.01706	.01248	.00943	.00738	.00590	.00401	.00290	.00219
2.5	.05576	.03426	.03489	.03455	.03360	.03211	.03150	.02953	.02599	.01945	.01447	.01096	.00850	.00674	.00546	.00378	.00276	.00210
3	.02566	.02519	.02470	.02491	.02444	.02389	.02330	.02216	.02007	.01585	.01230	.00962	.00763	.00617	.00505	.00355	.00263	.00201
4	.01493	.01452	.01495	.01526	.01446	.01418	.01395	.01356	.01281	.01084	.00900	.00742	.00612	.00511	.00431	.00313	.00237	.00185
5	.00971	.00927				.00929			.00873	.00774	.00673	.00579	.00495	.00425	.00364	.00275	.00213	.00168
6	.00680					.00632			.00629	.00574	.00517	.00457	.00404	.00354	.00309	.00241	.00192	.00154
7	.00503					.00493			.00466	.00438	.00404	.00370	.00330	.00296	.00264	.00213	.00172	.00140
8	.00386					.00377			.00354	.00344	.00325	.00297	.00273	.00250	.00228	.00185	.00155	.00127
9	.00306					.00227			.00275	.00273	.00264	.00246	.00229	.00212	.00194	.00163	.00139	.00116
10								.00210	.00220	.00225	.00221	.00203	.00200	.00181	.00171			

TABLE 7

Function "H'

(after Ahlvin and Ulery, 1962)

z/a \ r/a	0	0.2	0.4	0.6	0.8	1	1.2	1.5	2	3	4	5	6	7	8	10	12	14
0	2.0	1.97987	1.91751	1.80575	1.62553	1.27319	.93676	.71185	.51671	.33815	.25200	.20045	.16626	.14315	.12576	.09918	.08346	.07023
0.1	1.80998	1.79018	1.72886	1.61961	1.44711	1.18107	.92670	.70888	.51627	.33794	.25184	.20081						
0.2	1.63961	1.62068	1.56242	1.46001	1.30614	1.09996	.90098	.70074	.51382	.33726	.25162	.20072	.16688	.14288	.12512			
0.3	1.48806	1.47044	1.40979	1.32442	1.19210	1.02740	.86726	.68823	.50966	.33638	.25124							
0.4	1.35407	1.33802	1.28963	1.20822	1.09555	.96202	.83042	.67238	.50412									
0.5	1.23607	1.22176	1.17894	1.10830	1.01312	.90298	.79308	.65429	.49728	.33293	.24996	.19982	.16668	.14273	.12493	.09996	.08295	.07123
0.6	1.13238	1.11998	1.08350	1.02154	.94120	.84917	.75653	.63469										
0.7	1.04131	1.03037	.99794	.91049	.87742	.80030	.72143	.61442	.48061									
0.8	.96125	.95175	.92386	.87928	.82136	.75571	.68809	.59398										
0.9	.89072	.88251	.85856	.82616	.77950	.71495	.65677	.57361										
1	.82843	.85005	.80465	.76809	.72587	.67769	.62701	.55364	.45122	.31877	.24386	.19673	.16516	.14182	.12394	.09952	.08292	07104
1.2	.72410	.71882	.70370	.67937	.64814	.61187	.57329	.51552	.43013	.31162	.24070	.19520	.16369	.14099	.12350			
1.5	.60555	.60233	.57246	.57633	.55559	.53138	.50496	.46379	.39872	.29945	.23495	.19053	.16199	.14058	.12281	.09876	.08270	.07064
2	.47214	.47022	.44512	.45656	.44502	.43202	.41702	.39242	.35054	.27740	.22418	.18618	.15846	.13762	.12124	.09792	.08196	.07026
2.5	.38518	.38403	.38098	.37608	.36940	.36155	.35243	.33698	.30913	.25550	.21208	.17898	.15395	.13463	.11928	.09700	.08115	.06980
3	.32457	.32403	.32184	.31887	.31464	.30969	.30381	.29364	.27453	.23487	.19977	.17154	.14919	.13119	.11694	.09558	.08061	.06897
4	.24620	.24588	.24820	.25128	.24168	.23932	.23668	.23164	.22188	.19908	.17640	.15596	.13864	.12396	.11172	.09300	.07864	.06848
5	.10805	.19785				.19455			.18450	.17080	.15575	.14130	.12785	.11615	.10585	.08915	.07675	.06695
6	.16554					.16326			.15750	.14868	.13842	.12792	.11778	.10836	.09990	.08562	.07452	.06522
7	.14217					.14077			.13699	.13097	.12404	.11620	.10843	.10101	.09387	.08197	.07210	.06377
8	.12448					.12352			.12112	.11680	.11176	.10600	.09936	.09460	.08848	.07800	.06928	.06200
9	.11079					.10989			.10854	.10548	.10161	.09702	.09234	.08784	.08298	.07407	.06678	.05976
10								.09900	.09820	.09510	.09290	.08980	.08300	.08180	.07710			

For a rectangular area (Fig. 8-19)

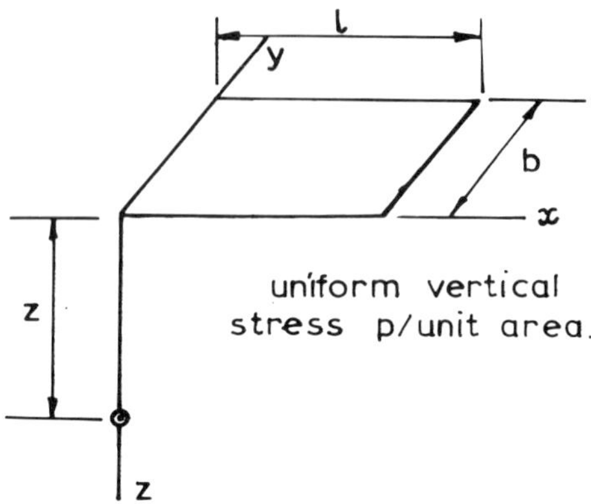

Fig. 8-19. Distributed load on surface of semi-infinite elastic solid – Notation for rectangular area.

Displacement beneath the corners of a rectangle at depth z,

$$\rho_z = \frac{pb}{E}(1-\nu^2)\left(A - \frac{1-2\nu}{1-\nu}B\right) \tag{8.13}$$

where

$$A = \frac{1}{2\pi}\left[ln \frac{\sqrt{1+m_1^2+n_1^2}+m_1}{\sqrt{1+m_1^2+n_1^2}-m_1} + m_1\, ln \frac{\sqrt{1+m_1^2+n_1^2}+1}{\sqrt{1+m_1^2+n_1^2}-1}\right]$$

$$B = \frac{n_1}{2\pi}\tan^{-1}\frac{m_1}{n_1\sqrt{1+m_1^2+n_1^2}}$$

$m_1 = l/b$ and $n_1 = z/b$.

Explicit expressions and influence factors for the vertical displacement at the surface ($z=0$) for four points (O, M, C, N) shown in Fig. 8-20 have been calculated by GIROUD (1968).

For all points

$$\rho_z = \frac{(1-\nu^2)}{E}pbI \tag{8.14}$$

where I = influence factor. Its value for different l/b ratios and the 4 different points are given in Table 8.

In the derivation of the above equations, it is assumed that load is uniformly distributed and that the sub-grade is homogeneous and isotropic. In practice, however, these conditions are not met. The rock underneath the test plate may be cross-anisotropic (e.g. sedimentary rocks) or completely anisotropic.

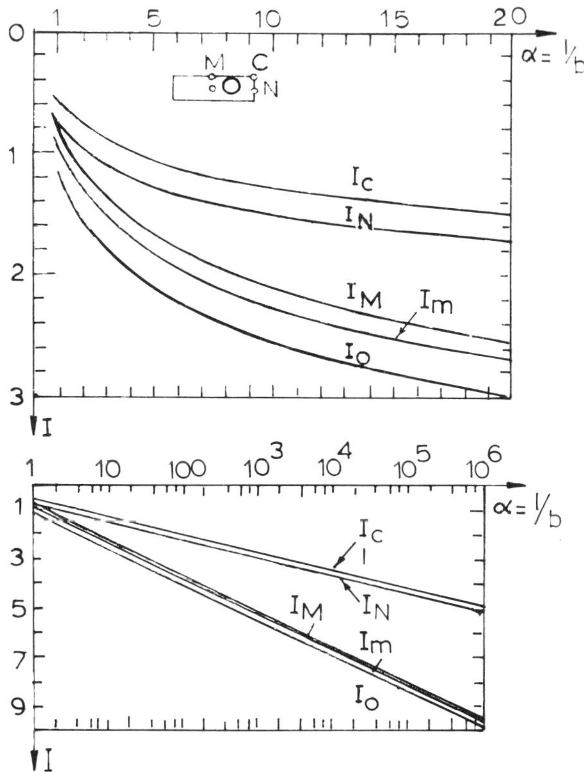

Fig. 8-20. Influence factors for vertical surface displacement beneath a rectangle (I_m = mean) (after GIROUD, 1968).

Similarly, the load distribution beneath the loading surface may not be uniform (as already discussed in section 2.2., Vol. I). This results in errors in the computed value of the modulus of deformation.

The influence of shape of the load distribution on the measured displacement for a circular area is given in Table 9. A comparison of the deformation values under different conditions shows that the displacements at different points are very considerable and are dependent upon the load distribution. In most of the cases, in actual testing, either a rigid plate (case IV) or flat jack (case I) is used. The mean displacement values in these two cases will not differ more than about 7 to 8%. When the loading plate allows lateral displacement of the rock layer direct in contact with it and at the same time maintains its rigidity, an extreme condition of parabolic loading with zero load at the centre and maximum at the edge may occur giving extreme conditions. It is therefore important that proper care be taken to make sure that such extreme and unpredictable conditions are avoided.

TABLE 8

Influence Factors *I* for Vertical Surface Displacement beneath Rectangle

(after GIROUD, 1968)

$\alpha=\frac{l}{b}$	I_C	I_M	I_N	I_O	I_m	$\alpha=\frac{l}{b}$	I_C	I_M	I_N	I_O	I_m
1	0.561	0.766	0.766	1.122	0.946	15	1.401	2.362	1.621	2.802	2.498
1.1	0.588	0.810	0.795	1.176	0.992	20	1.493	2.544	1.713	2.985	2.677
1.2	0.613	0.852	0.822	1.226	1.035	25	1.564	2.686	1.784	3.127	2.817
1.3	0.636	0.892	0.847	1.273	1.075	30	1.622	2.802	1.842	3.243	2.932
1.4	0.658	0.930	0.870	1.317	1.112	40	1.713	2.985	1.934	3.426	3.113
1.5	0.679	0.966	0.892	1.358	1.148	50	1.784	3.127	2.005	3.568	3.254
1.6	0.698	1.000	0.912	1.396	1.181	60	1.842	3.243	2.063	3.684	3.370
1.7	0.716	1.033	0.931	1.433	1.213	70	1.891	3.341	2.112	3.783	3.467
1.8	0.734	1.064	0.949	1.467	1.244	80	1.934	3.426	2.154	3.868	3.552
1.9	0.750	1.094	0.966	1.500	1.273	90	1.971	3.501	2.192	3.943	3.627
2	0.766	1.122	0.982	1.532	1.300	100	2.005	3.568	2.225	4.010	3.693
2.2	0.795	1.176	1.012	1.590	1.353	200	2.225	4.010	2.446	4.451	4.134
2.4	0.822	1.226	1.039	1.644	1.401	300	2.355	4.268	2.575	4.709	4.391
2.5	0.835	1.250	1.052	1.669	1.424	400	2.446	4.451	2.667	4.892	4.574
3	0.892	1.358	1.110	1.783	1.527	500	2.517	4.593	2.738	5.034	4.717
3.5	0.940	1.450	1.159	1.880	1.616	600	2.575	4.709	2.796	5.150	4.833
4	0.982	1.532	1.201	1.964	1.694	700	2.624	4.807	2.845	5.248	4.931
4.5	1.019	1.604	1.239	2.038	1.763	800	2.667	4.892	2.887	5.333	5.015
5	1.052	1.669	1.272	2.105	1.826	900	2.704	4.967	2.925	5.408	5.092
6	1.110	1.783	1.330	2.220	1.935	10^3	2.738	5.034	2.958	5.476	5.158
7	1.159	1.880	1.379	2.318	2.028	10^4	3.471	6.500	3.691	6.941	6.623
8	1.201	1.964	1.422	2.403	2.110	10^5	4.204	7.966	4.424	8.407	8.089
9	1.239	2.038	1.459	2.477	2.182	10^6	4.937	9.432	5.157	9.874	9.555
10	1.272	2.105	1.493	2.544	2.246	∞	∞	∞	∞	∞	∞

The influence of the cross anisotropy on the distribution of vertical displacement for a uniformly loaded circular area is given in Fig. 8-21. The error in the deformation modulus calculated by measuring deformation values are dependent upon the deformational anisotropy (E_h/E_v-ratio of horizontal to vertical deformation modulus values) and POISSON's ratio values. The ratio in modulus of deformation for rocks in different directions usually lies in the region of 1 to 2.5 and as such the error in the modulus value in the direction of loading due to cross anisotropy may be of the order of 20 to 50%. GERRARD and WARDLE (1973a, b) have discussed the case of cross anisotropy under a variety of loading systems on the deformation of the surface and deeper lying points.

The results will be under-estimated when $E_h/E_v \gg G_h/G_v$ and over-estimated when $E_h/E_v \ll G_h/G_v$. These ratios effect the surface displacement much more than any variation in load distribution under the plate. The errors are lower for materials having high compressibility, but larger for incompressible materials (GERRARD, DAVIS and WARDLE, 1972). ZIENKIEWICZ and STAGG (1967) have discussed cross anisotropy and how to account for it in plate bearing tests.

TABLE 9

Influence Factor for Calculation of Displacement for Different Load Distribution Conditions under a Circular Area

(after HAST, 1943)

Load Distribution	Influence Factor, I		
	ρ_o	ρ_a	ρ_m
Case I Uniformly distributed	1.57	1.0	1.33
Case II Parabolically distributed; zero at the edge and maximum at the centre	2.09	0.89	1.42
Case III Parabolically distributed; zero in the centre and maximum at the edge	1.05	1.11	1.24
Case IV Rigid loading plate	1.23	1.23	1.23

ρ_o = displacement at the centre of the loaded area at zero depth.

ρ_a = displacement at the edge of the loaded area at zero depth.

ρ_m = mean displacement of the loaded surface at zero depth.

$$\rho = \frac{4(1-\nu^2)I}{E \cdot \pi^2 \cdot a}$$

The deformations at depths greater than the radius of the loading plate are not influenced by the anisotropy of the material or the uniformity of load.

The problem becomes more complicated with layered systems as very commonly occurs in actual practice. When the thickness of the various layers is small compared to the loading diameter, the modulus values evaluated are much less effected (SALAMON, 1968). When the thickness of the beds is not a fraction of the loaded diameter, a number of solutions have been proposed (STEINBRENNER, 1934; PALMER and BARBER, 1940; ODEMARK, 1949; VESIC, 1963; UESHITA and MEYERHOF, 1967). POULOS and DAVIS (1974) have described methods for single, double and multilayered systems. The equivalent or resultant modulus of a two layer system (homogeneous isotropic linear elastic layers) is dependent upon the elastic parameters E_1, ν_1 and E_2, ν_2. The influence factors for the vertical surface displacement at the centre of the circular area are shown in Fig. 8-22 for $\nu_1 = 0.2$, $\nu_2 = 0.4$ (BURMISTER, 1962);

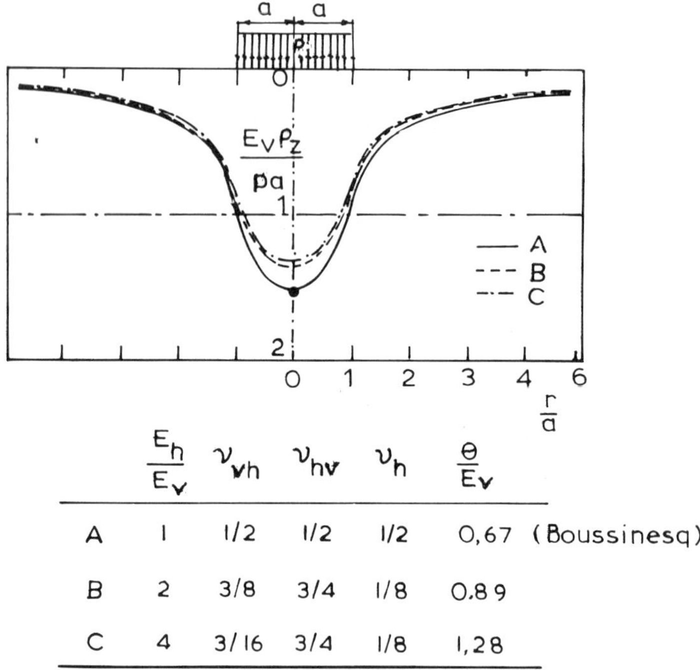

	$\frac{E_h}{E_v}$	ν_{vh}	ν_{hv}	ν_h	$\frac{\theta}{E_v}$	
A	1	1/2	1/2	1/2	0,67	(Boussinesq)
B	2	3/8	3/4	1/8	0.89	
C	4	3/16	3/4	1/8	1,28	

Fig. 8-21. Distributions of vertical displacement of surface (after KONING, 1960).

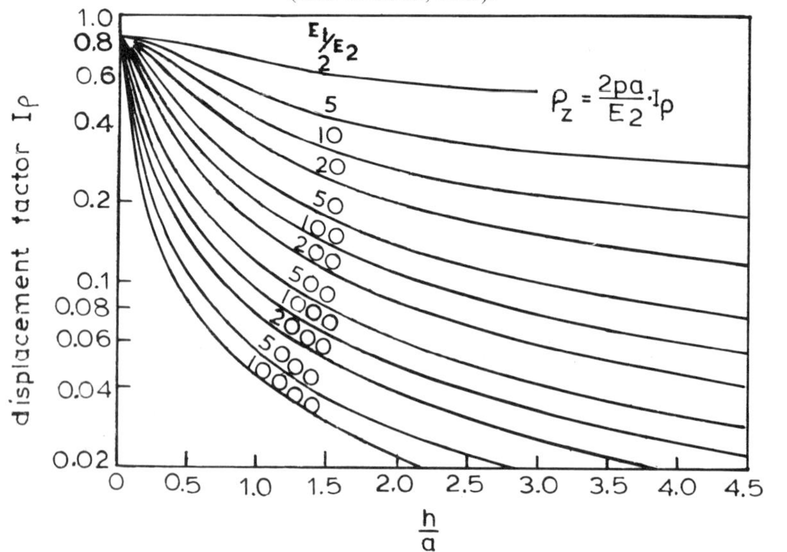

Fig. 8-22. BURMISTER layered system theory. Vertical displacement at centre of circle. $\upsilon_1 = 0.2$; $\upsilon_2 = 0.4$ (after BURMISTER, 1962).

in Fig. 8-23 for $\nu_1 = \nu_2 = 0.5$ (BURMISTER, 1945) and in Fig. 8-24 for $\nu_1 = \nu_2 = 0.35$ (THENN DE BARROS, 1966). An alternative interpretation of Fig. 8-23 has been given by UESHITA and MEYERHOF (1967) where the equivalent YOUNG's modulus E_e is plotted for different values of h/a and E_1/E_2 (Fig. 8-25).

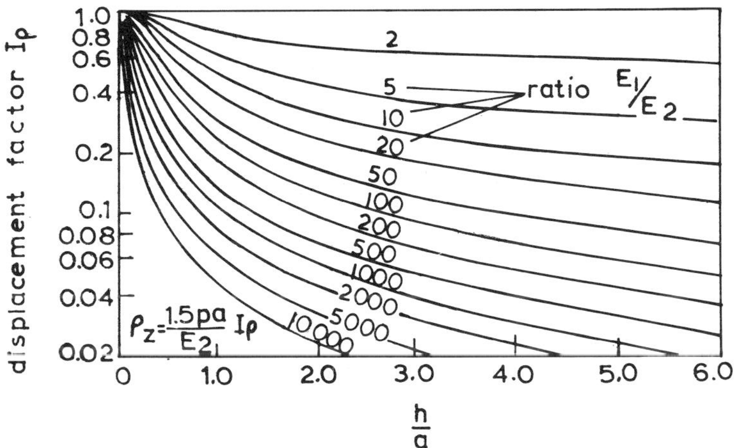

Fig. 8-23. BURMISTER layered system theory. Vertical displacement at centre of circle. $\upsilon_1 = \upsilon_2 = 0.5$ (after BURMISTER, 1945).

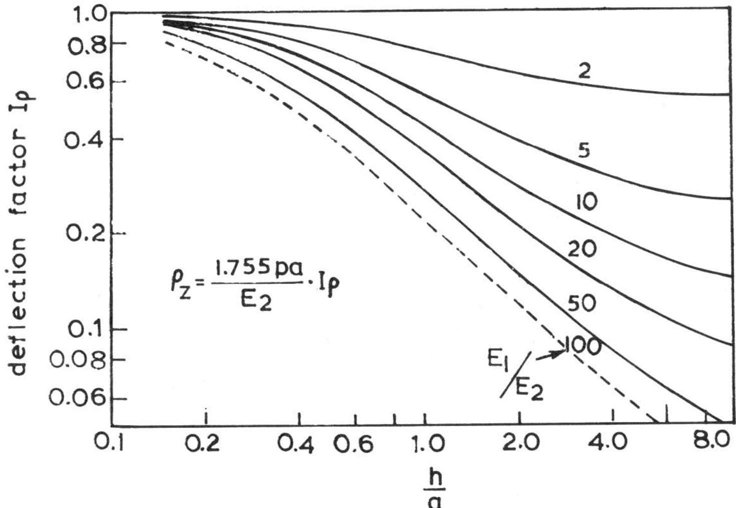

Fig. 8-24. Vertical displacement at centre of circle. $\upsilon_1 = \upsilon_2 = 0.35$ (after THENN DE BARROS, 1966).

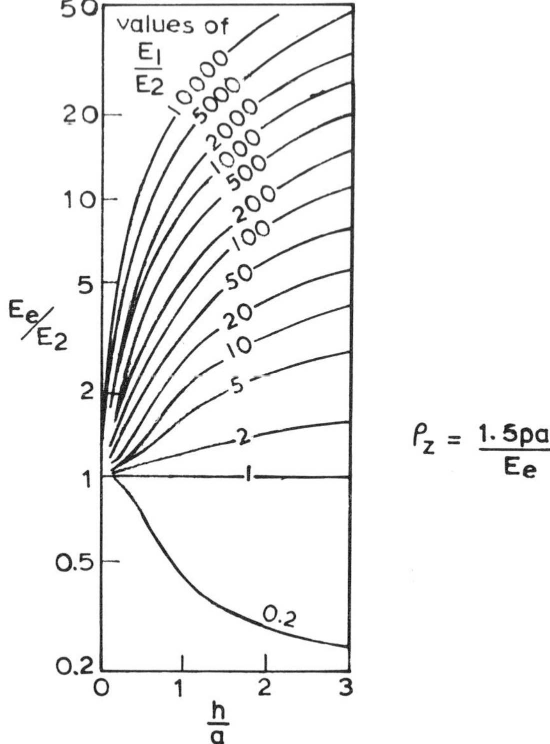

Fig. 8-25. Equivalent modulus E_e of two-layer system. $\upsilon_1 = \upsilon_2 = 0.5$ (after UESHITA and MEYERHOF, 1967).

This concept can be easily extended to more than a two layer system. For example, for a three layer system, the upper two layers are replaced by a single layer of thickness $(h_1 + h_2)$ having an equivalent modulus of

$$E_e = \left[\frac{h_1 \sqrt[3]{E_1} + h_2 \sqrt[3]{E_2}}{h_1 + h_2} \right]^3 \tag{8.15}$$

A three layer system is thus reduced to a two layer system and the modulus values may be appropriately corrected using the displacements from Figs. 8-22, 8-23 and 8-24. There seems to be no reason why this approach cannot be extended to more than three layers.

An approximate solution has also been proposed by RAPPOPORT (1970) applicable for thin layers when the error seems to be under 10%.

When the top layers of a highly compressible low rigidity material are underlain by a very rigid layer with a rough contact between the two layers such that no displacement occurs at the interface, the case of a finite layer is applicable (Fig. 8-26). The relationship and influence factor for vertical displacement are given in Tables 10 and 11. Detailed analyses are given by MILOVIC (1970) and TSYTOVICH et al (1970). TSYTOVICH et al (1970) believe that the interpretation of in situ test results using linearly deformed layer of a finite depth is more universal as compared to the semi-half space method and the results are in good agreement with seismo-acoustic measurements.

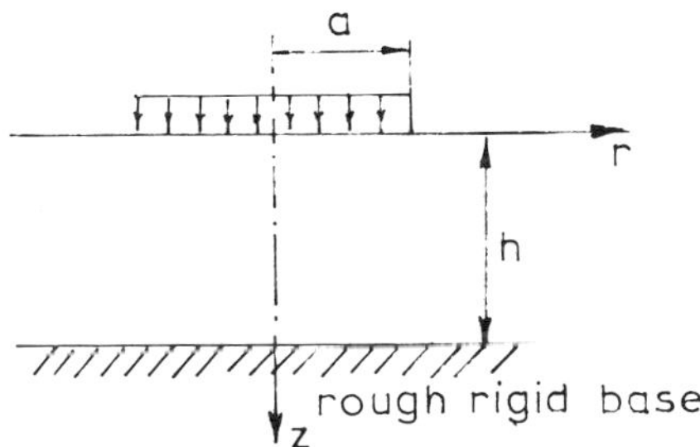

Fig. 8-26. Case of a finite layer.

TABLE 10

Influence Factors for Vertical Displacement at Edge of Circular Area

(after EGOROV, 1958)

h/a	I_e	
	Smooth interface (all ν)	Adhesive interface ($\nu = 0.3$ only)
0.2	0.005	0.04
0.5	0.12	0.10
1	0.23	0.20
2	0.38	0.34
3	0.45	0.42
5	0.52	0.50
7	0.56	0.54
10	0.58	0.57
∞	0.64	0.64

Edge displacement $\rho_e = \dfrac{2pa(1-\nu^2)I_e}{E}$

TABLE 11

Influence Factors I_ρ for Vertical Surface Displacement Within Circular Area

(after MILOVIC, 1970)

(Rough Rigid Base)

		r/a					
ν	*h/a*	0	0.2	0.4	0.6	0.8	1.0
0.15	1	0.464	0.458	0.441	0.408	0.348	0.208
	2	0.684	0.674	0.645	0.593	0.509	0.348
	4	0.811	0.800	0.768	0.710	0.619	0.463
	6	0.839	0.827	0.794	0.736	0.646	0.501
0.30	1	0.397	0.392	0.379	0.351	0.301	0.173
	2	0.613	0.604	0.578	0.531	0.456	0.305
	4	0.740	0.732	0.703	0.651	0.568	0.420
	6	0.770	0.762	0.733	0.681	0.597	0.458
0.45	1	0.278	0.276	0.267	0.250	0.213	0.109
	2	0.489	0.482	0.461	0.422	0.361	0.229
	4	0.612	0.608	0.585	0.541	0.472	0.340
	6	0.637	0.635	0.612	0.568	0.499	0.374

$$\rho_z = \frac{2paI_\rho}{E}$$

It shall be noted that in plate loading test, only one of the constants (either E or ν) can be evaluated and require the knowledge of the other. For example if the value of ν is not accurately known, the value of E so calculated will be very much unreliable. The displacement values at the surface are very much affected by changes in ν. The influence is considerably less for low values of ν in the range of 0 to 0.2. It increases rapidly as ν increases to 0.5.

8.5.1.2. Testing technique

The testing technique consists of loading an especially prepared surface, wall, roof or floor of any excavation (tunnels or open cuts) with the help of high capacity jacks. The shape of the area is usually circular or square and its linear dimensions are dependent upon variability of rock. In homogeneous rocks, the results from small diameter plates may be quite acceptable, but in rocks having high variability, loading area required may be extremely large and sometimes beyond the practical testing limits. In practice, the loading plate area usually does not exceed 1 m^2 (10.8 ft^2). GICOT (1948) and TALOBRE (1957, 1961) used 9–10 in (22.9–25.4 cm) diameter loading plates. U.S. Bureau of Reclamation (1965) used 24 in. (61.0 cm) diameter plates. Shannon and Wilson, Inc. (1964) used 24 in. (61.0 cm) and 34 in. (86.4 cm) diameter plates whilst Rocha, Serafim and Da Silveira (1955) used 1 m^2 (10.8 ft^2). Serafim and Guerreiro (1966) used 2 m^2 (21.6 ft^2) loading surfaces in the testing of foundation rocks. The trend is to load larger surface areas 6–10 m^2 (65–108 ft^2) and the load requirement increasing to 2000–5000 tonnef (1969–4921 tonf). The best method is to use loading plates of different diameters and calculate the deformation modulus for each plate separately. These results can then be extrapolated to arrive at an acceptable value. Even when it is not possible to extrapolate the results, it gives at least an indication of the uncertainty involved and the variability of the material. In other cases, the volume of rock affected must be at least 150 times the average size of a unit block.

Depending upon the joint density and homogeneity, size effect may be absent in certain rocks. Jiminez-Salas and Uriel (1964) found no change in the shearing resistance of lignites for the plate dimensions from 0.5 m × 0.5 m to 4.0 m × 4.0 m (1.6 ft × 1.6 ft to 13.1 ft × 13.1 ft). The results of tests by Bollo, Navalon and Sanz-Saracho (1966) on granite and Silurian schist using flat jacks are given in Table 12. The influence of size here is absent, but it looks that relaxation of surface has played a role in this case. The increase in modulus with size is due to the stress entering deeper into a more firmer less relaxed rock. Talobre, Fox and Meyer's (1964) tests on lime conglomerates using pressure chamber method with tunnel diameter 1.7 m (5.18 ft) and 5.0 m (15.24 ft) length gave modulus values quite close to those obtained from plate loading tests with 254 mm (10 in) diameter plate. Grishin et al (1961) and Evdokimov (1964) concluded that plate size has definite influence on the modulus values for limestone rocks. Flat jack method (Rocha, 1969) used in limestone rocks gave values which decreased with increase in size as it was varied from 1 m^2 to 3 m^2 (10.8 ft^2 to 32.3 ft^2).

TABLE 12

Influence of Plate Size on Modulus of Rocks (MPa)

(after Bollo, Navalon and Sanz-Saracho, 1966)

Rock Type	Rigid Plate 280 mm dia.		Flexible Plate 1 m^2		Flexible Plate 4 m^2	
	E_d	E_e	E_d	E_e	E_d	E_e
Granite	7420	12650	—	—	13720	19280
	—	—	12490	30790	17630	35950
	—	—	18620	32730	19660	37490
Silurian	2620	5670	—	—	3610	7950
Schist	2780	3940	—	—	3260	6800
	—	—	8760	18650	9240	20120

Displacements measured away from the plate, whilst necessarily being smaller, are more representative of the rock mass. It may be valuable to conduct a larger number of tests with relatively smaller size plate, than a few tests with a larger plate and statistically analyse the results. As the size of the loading plate increases it can no longer be assumed that plate is rigid and hence stress distribution under the plate may not be exactly known. Stress distribution is a function of the deformation modulus of rock (BOROWICKA, 1936), thickness and diameter of the loading plate etc. For high deformation modulus rock, stiffer and heavier plates are required. Stiffness of the loading plate should be at least an order of magnitude higher than that of the rock.

Site preparation for testing is important. When tests are to be conducted in underground galleries, the site should be at least 10 m (33 ft) away from the tunnel entrance. All loose rock that might be produced as a result of the blasting operations during the drivage of the tunnel must be removed and the testing surface made plane by chipping until surface asperities are within 5 cm (2 in). The depth to which the damaged rock is to be removed depends upon the nature of the rock, method of drivage and depth from the surface. It may vary from several centimetres to metres (inches to yards). All work near the testing site must follow hand mining and if blasting is required, it must be limited to at least 2 metres (6.6 ft) from the actual testing site.

The area surrounding the loaded surface must be a plane of sufficiently large dimensions to justify the assumptions made in the BOUSSINESQ theory, i.e. the load is acting on the plane boundary of an infinite sphere. This area should have a flat surface at least 3–4 times the radius of the loaded surface.

The loading pad may be of circular or rectangular shape. It usually consists of mortar pads with steel sheets bonded to them. A very rigid toughened steel plate 2.5 to 5.0 cm (1 to 2 in) thick usually suffices and is placed over it forming the loading surface.

In many cases, the rock surface may be prepared by concreting it to obtain a smooth surface and the loading jack is directly placed on it. The jack pistons are sufficiently stiff, but the area loaded by them is very limited.

The loading units usually consist of large capacity jacks 100–300 tonnef (98–295 tonf) capacity (ROCHA et al, 1955) depending upon the area to be loaded and the maximum pressure to be generated on the surface. Load as high as 720 tonnef (709 tonf) has been applied over an area of 1.2 m^2 (12.9 ft^2) (STUCKY, 1953). The maximum pressure is determined by the ultimate stress to which the rock is to be subjected. It is usually limited to 2.8–3.4 MPa (400–500 lbf/in^2).

The maximum number of jacks that can be used depends upon the size of the loading area and the size of the jack head, but it should be kept in view that sufficient deformation is produced which can be measured with reasonable accuracy.

The appropriate rate of loading has not been clearly established as yet. The loading rate should be such that it does not produce excessive pore water pressure during loading of rock.

Displacements to be measured range from 0.25–10 mm (0.01 to 0.4 in) and can be measured with the use of dial gauges reading to 0.001 mm (0.0001 in) and having a range of 10 mm (0.5 in). L.V.D.T.s may be used for remote recording of displacement at several points on the loading plate or at several other points away from the loading plate on the surface or in rock at different depths. The reference point for the deformation measuring devices should be so located that it is not effected by loading the rock surface. The anchor points in boreholes should be located in competent rock. Measuring displacement against the body of the jacks or against some frame anchored at points 3–4 times the plate diameter may not meet this requirement. Fig. 8-27 shows the deformations from a rigid plate (punch) and uniform pressure tests. The relative deformation of points away from the plate with respect to the centre point drops off more rapidly for a punch test than for a uniform pressure test. Even in a punch test, the displacement at 3 times the diameter equals to about 11% of penetration of the punch. A good solution of the problem of fixing reference points is to measure deformation at each point in relation to fixed points of a frame securely attached to the surfaces perpendicular to the loaded surface.

CARLSON joint meters (CLARK, 1965) or other extensometers (DODDS, 1974) can be used for measurement of displacements at various depths along the central line or at other points.

Besides measuring the deformation of the plate itself and at several points below it, it is also advisable to measure deformation away from the plate. Deformation measured away from the loading area gives a better measure of the nonlinear effects and is not influenced by the local circumstances.

The displacements away from the loaded area and at depth are considerably smaller and hence require greater care. These measurements provide a check against any gross errors of measurements. These are less sensitive to variations in pressure distribution than displacements directly under within the loaded area. These also allow a better assessment of the properties at depth as the displacements outside the loaded area are influenced to a much greater relative extent by the behaviour of rock well below the immediate contact with the loading plate. For a rigid circular loaded plate the displacements under the plate drop off very rapidly, the displacements away continue very much the same to much greater depths. In Fig. 8-28 ρ_0 represents the displacement along the axis below the plate and ρ_{3a} along a line normal to the axis at a distance of $3a$ from the plate axis. The dashed lines show vertical stress distribution below the plate along its vertical axis. It is clear that displacements at the surface are affected to a much larger degree by the surface rock (80% of displacement of plate is due to rock within $z/a < 4$). For a point outside the plate the depth contributing 80% of displacement is of the order of $20a$. These also permit to check the validity of the elastic assumption and permit calculation of the modulus values at depths.

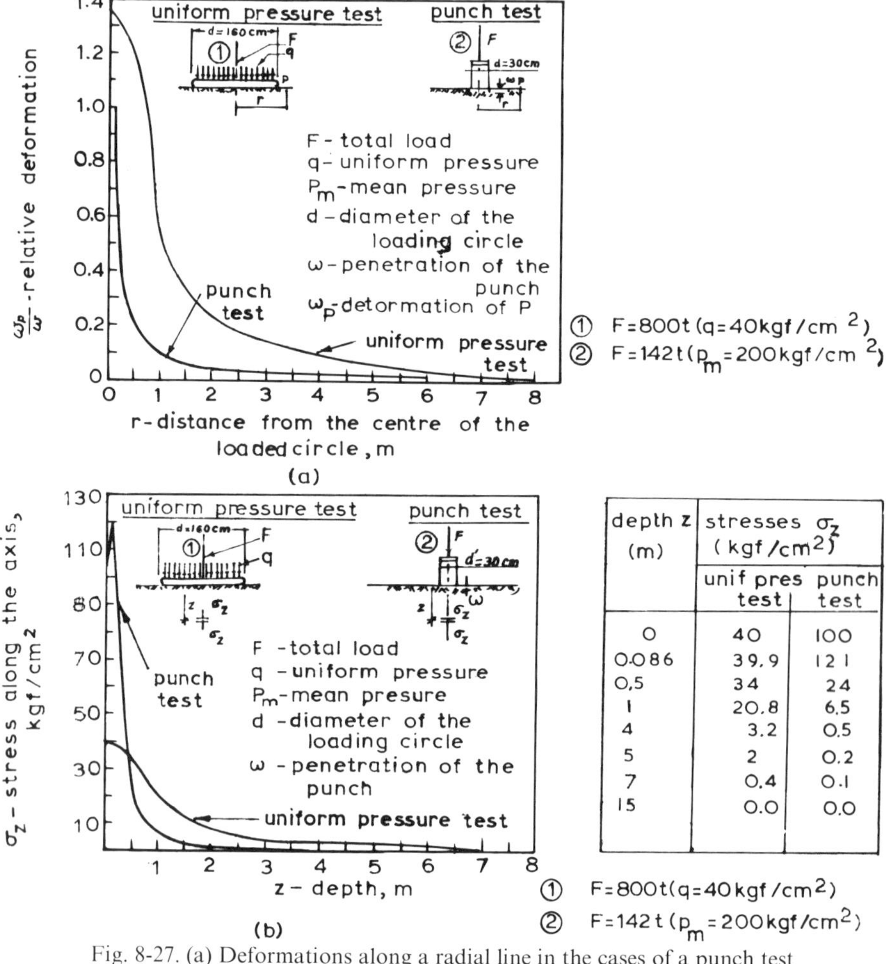

depth z (m)	stresses σ_z (kgf/cm²)	
	unif pres test	punch test
0	40	100
0.086	39.9	121
0.5	34	24
1	20.8	6.5
4	3.2	0.5
5	2	0.2
7	0.4	0.1
15	0.0	0.0

Fig. 8-27. (a) Deformations along a radial line in the cases of a punch test and of a uniform pressure test.
(b) Stresses along the axis of the loading surfaces acting on planes perpendicular to it, in the case of a punch test and of a uniform pressure test (after SERAFIM and GUERREIRO, 1966).

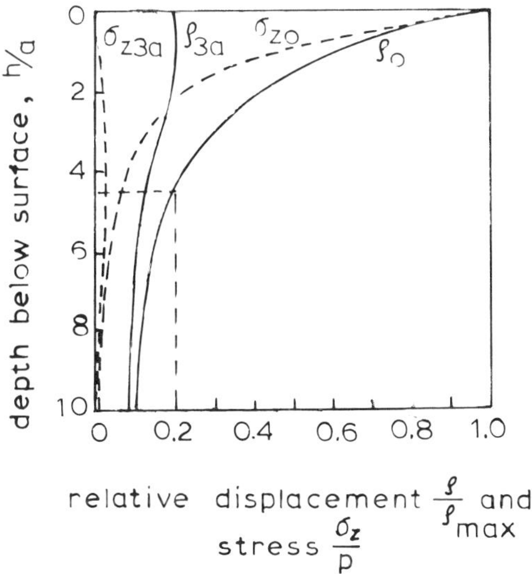

Fig. 8-28. Stress and normal displacement for a circular area, $\upsilon = 0.2$ (after STAGG, 1968).

It is a common practice to load specimens in plate bearing tests by stages. In general, the first cycle consists of about 25% of the maximum load; the rock is then unloaded and load increased in the second cycle to 50%; in the third cycle to 75% and in the fourth cycle to the maximum desired value or to failure. During each successive cycle, many times the load is not brought down to zero. However, it is advisable to bring the load to zero at second and third cycles at least as secant values at different loads both on unloading and loading phases of each cycle may be required to be determined.

In cyclic loading, load is usually kept constant for 30–60 minutes at the top of each loading cycle even when no creep tests are intended. It gives an idea if the possible maximum load has been achieved or exceeded. At least three cycles are repeated at the maximum load.

The following test procedure is recommended for plate bearing test.

(1) Depending on the rock type, the maximum load applied should be reached in 4 to 5 cycles of at least 5–10 minutes duration.

(2) The maximum load (on loading) and zero load (on unloading) in each cycle should be maintained until the deformation rate at the centre is less than 0.025 mm/hr (0.001 in/hr) taking readings at intervals of 1 minute to 15 minutes depending upon deformation rate (COATES and GYENGE, 1965).

(3) Creep of the rock at maximum pressure should be measured by keeping the pressure constant for 2–3 days.

(4) In wet rocks, pore pressure should be continuously monitored and rate of loading so adjusted that excessive pore pressure build up is avoided.

The number of tests to be conducted is dependent upon the geological structure, its variation from place to place and its heterogeneity. The larger the number, the more reliable shall be the results. In general, 3–4 tests are performed at each site and this number should be considered adequate.

The loading equipment should have the following features:

(1) It should be light and easily installed. The weight of the individual components should not exceed 80 kgf (175 lbf) so that these can be easily handled.

(2) The equipment should be capable of giving enough deformation to the rock under test. Highly deformable (weathered or highly jointed) rocks require testing equipment capable of large deformation 50–75 mm (2–3 in).

(3) The overall height of the equipment should be adjustable to operate under varying conditions.

(4) The load applying capability of the equipment should be higher than the maximum load to which the rock is likely to be subjected. Provisions should be available for maintaining constant pressures over long time (several hours or even days).

A number of loading systems have been used. A typical loading arrangement consists of a single or double jack (Wallace, Slebir and Anderson, 1969). Though rigid plates have been used in a number of field tests their use is limited only to cases where the loading area is small (maximum up to 30–40 cm (12–16 in) diameter). For large loading areas Freyssi flat jacks are commonly employed. These permit uniform loading and do not require a high rigidity loading plate which is likely to be damaged due to non-uniform deformation of the loaded area particularly when it is large. There is no limit to the number of flat jacks that can be used. Their number is dependent upon the size of the area to be loaded. The shape may be circular, rectangular or triangular, but for modulus studies, circular sections are most common. The equipment used by U.S. Bureau of Reclamation which has been proved very successful in a large number of tests is described below.

A general arrangement of a uniaxial jacking test equipment is given in Fig. 8-29. The system consists of flat jacks and four aluminium columns (25 cm (10 in) diameter) provided with adjustable screws to accommodate for varying height of the excavation. The circular flat jack used here essentially consists of two metal sheets brazed together at the edges and capable of holding a fluid under pressure transmitting the pressure to a test surface. The flat jack has rounded edges to permit greater extension as the test surface deforms. The external diameter of the flat jack is 87 cm (34.3 in) with a 5.1 cm (2 in) hole in the centre and its expansion is limited to 2.54 cm (1 in). For competent rocks, one flat jack on each end of the aluminium columns is sufficient to maintain the test load. In highly jointed and deformable rocks, more than one flat jack may be used to obtain greater deformation.

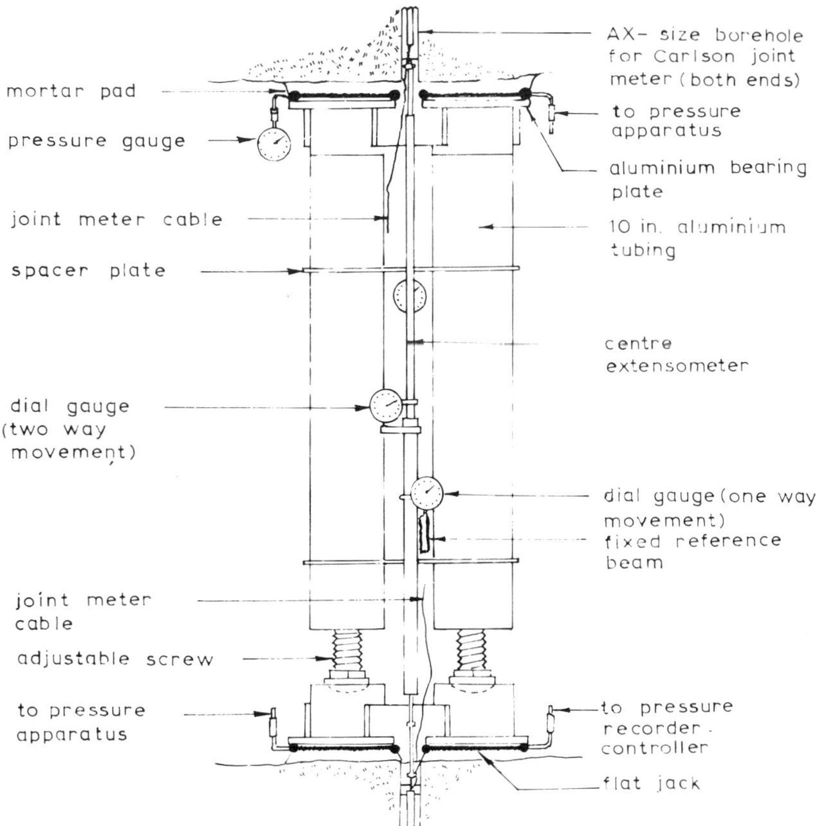

Fig. 8-29. U.S.B.R. plate loading system (after WALLACE et al, 1969).

Load to the rock surface is transmitted through a 91 cm (36 in) diameter concrete pad about 12.7 cm (5 in) thick to compensate for unevenness of rock surface. The jacks are calibrated individually and capable of applying maximum pressure of 6.9 MPa (1000 lbf/in^2) (70 kgf/cm^2) to the rock surface though the pressure in the system is about 10.3 MPa (1500 lbf/in^2) (105 kgf/cm^2). Pressure is maintained using an electrically operated pump and recorded on a recorder-controller. As pressure drops due to movement of rock, the controller activates the electric pump.

Rock deformation is measured in the centre at different depths using a 7 point retrievable extensometer (ROUSE and WALLACE, 1966) installed in a 6.06 m (20 ft) NX borehole. Seven L.V.D.T.s sense the deformation between the collar and the seven expandable anchors fastened within the hole (Fig. 8-48). Measurements are also made of the tunnel diameter midway between the jack using a tunnel diameter gauge. The tunnel diameter gauge consists of invar rod suspended on one side from the invert and the other side from the arch with a two way dial gauge between them to measure the relative movement.

The total shipping weight of the equipment is about 1000 lbf (454 kgf). Two components of the equipment weigh 70 lbf (32 kgf) each but the remaining components each weigh less than 40 lbf (18 kgf).

Equipment can be used in any direction in a tunnel or in a trench.

A very similar test arrangement was used by SHANNON and WILSON, Inc. (1964) at Dworshak dam site while the equipment described above was used at Morrow Point dam site. The characteristic features of the two tests are compared in Table 13. The comparison shows a number of marked differences in test procedure.

8.5.1.3. Testing in trenches or open pits

When tests are required to be conducted in trenches with loading axis horizontal, the loading jacks can be suspended from beam placed across the trench. The two walls of the trench provide the reaction necessary as in a tunnel.

When loading direction is vertical, the reaction has got to be supplied by specially prepared structure. The usual method for applying loads is by using a loading frame anchored to the ground with grouted anchors 25 mm (1 in) diameter 1 to 2 m (3.3 to 6.6 ft) in length (Fig. 8-30). Extra weight can be brought on using sand filled gunny bags. For greater loads, specially designed ground anchors of longer length can be used (Section 8.5.2.).

Deformation is measured against a datum bar suitably anchored into the wall of the trench or a frame supported at points well away from the loaded area.

Another easy method of obtaining the reaction is using the cable loading technique described in Section 8.5.2. The method is simpler and less costly in its preparation.

8.5.1.4. Interpretation of plate bearing test

The basic diagram obtained in a plate loading test for any measurement point is given in Fig. 8-31. The curve OA_1 corresponds to the loading and A_1E_1 corresponds to the unloading of the specimen in the first cycle. These are repeated in the subsequent cycles. If the modulus values are calculated taking pressure P_1 or P_2 and displacement D_1 or D_2, this is called deformation modulus (E_d). If only the deformation values ($D_1 E_1$) and ($D_2 E_2$) are taken into account, the modulus value so obtained is called the modulus of elasticity (E_e). For displacements at extremely small load (corresponding to tangent OC in Fig. 8-31) the modulus is called as the initial modulus E_o (modulus at zero load).

TABLE 13

Comparison of Two Recent Field-Jacking Tests

(after U.S. Bureau of Reclamation, 1965)

	Morrow Point Dam	Dworshak Dam
Total load	200 tons	425 tons
Size of shoe	24 in diameter, round	34 in diameter
Unit load	995 lbf/in^2	1000 lbf/in^2
Equipment	hydraulic jacks	34 in diameter Freyssi flat jacks on grout and adjustable struts
Direction	horizontal and vertical	horizontal and vertical
Rock surface	below blast damage	below blast damage. Less than 2 in deviation of surface
Gauge	rock deformation, Carlson	Carlson
Length	15 ft 3 in	15 to 18 ft
Grout	portland cement, 24 hr set	—
Surface deformation	micrometer extensometer between steel anchors	dial gauge extensometer 0.0001 in reading
Temperature control	none given	0.5 F, corrections for more than 1 F
Pressure control	pressure recorder and regulator	preset-electronic control
Testing	200 lbf/in^2, 6 days 0 lbf/in^2, 1 day 400 lbf/in^2, 6 days 0 lbf/in^2, 1 day 600 lbf/in^2, 6 days 0 lbf/in^2, 1 day	5 cycles, 5 to 10 min, 250 750, 1000 lbf/in^2
Strain rate	none given	less than 0.001 in/hr
Creep	not measured	checked with 2 to 3 day max load

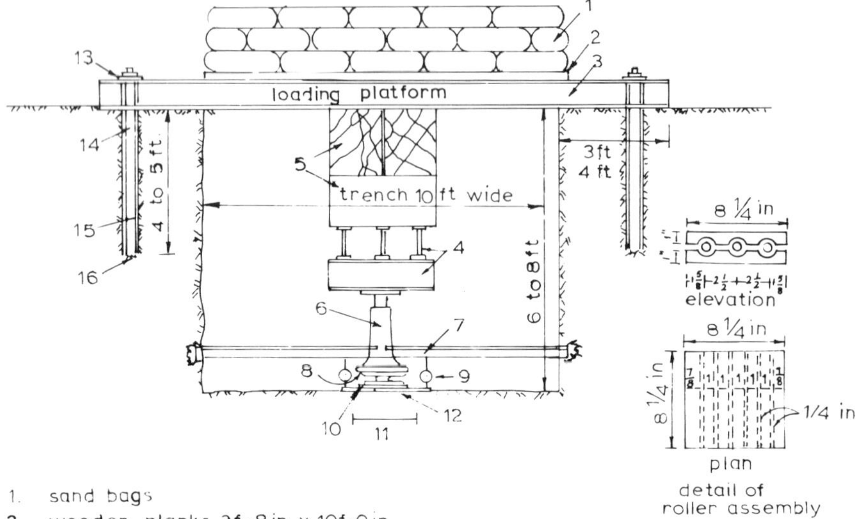

1. sand bags
2. wooden planks 3f-8in x 10f-0in
3. R.S.J. 225 mm x 110 mm @ 3.12 kgs/m
4. R.S.J's 225 mm x 110 mm @ 3.12 kgs/m
5. timber packing
6. hydraulic jack 200 tons capacity
   pressure gauge separate pump arrangement
7. datum bar suitably anchored
8. roller assembly
9. dial gauges
10. 1 in. thick steel plate 12 in. dia.
11. surface ground level
12. 1/4 in. thick cement plaster to make the surface even.
13. 3/4 in steel plate 3ft 8 in wide 5ft long
14. drill hole cement grouted.
15. pinning rod 1 in. dia.
16. split end and wedge arrangement for anchoring

Fig. 8-30. In situ vertical load tests in trenches.

The ratio OE/OD, i.e. ratio between plastic deformation and total deformation is considered by some American geologists as a characteristic of the rock mass. French geologists, however, consider ED/OE to be significant. The Austrians and Germans consider the ratio E_e/E_d to be significant. Basically, there is no difference between these, except that ED/OE gives a direct relationship easily interpretable as the ratio between the elastic strain to total strain.

While reporting plate loading tests, the various values such as E_o, E_d, E_e, (ED/OE) should be reported along with the test dimensions and pressures at which test is carried out. The values should be reported for each cycle. For the same cycle it is often necessary to determine the deformation modulus values at different loads particularly when the behaviour of rock is nonlinear. The concept of equivalent modulus could be used in such cases. The equivalent modulus E_p for a given pressure p (Fig. 8-32) in this case means the modulus of elasticity of an elastic medium which under the same test conditions represents the same maximum displacement δ. The ratio $E_p/E_o = \tan \alpha / \tan \alpha_o$ is an index of nonlinearity up to the pressure p.

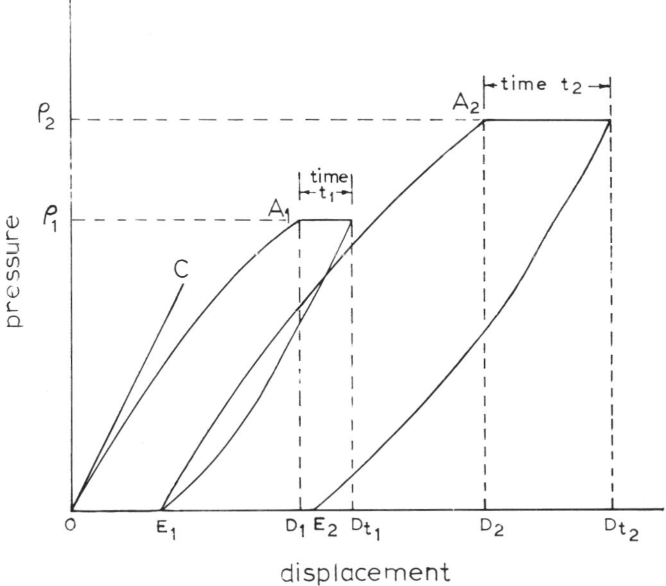

Fig. 8-31. Pressure displacement curve in a plate bearing test.

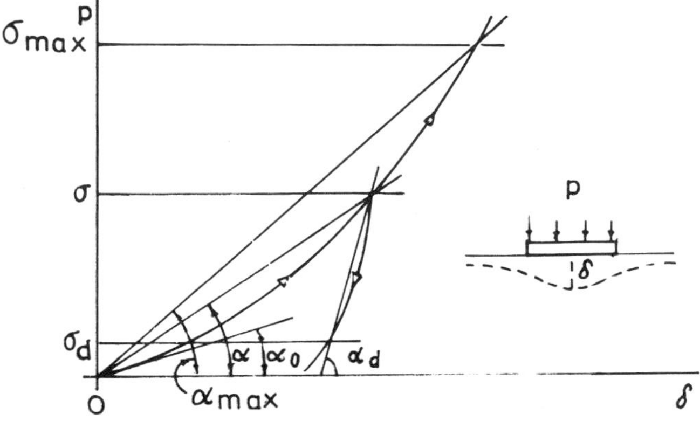

Fig. 8-32. Equivalent modulus of elasticity.

A number of difficulties arises in the estimation of the modulus values from plate loading tests. These difficulties are partially related to the conditions in which the tests are conducted and partially due to the geological structure. Some important points are discussed below.

1. In most of the cases, plate bearing tests are conducted in galleries or in trenches. The results are evaluated assuming mathematical theory for a load applied to the boundary of a semi-infinite solid. As there is little physical resemblance, a potential source of error exists. Unless open cracks in the tunnel arch, invert, or walls allow free movement, deformation of the loaded area is resisted. When a circular gallery is loaded with a uniform load extending an angle 2β at its centre, the radial displacements can be given by (MUSKHELISHVILI, 1953, JAEGER and COOK, 1969)

$$\rho_r = \frac{ap}{\pi E}\left(1+\nu\right)\left[2\beta + \sum_{n=1}^{\infty} \frac{1}{n}\left(\frac{3-4\nu}{2n-1} + \frac{1}{2n+1}\right) \cos 2n\beta \sin 2n\theta\right] \quad (8.16)$$

where ρ_r = radial displacement at the surface of the gallery and
a = radius of the gallery.

These displacements differ from that of a half-space loaded at its surface (compare with Eqs. 8.11 and 8.14). BOLLO et al (1966) have compared the case for a rectangular plate of 1.3 m × 0.7 m (4.3 ft × 2.3 ft) placed in a circular opening of 2.8 m (8.53 ft) in diameter. The longer dimension was placed parallel to the gallery axis. The modulus of deformation calculated using half-space gave value of $E = 73.1$ tonne/cm^2 (7.31 GPa) and for the infinite long loading in a circular opening gave a value of 82.5 tonne/cm^2 (8.25 GPa) an error of about 9% ($\nu = 0.3$, $\beta = 14°$). The errors decrease with decrease in the values of ν and β.

The error can be further reduced by having a flat surface around the loaded area at least equal to the loaded area.

The assumption of linear elastic isotropic behaviour of rock is far from true. Any type of nonlinear behaviour can be obtained either due to anisotropy or due to non-linearity of the material. Hence, the deformation curve does not have any unique solution.

2. In many cases, creep is measured using plate loading tests when load is kept constant from several minutes to days. This information is invariably limited to measurement of primary creep which is dependent upon stress field, temperature, moisture, etc. Information on creep based upon surface measurements could be useful only if it is at least correlated with the stress levels to be expected in prototype. To obtain sufficient creep information, load should be kept constant for a period of 3–30 days (depending upon rock type) and

deformation recorded for various time intervals. It has been found that the duration of development of deformation for certain rocks may be as long as 400 hours and creep deformation may be 100–200 percent of instantaneous deformation (TSYTOVICH et al, 1970).

The results of creep tests are quite often represented in the form of coefficients

$$\left(\frac{E_o - E_\infty}{E_o} \times 100\right) \text{ and } \left(\frac{E_\infty}{E_o} \times 100\right)$$

Table 14 shows the results for two different types of rocks conducted by LNEC (Portugal). The results show a considerable reduction in value of modulus of elasticity which can adversely effect the behaviour of a structure.

As already indicated, the deformation of the surface is greatly influenced by the properties of the material close to the loaded surface. The upper surface, in many cases, may be damaged due to the effects of excavation and surface preparation. Engineering judgement is essential to temper the results. Fig. 8-33 shows the type of curve commonly obtained with a damaged rock. The concavity towards the stress axis is due to compaction of fractured rock. This concavity is not present in the second and subsequent cycles which indicates that damage to the rock is only local and surficial. The high stresses near the loading surface compact the rock and its influence greatly decreases in the subsequent cycles. As the modulus changes, the difference between third and fourth cycle is not very much. The values determined from these cycles give the modulus value for the undamaged rock. However, not in all cases is this phenomenon transient. Fig. 8-34 shows high inelastic strain that persists in all the cycles indicating deeper lying less rigid layer. On the other hand Fig. 8-35 shows no elasticity even though it has high modulus value. The concave portion is only in the beginning and is the result of damage where a few large widely spaced cracks have been opened due to blasting.

Calculation of modulus values separately for different load values from surface deflections permits to obtain information about the moduli of the different layers. The values obtained at initial loads are very much due to the compression of the upper layers immediately beneath the plate and the contribution by the lower layers increases as the load increases. The overall contribution to surface displacement is limited to depth of about twice the plate diameter. Testing using different diameter plates can help to gain an insight into the modulus of deeper lying layers without measurement of displacements at depths.

3. High shear stresses develop beneath the loaded surface which under cyclic and sustained loading may result in high deformation of the rock due to shear strain. This will tend to lower the computed modulus. The shear strains,

TABLE 14
Influence of Time on Deformability
(after SILVEIRA, 1966 b)

Rock type	Site	Direction of the forces	E_o (kgf/cm^2)			$\frac{E_o - E_\infty}{E_o} \times 100$			$\frac{E_\infty}{E_o} \times 100$			Number of tests
			Min.	Max.	Average	Min.	Max.	Average	Min.	Max.	Average	
Schist	Alcantara	⊥	5,000	123,000	34,000	2	33	13	67	98	87	16
		‖	11,000	227,000	72,000	5	45	15	55	95	85	13
	Caneiro	⊥	6,000	153,000	47,000	8	55	32	45	92	68	10
		‖	106,000	117,000	112,000	22	23	22.5	77	78	77.5	2
	Dao	⊥	27,000	71,000	47,000	32	47	40	53	68	60	10
		‖	398,000	462,000	430,000	3	6	5	94	97	95	2
Schist with layers of graywacke	Aguieira	⊥	11,000	35,000	17,000	20	40	29	60	80	71	5
		‖	33,000	254,000	89,000	20	52	33	38	80	67	5
Graywacke	Aguieira	⊥	20,000	29,000	25,000	12	33	20	67	88	80	4
		‖	39,000	128,000	81,000	15	45	29	55	85	71	6
	Caneiro	⊥	35,000	69,000	52,000	40	55	47	45	60	53	2
Phyllite	Aguieira	⊥	8,000	9,000	8,500	30	32	31	68	70	69	2
		‖	23,000	24,000	23,500	13	18	16	82	87	84	2
Shale	Bemposta	–	8,000	141,000	33,000	6	42	15	58	94	85	22
Granite	Alvarenga	–	12,000	584,000	238,000	6	60	30	40	94	70	18
	Vilarinho	–	6,000	137,000	47,000	5	56	19	44	95	81	22
	Alto Lindoso	–	69,000	670,000	280,000	5	38	20	62	95	80	4
Conglomerate	Cachi	–	19,000	56,000	31,000	7	15	11	85	93	89	7

⊥ – normal to schistosity
‖ – parallel to schistosity
E_o – instantaneous modulus of deformability in the sites of creep tests
E_∞ – modulus of deformability assuming total deformation (initial + creep)

however, are important in closely jointed or altered rock and where the plate size is small. The plate jacking test also suffers from the size effect discussed earlier.

4. The distribution of stress in a jointed rock mass is very much dependent upon joint orientation (KRSMANOVIC and MILIC, 1964; MALINA, 1969). When the joint set is parallel to the direction of loading, the diffusion of stresses does not take place resulting in the concentration of stresses under the plate axis. This will give higher deformation and result in lower calculated modulus values. On the other hand when a joint set is inclined to the axis of loading, the stress bulb may split into two giving very low deformation along the load axis resulting in high calculated modulus values. In both these cases, it is better to use mean deformation of the full plate in the calculation of moduli.

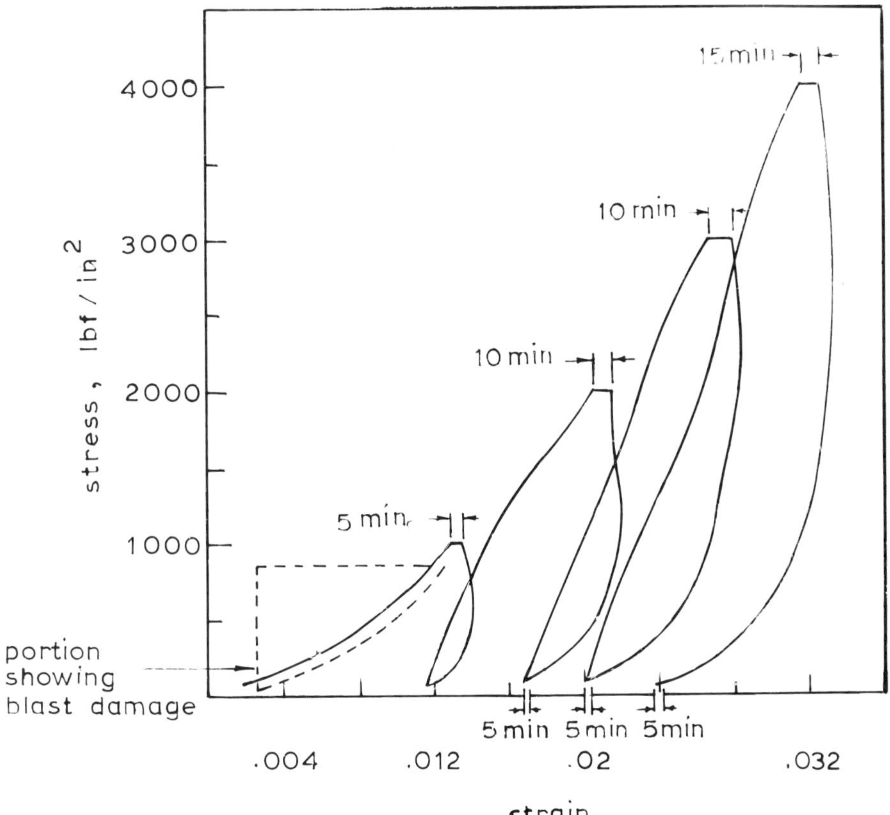

Fig. 8-33. Typical curve showing blast damage (after DODDS, 1974).

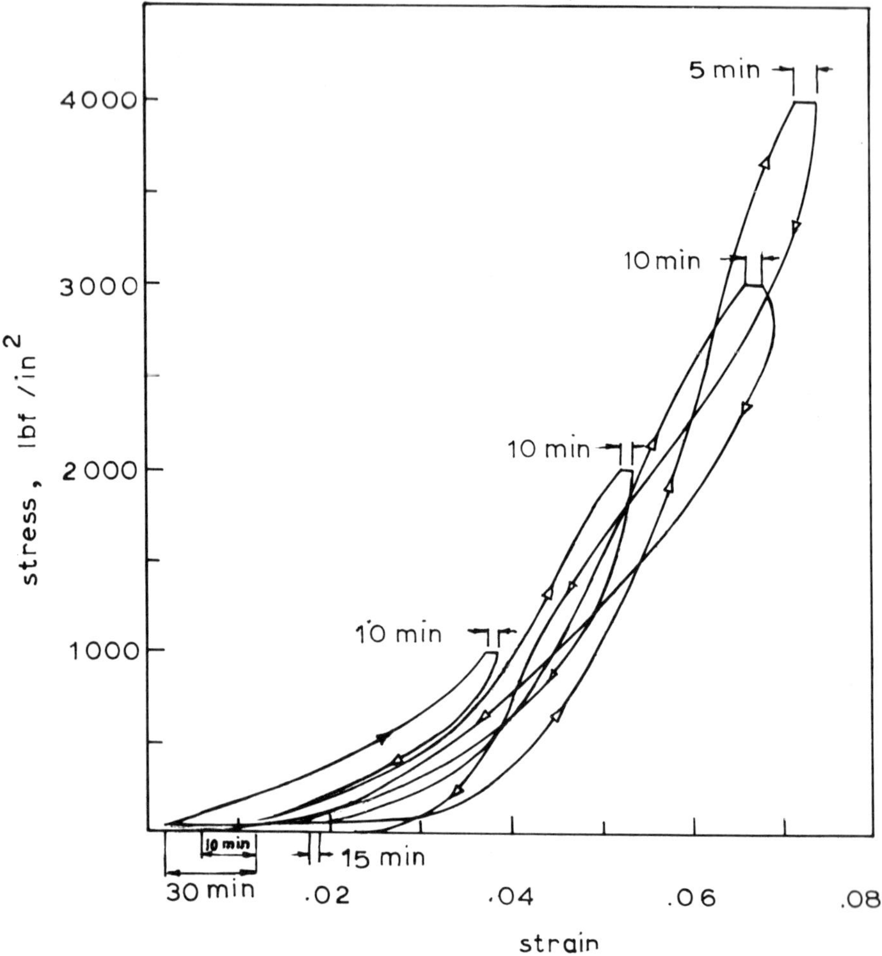

Fig. 8-34. Typical curve showing deep layer of stress relieved rock (after DODDS, 1974).

A comparison of the surface settlement curve with the theoretical curve sometimes indicates that the magnitude of the settlement decreases rapidly with increase in distance from the edge of the plate than that predicted by elastic theory of homogeneous isotropic medium. Such an indication points clearly to the different state of stress distribution. It could either be due to presence of structural discontinuities or due to the large variations in the modulus value of the top layer compared to the bottom layer (TSYTOVICH et al, 1970).

It is very often found that the modulus values obtained from the loading of a tunnel in two opposite directions (roof and floor) are different, though the rock may seem to be the same. When these different results are obtained, it is important to examine the cause of the anomaly. In many cases it may be due to the direction of loading of the foliations or due to partial decompression. Fig. 8-36 represents the results of tests at a number of dam sites on granites, schists and quartzites. It is seen that dispersion of the values of E_1 and E_r is comparatively symmetric while marked asymmetry is present in the values of E_t and E_b probably due to unequal decompression in roof and floor.

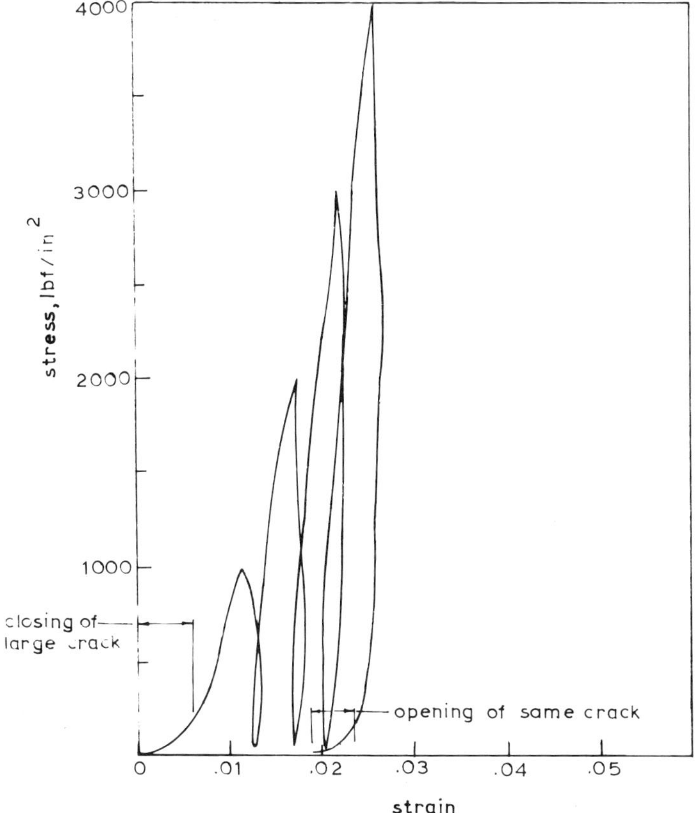

Fig. 8-35. Typical curve showing blast damage (after DODDS, 1974).

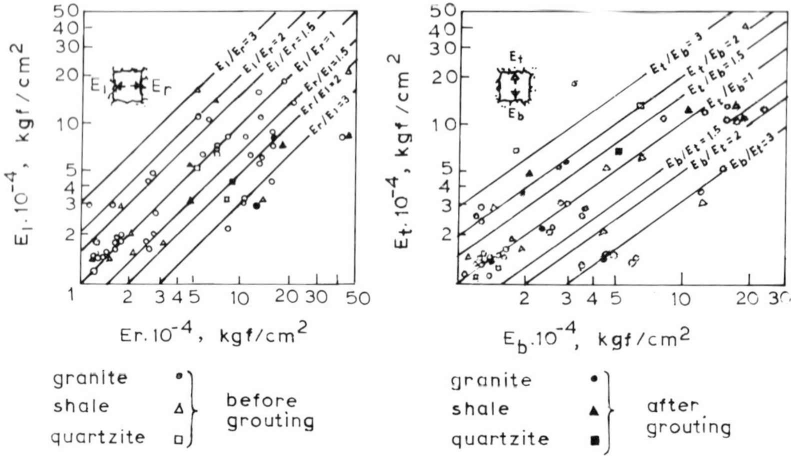

Fig. 8-36. Heterogeneity of foundation rocks (after SILVEIRA, 1966a).

8.5.2. Modifications of Plate Bearing Test

Many variations of the plate bearing test have been used by several investigators. Some more common ones are described here.

8.5.2.1. Compression in narrow slits

The method was first used in Yugoslavia (KUJUNDZIC, 1963) and has been subsequently developed in Portugal (ROCHA, 1969).

The method consists of placing a hydraulic flat jack (FREYSSINET jack) of 2 m diameter (or any other size) in a slit and the empty space is filled with concrete (Fig. 8-37). The slit may be made in any direction, vertical, horizontal or in any other direction. The thickness of the slit should be as small as possible to decrease the thickness of the concrete layer and to avoid its effect.

The deformation is measured by measuring the volumetric change of the pressurising flat jack. Once the jack has been placed, it is connected to a U-tube filled with water connected at one end with the flat jack and the other end to an air reservoir. The initial reading when the pressure in the flat jack is equal to the atmospheric pressure is taken. The compressed air reservoir valve is then opened to apply pressure to the jack. The pressure is read at the manometer and the change in volume of the flat jack and the pipes is determined by reading the displacement of the water level in the U-tube. The difference in level in the U-tube multiplied by the area of cross-section gives the volume change of the flat jack and the pipes connecting the U-tube to

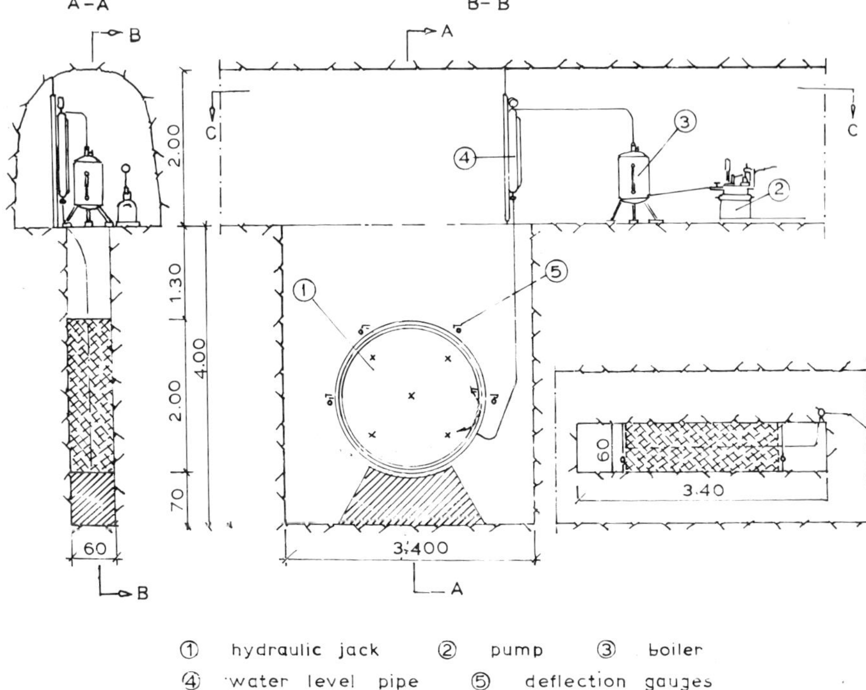

Fig. 8-37. The device for "hydraulic jack" test (after KUJUNDZIC et al, 1964).

the jack. The average deformation of the loaded surface of the rock in the slit is given by

$$\delta_{av} = \frac{fh}{2F} \tag{8.17}$$

where δ_{av} = mean deformation of the rock surface, cm
f = cross-section of the pipe, cm^2
h = difference of water level in the pipe, cm and
F = hydraulic jack surface, cm^2

Different diameter glass tubes can be selected to accommodate high or low volume changes and increase accuracy of readings. When the modulus value of rock is high, the error due to compressibility of concrete, water or hydraulic medium may be quite high and should be taken into account. The Eq. 8.17 can be written as

$$\delta_{av} = \frac{f(h - \Delta h)}{2F} - \Delta\delta_c \tag{8.18}$$

where Δh = correction due to compressibility of water and
$\Delta\delta_c$ = correction for deformation of concrete.

Compressibility of concrete is given by

$$\Delta\delta_c = \frac{pt}{E_c} \tag{8.19}$$

where p = pressure
t = thickness of concrete grout and
E_c = modulus of deformation of concrete.

The modulus value is computed from the average deformation,

$$E = \frac{0.27\,P(1-\nu^2)}{a\delta_{av}} \tag{8.20}$$

where P = total load
ν = POISSON's ratio of the rock
δ_{av} = average deformation and
a = radius of the jack.

The LNEC, Lisbon, Portugal have developed the technique further. The slit can be cut with a diamond slotting machine with a diameter d to a depth h and thickness t (Fig. 8-38). A flat jack of thickness t and height $(h-a)$ is introduced into the slit. Pressure p is applied and the deformation of the flat jack is measured at different points on the flat jack using strain-gauged deformeters inserted in the jack. These permit measurement of deformation at different points of the rectangular flat jack. Average deformation for the rock surface can be calculated by measuring the volume changes as described

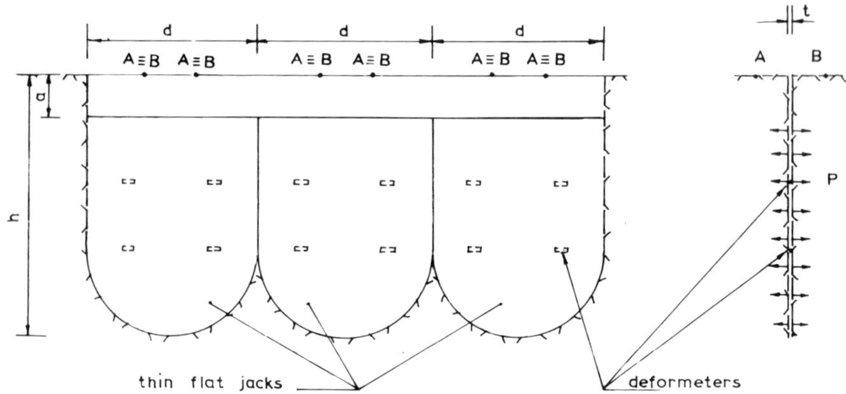

Fig. 8-38. Determination of deformability using flat jacks (after LNEC, 1970).

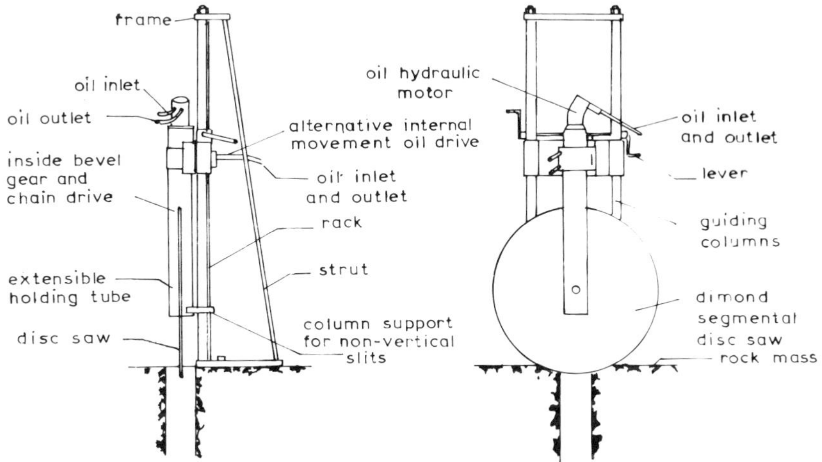

Fig. 8-39. Slit cutting machine.
(after LNEC, 1970).

earlier. The slit cutting machine (Figs. 8-39 and 8-40) essentially consists of a 1 m (3.3 ft) diameter, 7 mm (0.28 in) thick diamond disc mounted at the end of a pipe. The power transmission unit is mounted inside the pipe. By drilling a hole of 170 mm (6.7 in) diameter, at the test site, the pipe can freely move into it permitting the placing of the flat jack at any desired depth beneath the surface. Fig. 8-41 shows a flat jack being introduced into the slit where two others are already in position.

The size of the loading surface can be varied by using a larger flat jack or a number of flat jacks placed side by side. Test areas up to 4 m^2 (43 ft^2) have been used (FERNANDEZ-BOLLO, 1968).

The analysis of results obtained using flat jack method in a narrow slit is not simple. From measurements of deformation of points placed near the centre of the flat jack, consistently lower values of deformation modulus are obtained (ROCHA, 1969). This is due to the restraining influence of the slit ends. It is suggested that slit diameter should be 2–3 times the flat jack diameter.

The closeness of the surface to the flat jack results in loading a near quarter space rather than a half space as envisaged by BOUSSINESQ solution. Unless the distance from the surface is large, the modulus value will be lower. The modulus values determined from measurements at points placed away from the surface will give a better approximation provided the slit dimensions are large enough. A finite element solution could be used to determine the influence of loading on the test geometry.

Fig. 8-40. Slit cutting in the wall. (after LNEC, 1970).

In the absence of any finite element solution, the jacks can be calibrated using very large blocks of artificial material. A general relationship could be formulated where the deformation modulus can be given by

$$E = c\frac{p}{\rho} \tag{8.21}$$

where E = modulus of deformation

p = pressure

ρ = displacement of wall at a point and

c = constant, the value being dependent upon the point at which ρ is measured, size of flat jack, its relationship with slit dimension and nearness of surface.

Fig. 8-41. Test arrangement for deformability using flat jacks. Flat jack is being introduced into the slit (after LNEC, 1970).

Model tests have shown that when the size of the flat jack is equal to that of the slit, the displacement at the top edge of the quarter space is about 50% higher than at the position close to the bottom edge of the slot (ROCHA, 1969).

8.5.2.2. Cable jacking method

Cable jacking is a simple method of loading a surface in a trench or in an open pit where reaction is not available. The method was first suggested by JAEGER (1961) but was first tried by ZIENKIEWICZ and STAGG (1967). The method consists of anchoring a cable at a distance 6 to 10 times the bearing pad diameter in a bore drilled at the centre of the loading area. Load up to 1016 tonnef (1000 tonf) can be applied using a single cable and for larger areas, more than one cable can be used. The method also permits applying horizontal loads to the surface of the rock and thereby permits calculation of anisotropy in modulus values without having to conduct tests in two directions. A test arrangement using 4 cable loading pads is shown in Fig. 8-42a. A centre hole of 112 mm (4.4 in) diameter, 12 m (39.4 ft) deep, is drilled at each pad location and a multistrand cable is grouted. The cables can give a load in the range of 254–305 tonnef (250–300 tonf). A square reinforced loading block is cast in place on the top of which three 102 tonnef (100 tonf) jacks are mounted. The cable head consists of 19 mm (0.75 in) thick shell plate cylinder, 300 cm (118 in) high and filled with concrete resting on the jacks to which the cable is securely tightened. Horizontal loading is similarly applied using a 102 tonnef (100 tonf) one or more jacks placed between the two concrete blocks.

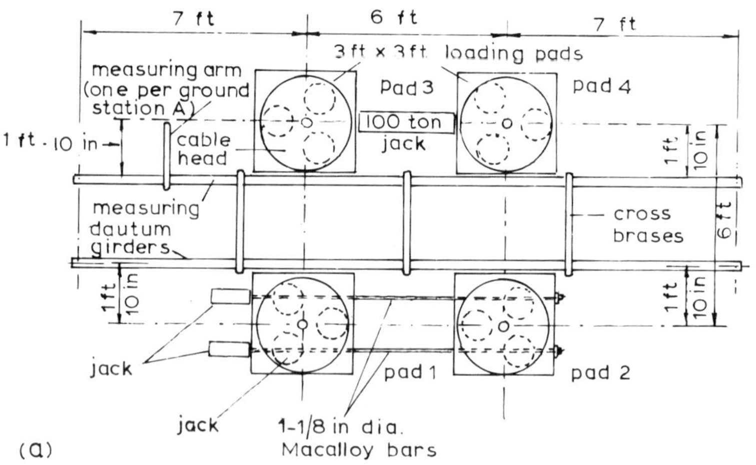

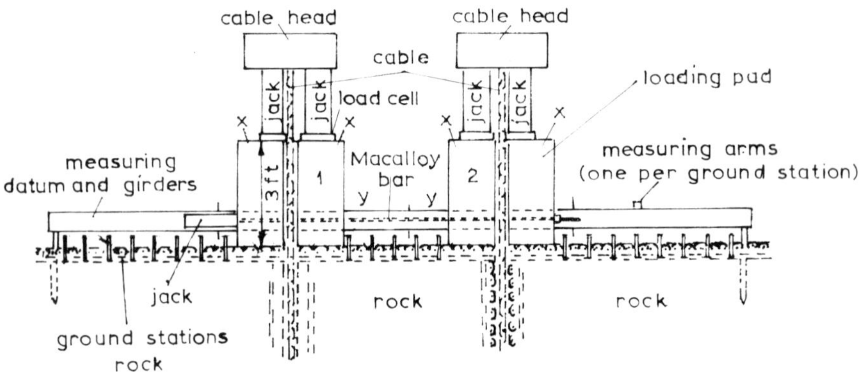

Fig. 8-42a. Cable jacking method with 4 loading pads
(a) Plan view of layout
(b) Section through holes 1 and 2
(after ZIENKIEWICZ and STAGG, 1967).

Fig. 8-42b. General view of cable loading setup with recording gear.

Fig. 8-42c. Close view of inclinometers and LVDTs (after LAMA et al, 1977).

Surface deformation can be measured from the measuring datum girders at a number of points placed on the concrete blocks and away from it in a gird pattern. When horizontal surface load is applied, horizontal deformations are also measured between different points using suitable extensometers. A general elastic solution for a half-space loaded by different types of loads is given by GERRARD and WARDLE (1973a, b). Figs. 8-42b and c show a view of cable loading equipment developed by CSIRO, Australia. The measurement of both surface points and at depth is done using LVDTs and inclinometers (LAMA et al, 1977).

8.5.2.3. Goffi's method

The method makes use of the deformation caused by the application of a concentrated load on the circular tunnel surface (GOFFI, 1963, 1966). The generatrix of a circular tunnel under an uniform linear radial load P undergoes deformation which is given by (Fig. 8-43a)

$$\begin{aligned} W_{\mathrm{A}} &= P\Omega_{\mathrm{A}} \\ W_{\mathrm{A}'} &= P\Omega_{\mathrm{A}'} \end{aligned} \tag{8.22}$$

where Ω_{A}, $\Omega_{\mathrm{A}'}$ are the areas of influence.

Analytically, the displacement W is a function of the modulus of elasticity of the rock surrounding the tunnel, E, the radius of the tunnel, a, and the load, P, such that

$$W = P\,f(E, a) \tag{8.23}$$

Equating (8.22) and (8.23) we get

$$P\,f(E, a) = P\Omega \tag{8.24}$$

from which the value of E can be determined as a function of the experimental data and geometric dimensions.

In practice, load is applied to a rectangular area of length b and width h subtending an angle 2ε at the centre of the tunnel (Fig. 8-43b). The modulus of elasticity can be given by (assuming $2\varepsilon \rightarrow o$),

$$E = \frac{P(1+\nu)}{2\pi\Omega_{\mathrm{A}}}\left[4(1-\nu)\,(1-\log\frac{h}{4a} - (1-2\nu)\frac{\pi h}{4a}\right] \tag{8.25}$$

where ν = POISSON's ratio
a = radius of the tunnel
P = total load and
h = width of loaded surface.

The method is quite reliable and accurate when rock is elastic, homogeneous and isotropic since it does not demand the geometry restrictions imposed by BOUSSINESQ's solution.

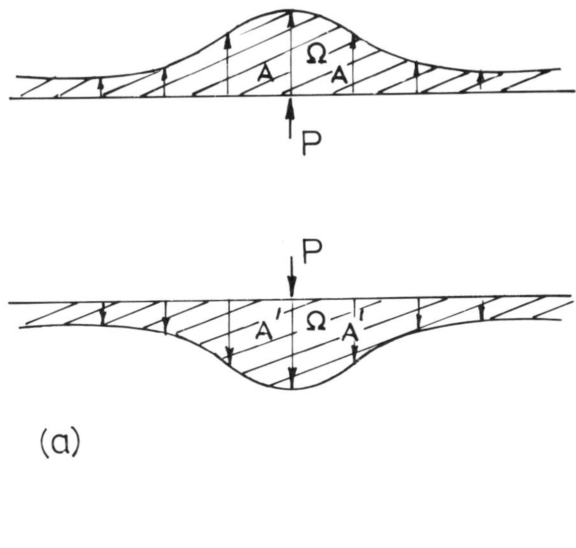

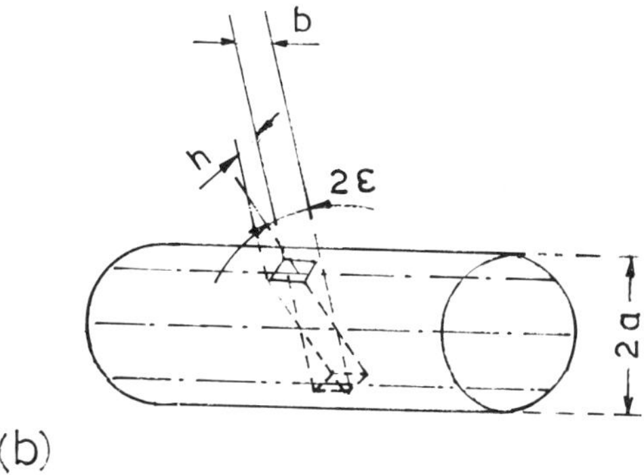

Fig. 8-43. Generatrix of a circular tunnel
(a) Under an uniform linear radial load P
(b) With load applied to a rectangualr area of length b and width h subtending an angle 2ε at the centre of the tunnel.

Though the value determined is dependent upon ν, the knowledge of which is necessary but any error in its estimation influences the value of E to a very small degree.

The use of this method gave satisfactory results in modulus determinations in Mont Blanc tunnel where 600 tonnef (591 tonf) jacks were used for the loading of the excavation generatrix.

8.6. Pressure Tunnel Test

8.6.1. Theoretical Basis

The method consists of applying uniformly distributed radial load to the surface of the tunnel while the diametral deformation of the tunnel boundary is measured. The calculations are based upon treating the tunnel as a thick-walled cylinder and assuming the external radius to be infinite. The stresses in the diametral plane at points sufficiently far removed from the edges of the loading area can be given by the relationship:

$$\sigma_r = \frac{pa^2}{r^2} \tag{8.26}$$

$$\sigma_t = -\frac{pa^2}{r^2} \tag{8.27}$$

where p = hydrostatic pressure
a = radius of the tunnel and
r = distance of the point from the centre of the tunnel.

If the deformation of the tunnel is measured at some distance from the tunnel wall, the modulus can be calculated using the relationship

$$E = \frac{pa(1-\nu)}{W_r} \tag{8.28}$$

where W_r = radial displacement due to internal pressure at the point r and
ν = POISSON's ratio for rock.

The above equation assumes an infinitely long gallery under uniform pressure. In practice, the loaded area is much smaller and the measured deformations are smaller than if the gallery was infinitely long. A coefficient ψ must be introduced into Eq. 8.28, the value of which is smaller than one. Eq. 8.28 therefore becomes

$$E = \frac{pa(1-\nu)\psi}{W_r} \tag{8.29}$$

WESTERGAARD (1941) and TRANTER (1946) have discussed the influence of finite loading. SEEBER (1960) has shown that deformation measured at the centre is about 93% of the ideal case. Decreasing the loaded length decreases displacements and inherent inaccuracies in the measurements increase. Deformation increases rapidly as the loaded length approaches and exceeds tunnel diameter (Fig. 8-44). Excessively large length increases cost and is not very helpful. The volume of the rock, effected around the tunnel, increases proportionally to the square of the tunnel radius and is advantageous. The amount of surface deflection as well as the difference between surface of

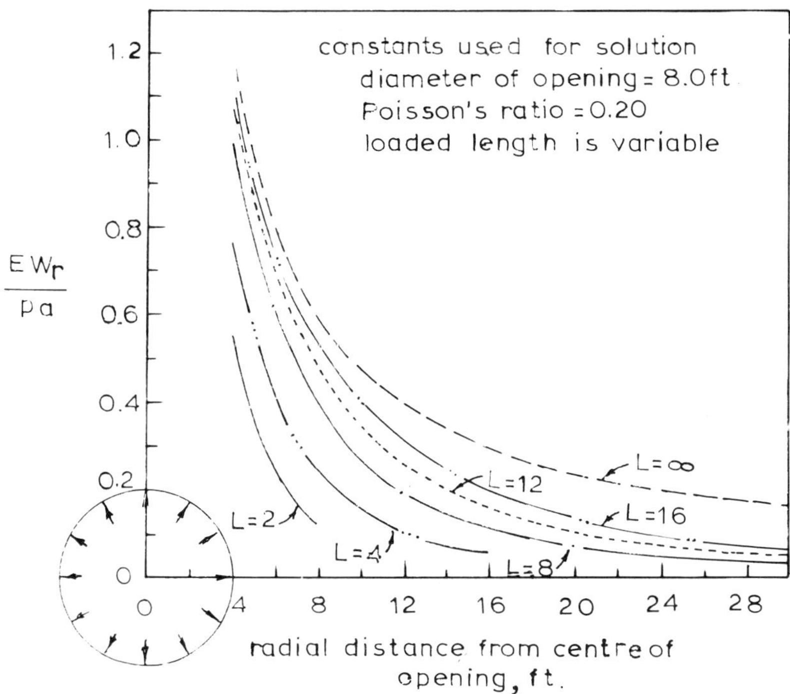

notation

E = modulus

w_r = radial displacement

p = intensity of applied radial load

a = radius of opening

l = loaded length, ft.

Fig. 8-44. Effect of loaded length on radial displacement (after MISTEREK, 1969).

opening and a deeper lying point 9 m (30 ft) away decreases rapidly at loaded length equals to diameter (Fig. 8-45). The knowledge of ν is again essential for calculation of E. Using an exact solution of TRANTER (1946), the influence of ν on $(E.W_r/pa)$ is shown in Fig. 8-46. It is clear that error in E due to inaccuracy in ν of ± 0.1 will be about 10%. POISSON's ratio values will normally be known more accurately than ± 0.1.

Optimum length of loaded surface should be approximately 5–6 times the diameter.

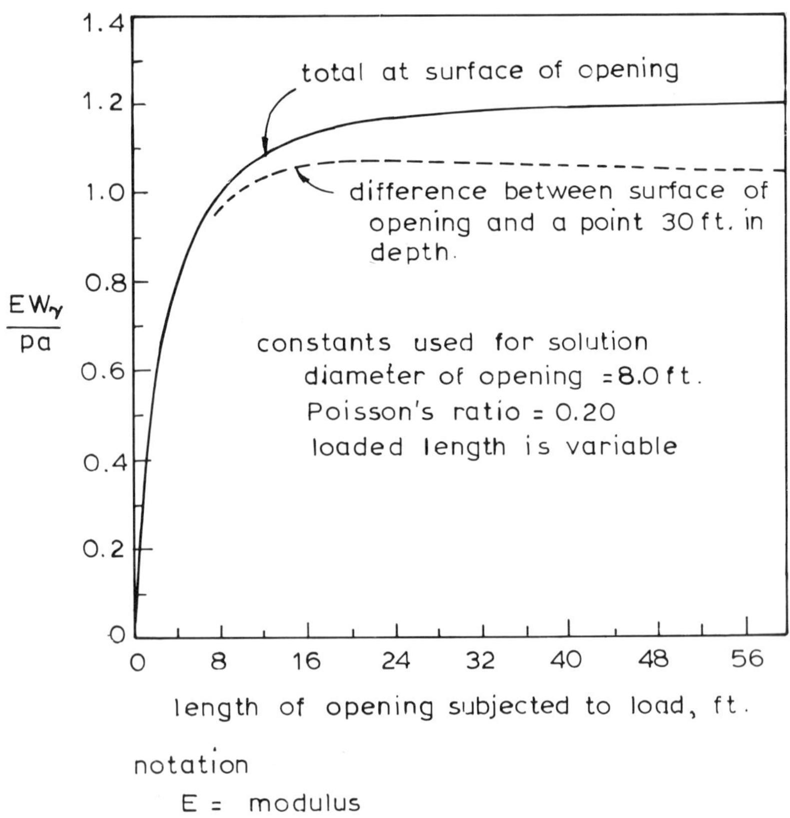

Fig. 8-45. Effect of loaded length on displacement at the surface and at some point away from the tunnel (after MISTEREK, 1969).

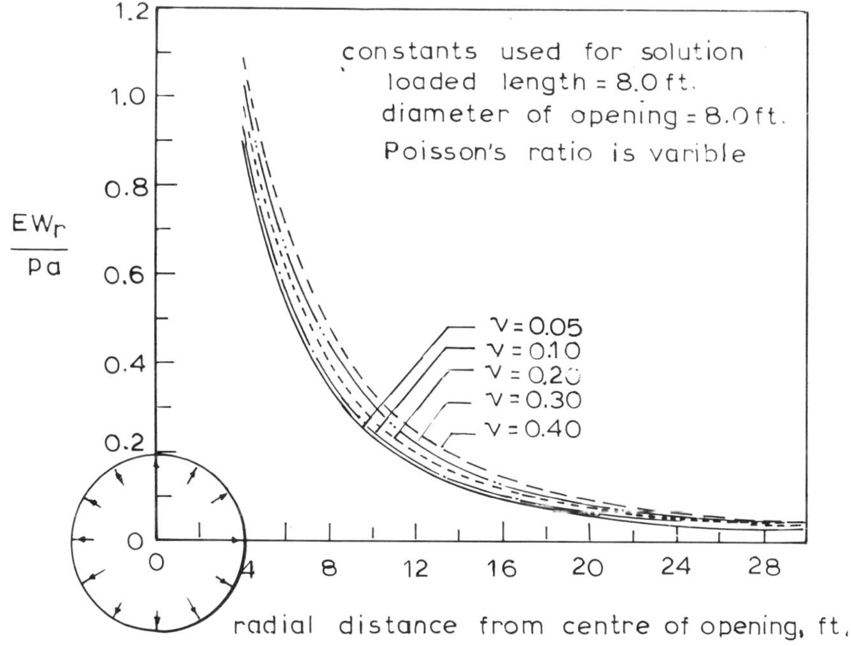

notation

E – modulus.

w_r – radial displacement

p – intensity of applied radial load

a – radius of opening

ν – Poisson's ratio

Fig. 8-46. Effect of POISSON's ratio on radial displacement (after MISTEREK, 1969).

8.6.2. Hydraulic Pressure Chamber Test

This is the oldest method for determination of the in situ modulus of rock mass and dates back to the beginning of the century. In this method, pressure is applied to the tunnel perimeter through a hydraulic medium.

The test is conducted in a section of the tunnel or exploratory adit which is plugged at one end with an impermeable dam, the other end being dead end or the tunnel may be plugged at two ends (Fig. 8-47). This section of the tunnel is lined with a nonpermeable lining usually rubber or PVC so as not to allow the hydraulic fluid (water) to come in contact with the rock wall and thereby alter its properties or develop hydrostatic pressure in the joints. To protect the lining from rupturing due to sharp edges and unsmooth surface of the tunnel rock, the test section is lined with thin reinforced concrete walling or gypsum, or plaster is placed in 4–8 segments round it which serves as a smooth back for the impermeable rubber ring.

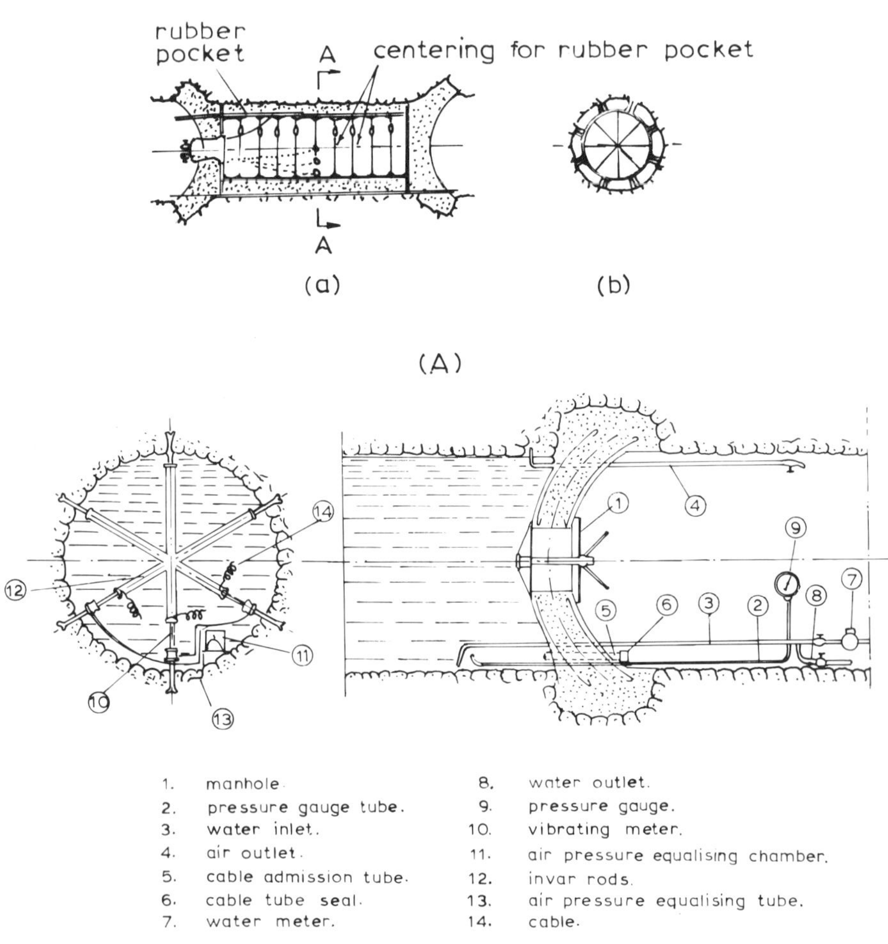

Fig. 8-47. (A) Standard arrangement for pressure tunnel test.
(a) Tunnel for rock test; (b) Section A-A
(after STAGG, 1968).
(B) Details of hydraulic pressure chamber test
(after ROCHA, SERAFIM and DA SILVEIRA, 1955).

Because of the necessity of closing the tunnel ends for pressure chamber test, the "Jaroslav Cerni" Institute, Belgrade, has developed a precast concrete wall of a conical shape. It is placed in gallery over a special concrete bed and is recoverable. This permits cutting down test cost and time for preparation of tests (KUJUNDZIC, 1963).

The length of the section under test must be large enough so as to make the error due to end effect negligible in the calculation of the value E. Tunnel diameter is usually limited to 1.5–2 m (4.9–6.6 ft).

The section is loaded by filling it with water and applying pressure with a pump. The technique of application of load is necessarily the same as in plate jacking method. The pressure applied may vary from 15–40 bars (1500–4000 kPa) (218–580 lbf/in^2). Deformation of the tunnel is measured by various types of extensometers, waterproof vibrating wire deflectometers, CARLSON joint meters or multipurpose strain meters. A 7 point retrievable extensometer (REX – 7P) developed by U.S. Bureau of Reclamation is shown in Fig. 8-48. The REX – 7P is installed in 6.1 m (20 ft) NX hole. Seven LVDTs are used to sense the deformation between the collar of each hole and the seven expandable anchors fastened in the hole. The LVDT sensing head is placed at collar of the hole and also contains a gauge point for the tunnel diameter. 8 mm (5/16 in) stainless steel metering rods connect each of the anchors to the LVDT head. The anchors consist of 3 legged "scissor jacks". A setting rod engages lugs on the anchor and tightens the anchor in position. Metering rods attached to the anchors pass through the holes provided the anchors placed before them. Anchors are tightened starting with the farthest most anchor. The extensometer is capable of measuring movements of 200 μ in. (0.005 mm). A number of such units (6–10) can be installed in the wall at different points in holes drilled perpendicular and radially outwards from the tunnel axis. Another 8 point measuring extensometer, very simple in design is described by BRUGMAN (1968).

Changes in tunnel diameter can be measured with spring loaded invar wires using LVDTs as sensing elements. It is important that all sensing elements are waterproof and designed to stand the hydraulic pressure to which they are to be subjected. Water temperature measurements are taken to correct for instrumental errors.

The value of E calculated from measurements at depth is more reliable since it refers to the conditions behind the decomposed and fractured zone. It is advisable to measure the radial displacement at various depths and use Eq. 8.29 rather than radial strain.

The advantage of pressure tunnel tests is that it affects a much larger rock mass and imposes a better defined stress state on the surrounding rock. On the other hand the method is very expensive. The method is applied only to

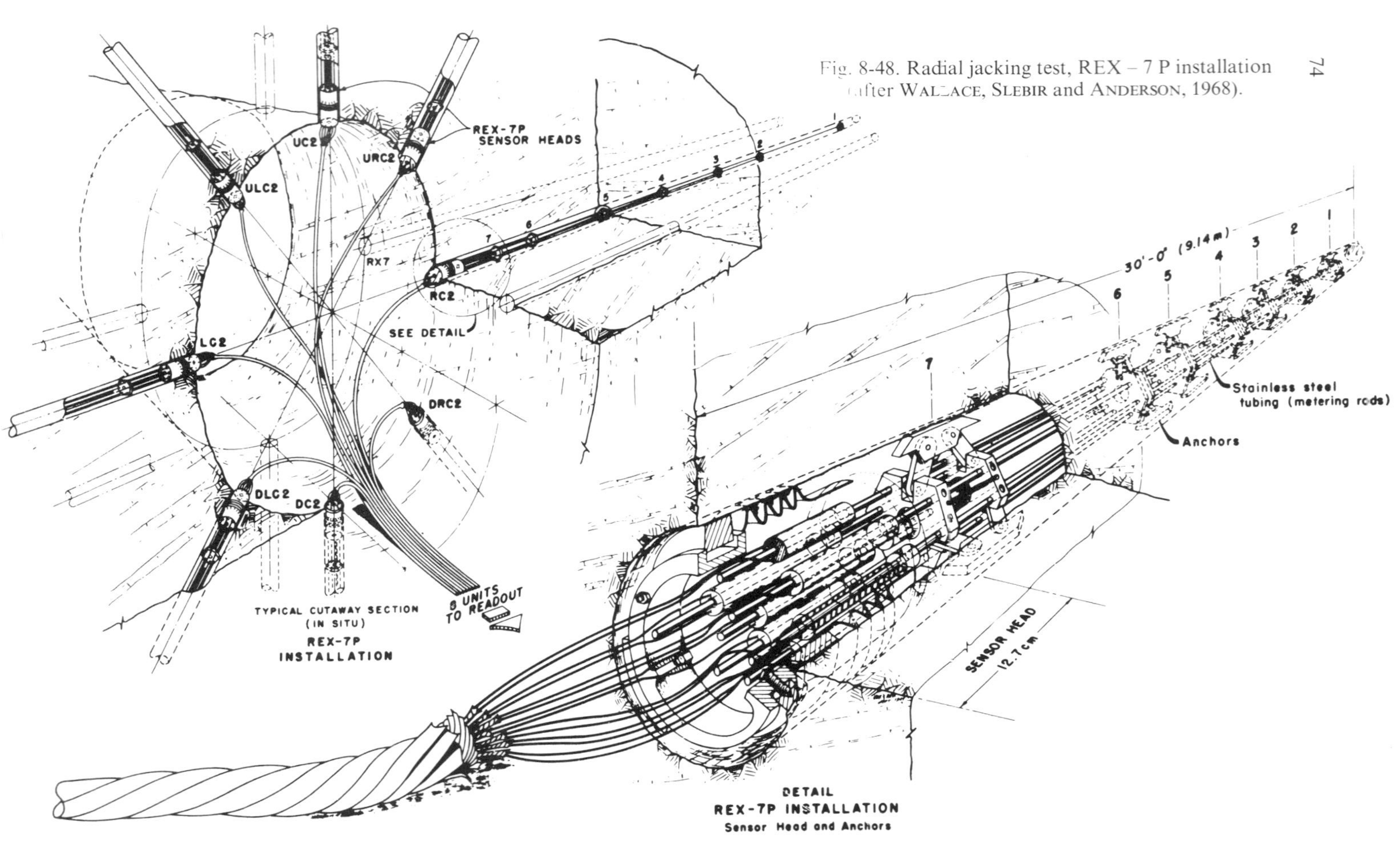

Fig. 8-48. Radial jacking test, REX – 7 P installation (after WALLACE, SLEBIR and ANDERSON, 1968).

major structures (dams, underground power houses etc.) and has mainly been used in pressure tunnel sites. Because of the costs involved the number of sites that can be tested are limited to 2 or 3 and hence it may be supplemented with small diameter jacking tests. The method has been most commonly employed in France and Italy (OBERTI, 1960) where it has been used for over 30 years.

There have been certain objections to the use of the method since the method induces tensile hoop stresses in the rock and at certain points, the tensile stress may exceed the compressive stress and cracks may open up giving lower value of strain modulus. The results are, however, directly useful for pressure tunnels and tunnel lining design or testing of tunnel liners (NONVEILLER, 1954; KUJUNDZIC, 1957).

8.6.3. Radial Jacking Test

Radial jacking test is a modification of the pressure chamber test where pressure is applied through a series of jacks placed close to each other. A number of loading techniques have been developed in different countries such as Yugoslavia, Austria, West Germany, Switzerland and U.S.A.

The basic features of the equipment consist of 16 flat jacks of length 2.0–2.4 m (6.5–8.0 ft) with width of 40–41 cm (16 in). The flat jacks are curved along their width and form segments of a circular ring of diameter about 2.2–2.4 m (7.2–8.0 ft). The tunnel surface to be loaded (Fig. 8-49) is given a shot-concrete lining of 10–13 mm (3/8–1/2 in) (sometimes up to 15 cm (6 in), LAUFFER and SEEBER, 1966) and the pressure jacks are placed against the concrete lining.

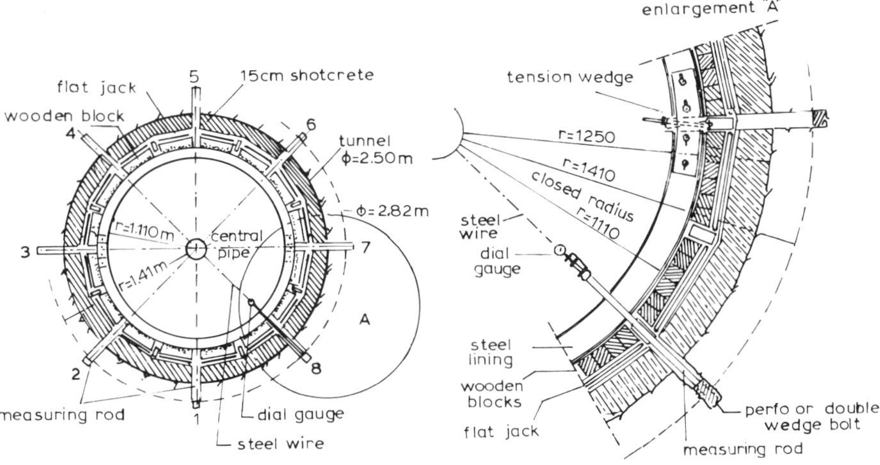

Fig. 8-49. Radial press TIWAG (after LAUFFER & SEEBER, 1966).

Chip wood blocks are placed between the flat jacks and steel linings. These blocks are shaped to fit the steel rings on the inner side and the flat jacks on the outer side. These serve as cushions, prevent damage to the jacks and distribute load to the steel sets uniformly.

The tunnel surface when prelined with shotcrete is specially prepared to have flat segments on which the flat jacks rest. This technique has been followed on the Continent very successfully.

In the U.S. Bureau of Reclamation practice, the inner steel ring and the jacks are installed before hand. The ring and flat jacks are held centrally in the bore of the tunnel by suspending them from the roof of the tunnel on a heavy pipe frame. The space between flat jacks and rocks is packed by pumping concrete (Figs. 8-50 and 8-51). The back surface of the jacks is protected by a thin PVC sheet against adhesion with concrete to allow easy recovery.

The steel ring sets are in 4 or 8 sections and are designed to withstand the maximum pressure that the jacks are expected to exert on the tunnel walls.

Displacements are measured using multiple point extensometers or dial gauges. Wooden plugs are provided before hand separating the various jacks through which radial holes are drilled for installation of extensometers.

Fig. 8-50. Hydraulic pump and hose manifolds through which pressures are applied to flat jacks, concrete and rock (after WALLACE, SLEBIR and ANDERSON, 1968).

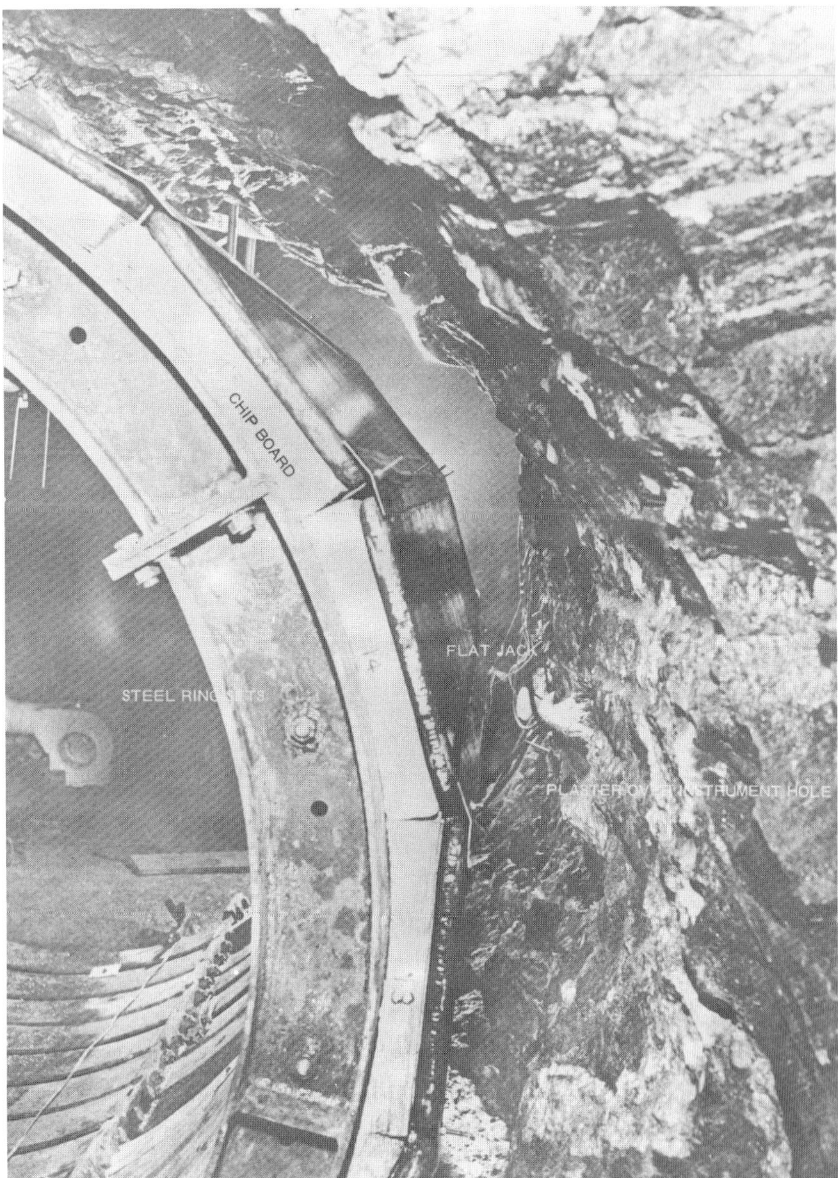

Fig. 8-51. Close-up of portion of radial jacking test.
Area between jacks and rock to be back-filled with concrete
(after WALLACE, SLEBIR and ANDERSON, 1968).

European techniques mostly use perfo or double wedge anchors for installation of measurement points for tunnel deformation, while the U.S. Bureau of

Reclamation uses REX-7 point extensometers as described earlier. TIWAG, Interfels or Yugoslavian techniques make use of a central pipe as a reference point against which deformation is measured (Fig. 8-52).

Fig. 8-52. Radial press TIWAG
(after Interfels, Salzburg).

8.6.4. Analysis of Results from Pressure Tunnel Tests

Pressure tunnel tests are very useful and have direct application when tunnel linings are tested which are to be subjected to water pressures (pressure tunnels in hydro-electric schemes). Measurement of tunnel surface deformation gives directly the values about the likely stresses to which the tunnel lining shall be subjected. In such cases, it is quite sufficient to make measurements of the tunnel surface. In Europe, radial jacking tests have mostly been carried out for the above purpose and that is why displacements at depths have been rarely measured.

On the other hand, in dam abutment analysis, the larger volume of rock mass is in view and it is important that measurements are carried out at greater depths. The U.S. Bureau of Reclamation designed radial jacks for this purpose and hence the gauging equipment for studies of rock moduli at depths away from the tunnel.

While measuring displacements, with respect to the tunnel axis (TIWAG system), it is important that the central pipe is fixed on supports away from the loaded diameter. Measurement points should be located not only within the loaded area but also away from the loaded area up to about 0.5–1.0 times the diameter. This also establishes whether the reference points were really away from the influence area.

While using the U.S. Bureau of Reclamation technique, there are no reference points, but the modulus values are calculated by knowing the difference in W_r values for different points at different depths. Anchors can however be fixed connecting the opposite ends of the tunnel surface with measuring rods and gauges passing through the tunnel centre and placed at 90° to each other. This allows measurement of change in tunnel diameter along two axes, but has the disadvantage that it presupposes symmetrical deformation around the central axis, which may be a gross error. Fig. 8-53 shows this point very clearly.

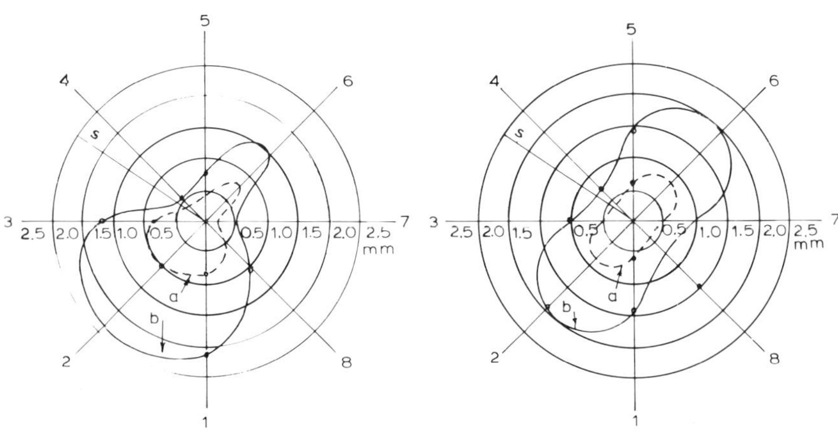

a — elastic deformations. b — total deformations.
s — position of the foliation plane of the rock.
1 to 8 – measuring points.

Fig. 8-53. Circumferential distribution of radial deformation in the test zone. (Left) True form of the curve when plotting the measured deformation of the radius. (Right) Form of the curve when plotting the half enlargements of diameter (after LAUFFER and SEEBER, 1961).

The presence of a fractured zone around the tunnel greatly influences the E values calculated from the tunnel surface displacement measurements. Effective modulus of the fractured and unfractured rock together, that would give the same deformation as the fractured rock around the tunnel can be calculated using the following relationship (SEEBER, 1960)

$$E_{\text{eff}} = \frac{pa}{W_{\text{a}}}\left[(1-\nu)\right]\left[\ln\frac{c}{a}\right] \tag{8.30}$$

where c = radius of fracture.

The value of c may be determined by observation through boreholes, or seismic surveys.

Measurement of displacement at different depths permits calculation of modulus of different layers away from the tunnel wall. A sudden change in the E value of rock indicates either a change in rock bed or change in its state. Geological information obtained from drill cores at site can help decide about this zone of fracturing c near the tunnel wall or presence of weaker beds away from the tunnel wall.

Modulus values calculated without taking into consideration the fractured zone may be 50–100% in error (SEEBER, 1960). Fig. 8-54 shows the influence

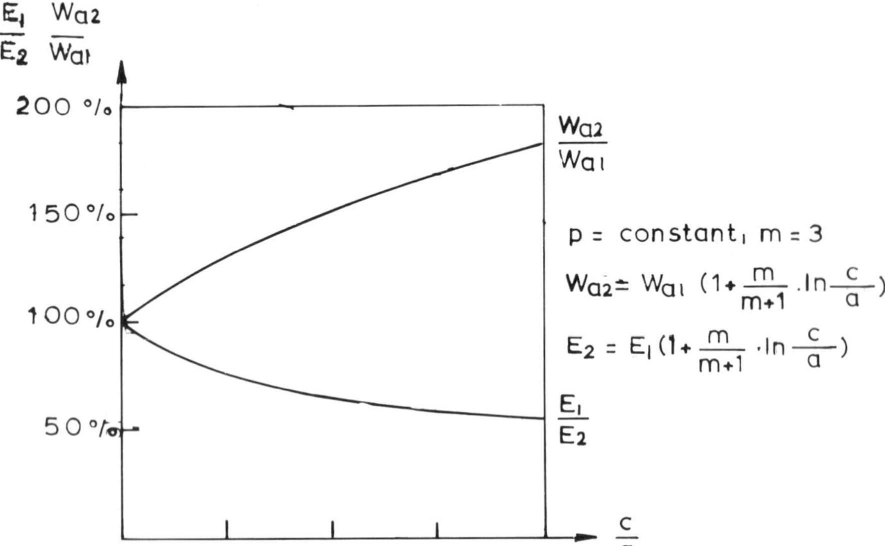

Fig. 8-54. Influence of fracture zone on measured deformation and calculated modulus. W_{a1}, E_1 for unfractured rock. W_{a2}, E_2 for fractured rock (after SEEBER, 1960).

of fracture zone on the calculated modulus values and measured displacements of tunnel surface.

The existing stress field which gets redistributed around the tunnel results in development of tensile stresses at some points in the tunnel walls and compressive stresses at other points. Superimposed upon this field is the stress field produced due to radial jacking or hydraulic pressure in the tunnel. Tangential tensile stresses tend to add into the tensile stress already in existence and may result in fracturing of the rock. Modulus value will greatly reduce as fracturing advances forward into depth. Creep measurements at constant pressures can help clarify this point, particularly if there is no big change in the rock types at different points around the tunnel. The knowledge of virgin stress field in the area can help in defining region of high tensile stresses. The results will be in error if the rock surrounding the tunnel is allowed to crack under imposed jack loads or pressure tests.

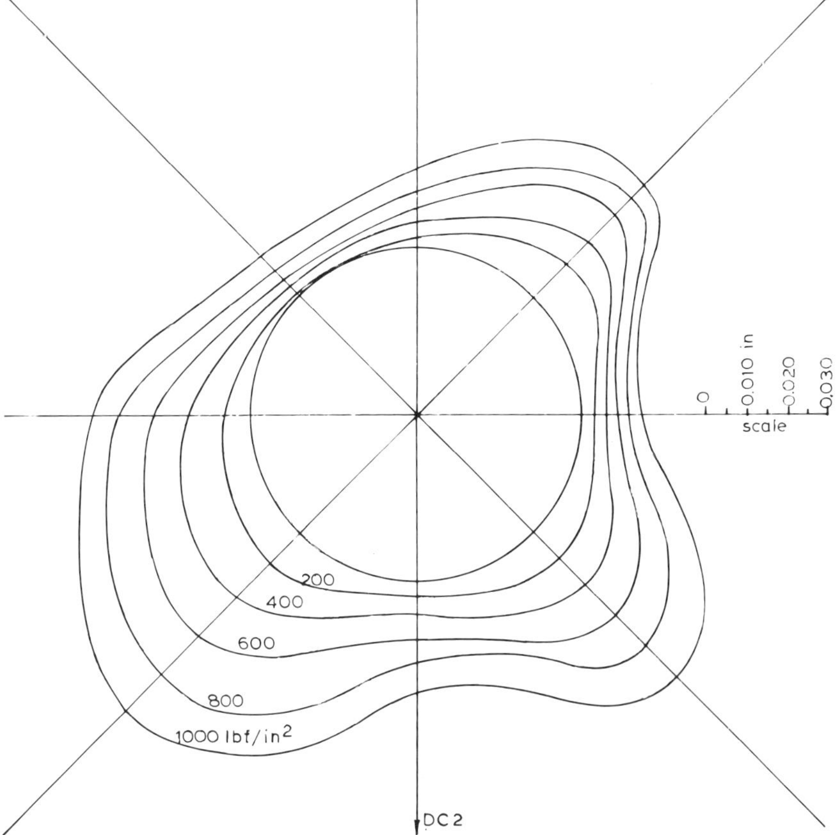

Fig. 8-55. Deformation around the centre section of the radial jacking test (after WALLACE, SLEBIR and ANDERSON, 1968).

The thickness of shotcrete lining between the flat jacks and the rock is important and influences the results. The lining should be as thin as possible and discontinuous along the length of the tunnel so that load is not taken up by this lining and thereby decreasing the load acting on the rock.

The modulus values so obtained can be plotted around the centre section of the tunnel and classified into different groups depending upon geological observations (Figs. 8-55 and 8-56). Both modulus values on loading (deformation modulus) and unloading (elastic modulus) can be represented. Values can also be represented at different confining pressures which permit to define the pressure at which fracturing occurs and the direction which is sensitive to these pressures. Increase in differential deformation at high pressures between individual points is an indication of fracturing.

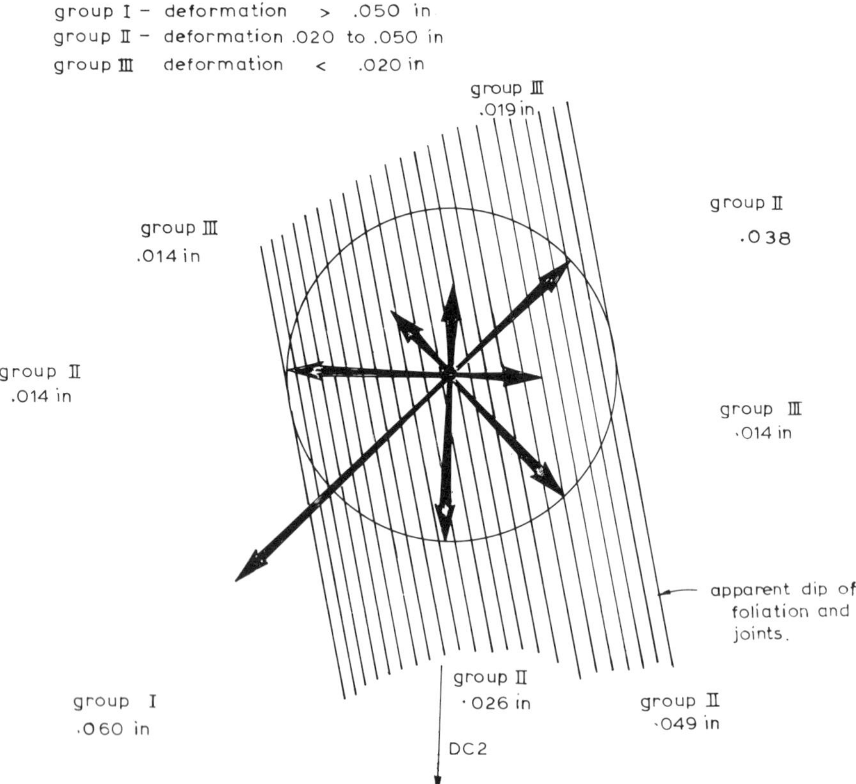

Fig. 8-56. Deformation grouping around the centre section of the radial jacking test. Based on joint openings, dip of foliation and joints and measured test values (after WALLACE, SLEBIR and ANDERSON, 1968).

Pressure tunnel tests offer many advantages over plate bearing tests because of the large volume of rock affected by them. The tests are costly to conduct and their application is reserved for those rock types which compose the biggest part of the foundation or sites of large structures which are apt to critically influence the prototype behaviour.

8.7. Borehole Tests

In the last 20 years, a number of instruments have been developed for measurement of rock deformability in small diameter boreholes. These instruments depend upon the application of pressure to the walls of a borehole and measurement of the radial response of the borehole wall. The instruments that apply a uniform internal pressure to the borehole walls are called dilatometers or pressuremeters. Those which apply force along a limited portion of the circumference of the borehole by forcing apart circular plates are called borehole jacks. Devices that force a small indenter into walls of a hole are called borehole penetrometers. Table 15 gives a summary of the three types of instruments used by various investigators.

8.7.1. Borehole Dilatometers

These are full circle radial expansion devices. Pressure is applied through an inflatable cylindrical probe placed down the borehole at desired depth. One of the earliest pressuremeters was developed by MÉNARD (1957, 1966). Since then a number of other pressuremeters have been developed (JANOD and MERMIN, 1954; KUJUNDZIC and STOJAKOVIC, 1964; COMES, 1965; ROCHA et al, 1966; TAKANO and SHIDOMOTO, 1966; Geoprobe Instrument, 1967; MCKINLAY and ANDERSON, 1975). The maximum contact pressure in these cells does not exceed 15 MPa (2200 lbf/in^2) (155 kgf/cm^2) and are hence suitable for softer or very highly weathered rocks.

The inflation pressure in some instruments is obtained through a gas cylinder (MÉNARD pressuremeter and Geoprobe) and change in diameter is calculated using the change in diameter during pressurising. In other instruments, pressure is applied using water or oil. The change in diameter is measured by either measuring the change in volume of the fluid (indirect method) and or by differential transformers (direct methods). Fig. 8-57 shows classification of the various dilatometers depending upon the method of measurement of diameter displacement. Dilatometers using indirect method of measurement of displacement are suitable for generally poor ground (soil or highly weathered rocks), but those using direct methods have a wider application. The use of differential transformers placed at different positions around the circumference also has the advantage that these permit measurement of anisotropy of rock.

TABLE 15 Devices for measuring rock deformability in boreholes

Type	Pressure condition	Interpretation formula	Name of device	Method of pressure application	Method of measuring deformation	Number of diameters measured	Diameter of borehole (mm)	(in)	Length of loaded area (mm)	(in)
Dilatometers	Uniform radial pressure all around the circumference of the borehole	$E = \frac{\Delta p}{\Delta(U_d/d)} K(1+\nu)$	MÉNARD pressure meter	gas pressure against water filling cell	volume change on expansion	integrated effect of all diameters	76 60 48 32	3.0 2.4 2.0 1.5	515 502 686 910	19.5 19 26 34.5
			Geoprobe instrument	ditto	ditto	ditto	76	3.0		?
			LNEC device	pump oil to expand cell	4 LVDTs	4	76	3.0	540	21.2
			JANOD-MERMIN device	ditto	3 LVDTs	3	168	6.6	770	30.4
			COMES' cell	ditto	3 LVDTs	3	160	6.3	1600	63.0
			tube deformeter	ditto	24 LVDTs	4	297	11.7	1300	51.2
			sounding dilatometer	ditto	2 "MCH" instruments	2	200 300	7.9 11.8	1000 1200	39.3 47.2
			pressiometer	ditto	4 LVDTs	2	46	1.8	680	26.8
			pressiometer	ditto	4 LVDTs	2	76	3.0	—	—
			dilatometer	ditto	8 Huggenberger	8	144	5.7	890	35.0
			elastometer 100		2 lever arms	2	62	2.4	520	20.5
			elastometer 200		1 LVDT	3	62	2.4	520	20.5
			lateral load tester	gas pressure	volume change	integrated effect	80	2.2	850	35.5
			pressure meter	gas pressure	volume change	integrated effect	EX, AX NX holes		762	30.0
Borehole jacks	unidirectional force of 2 stiff plates each contacting rock over an angle 2β		NX plate bearing test	pump oil to drive pistons	2 LVDTs	1	76	3.0	204	8.0
			centex cell (CEBTP device)	pump oil to drive a conical mandrel into a split cylinder	distance mandrel is driven measured inductively	1	76	3.0	306	12.0

Symbols
U_d = diametral displacement d = diameter of borehole p = applied pressure

Max. contract pressure bars	Can it be recovered after use?	Country of origin	Remarks	Reference
100	yes	France	considerable experience record	MÉNARD, 1957 and 1966
100	yes	?		Geoprobe Instrument
150	yes	Portugal		ROCHA et al, 1966
150	yes	France		JANOD & MERMIN, 1954
150	yes	France		COMES, 1965
40	yes	Japan		TAKANO & SHIDOMOTO, 1966
40 75	yes	Yugoslavia		KUJUNDZIC & STOJAKOVIC, 1964
200	yes	USSR		PRIGOZHIN, 1968
—	—	USSR	suitable for rock for E up to 10^5 MPa	PRIGOZHIN, 1968
100	yes	Spain	deformation accuracy 5×10^{-6} mm	ALAS, 1968
100	yes	Japan	for rock with compressive strength up to 10 MPa	Oyo Corporation, Japan
200	yes	Japan	for rock	Oyo Corporation, Japan
10–30	yes	Japan	for soils with compressive strength up to 1.0 MPa	Oyo Corporation, Japan
40	yes	USA	useful for $E = 6897 - 10342$ MPa	DIXON, 1969; HENDRON et al, 1969
640	yes	USA	$2\beta = 90°$	GOODMAN, 1966
?	not always in deep holes	France	$2\beta = 140°$	NOEL, 1963

TABLE 15 (continued)

Type	Pressure condition	Interpretation formula	Name of device	Method of pressure application	Method of measuring deformation	Number of diameters measured	Diameter of borehole (mm)	(in)	Length of loaded area (mm)	(in)
		$E=\frac{\Delta p}{\Delta(U_d/d)}K_1(v,\beta)$	Geoextensometer (New CEBTP device)	pump oil to drive pistons	2 LVDTs	1	76	3.0	306	12.0
			TALOBRE's jack	pump oil to drive pistons	?	1?	56	2.2	120	4.7
			German stress strain meter	spread wedges by driving a screw	"no contact" induction pick up	1	50	2.0	63	2.5
			PANEK's borehole flat jack	pump oil into a flat jack cemented in the borehole	not done at present could be measured by gauging across inside of flat jack	1	can be tailor-made for any length and diameter			
	radial pressure over opposed 2β sections	$E=\frac{\Delta p}{\Delta(U_d/d)}K_2(v,\beta)$	quadrantal jacks	pump oil into curved jacks in opposed quadrants	response of passive jacks in other two quadrants	integrated effect of all diameters	ditto			
			quadrantal jacks	ditto	2 LVDTs	2	76	3.0	280	11.0
Borehole Penetrometers	unidirectional pressure over a small area	empirical relationship	USBM penetrometer	pump oil to drive piston forcing indenting pin	dial gauge extensometer with cable drive	1	32	1.25	95	0.4
			HULT's device	pump oil to drive piston deforming proving ring	strain gauges on proving ring	1	32	1.25	~3	0.1
			DRYSELIUS' device (CTH 3)	pump oil to drive 3 pistons at 60°	strain gauges on cantilever elements	3	44		~5	~0.2

Max. contract pressure bars	Can it be recovered after use?	Country of origin	Remarks	Reference
340	yes	France	$2\beta = 143°$	ABSI and SEGUIN, 1967
?	yes	France	2β believed to be about 90°. Intended as stress meter, limited to shallow depth	TALOBRE, 1964
high	not always	Germany	slightly < 180°	MARTINI et al, 1964
300	no	USA	2β slightly < 180°, intended as stressmeter. Could be adapted but severe edge effects	PANEK & STOCK, 1964
300	yes	Australia and South Africa	$2\beta = 90°$ test never actually performed in situ. Could be adapted for shallow applications	JAEGER & COOK, 1963
350		Australia	$2\beta = 120°$	WOROTNICKI et al, 1976
very high	yes	USA	3/8″ diam. indenting pin forced into wall. Designed to measure roof bolt anchorage capability; limited depth	STEARS, 1965
very high	yes	Sweden	CA 1/8″ square piece forced into wall by deformation of proving ring. Designed as active stiff gauge for stress measurements	HULT, 1963
very high	yes	Sweden	3 pins forced into wall by oil pistons. Designed as active stiff gauge	DRYSELIUS, 1965

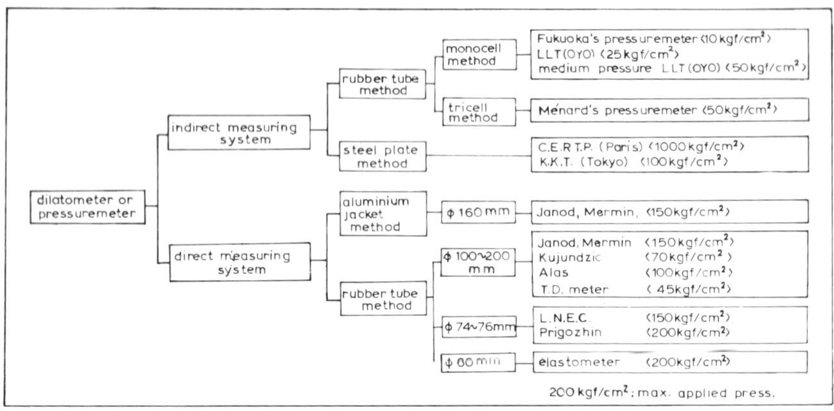

Fig. 8-57. Classification of dilatometers depending upon method of measurement of displacement.

Dilatometers which make use of indirect method require application of correction due to compressibility of fluid (water, gas or oil). All dilatometers require that there be no air trapped in the dilatometer or pressuring sleeve (except when otherwise required, e.g. OYO Elastometer 200). The error due to the changes in the thickness of the rubber sleeve needs to be taken into consideration, particularly for softer rocks. The compression behaviour of water, rubber and air is given schematically in Fig. 8-58. The errors arising due to rubber and oil/water are linear but due to air nonlinear. The reduction in volume due to water/oil is calculated as follows:

$$V_1 = kV\,dp \tag{8.31}$$

where V_1 = reduction in volume
k = compressibility
V = original volume and
dp = change in pressure

Assuming that the volume of the sleeve remains constant, change in the thickness of the rubber or neoprene sleeve can be obtained from the relationship (Fig. 8-59):

$$t = r - \sqrt{r^2 - \frac{S}{\pi}} \tag{8.32}$$

where t = thickness of the rubber sleeve
r = outside radius of the rubber sleeve and
S = cross-sectional area of the rubber tube.

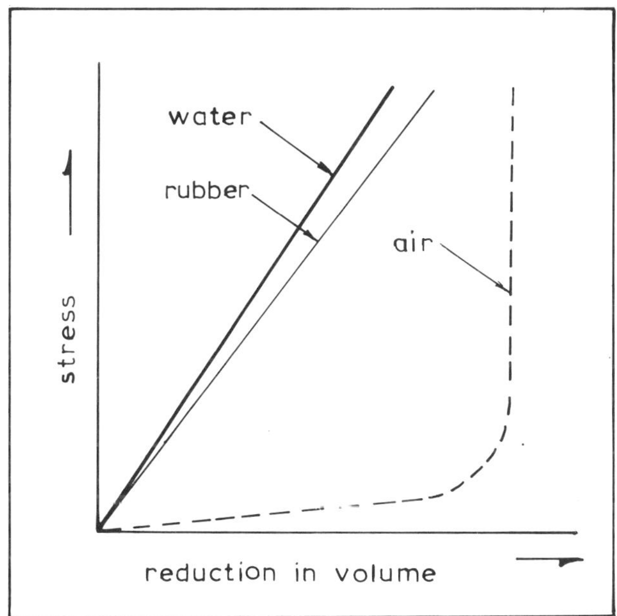

Fig. 8-58. Relations between stress and reduction in volume.

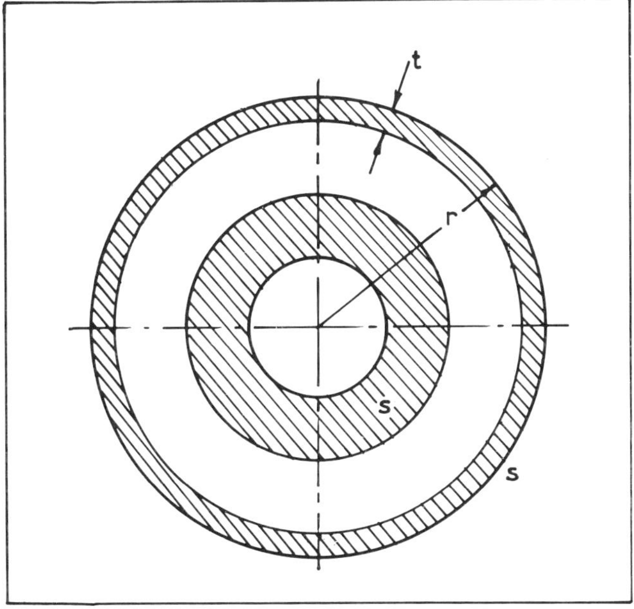

Fig. 8-59. Schematic diagram of change in thickness.

The diameter of boreholes required for the use of dilatometers varies under wide limits depending upon the various designs. The large diameter dilatometers have the advantage of affecting a larger volume of rock. The effective length of dilatometers varies between 200 mm (8 in) and 1600 mm (63 in) depending upon the diameter. Usually the ratio between length and diameter is greater than 4. Constructional details of some of the dilatometers are given below.

8.7.1.1. LNEC dilatometer

The instrument (Fig. 8-60) consists of stainless steel cylinder 540 mm (21 in) in length with an external diameter of 66 mm (2.6 in) and a wall thickness of 10 mm (0.4 in), wrapped in a rubber (neoprene) jacket 4 mm (0.16 in) thick. The external diameter of the instrument is therefore 74 mm (2.9 in), which corresponds to a clearance of 2 mm (0.079 in) with respect to a borehole 76 mm (3 in) in diameter. The liquid (water or oil) which applies the pressure on the side walls of the borehole is pumped into the space between the external surface of the cylinder and the internal face of the jacket. One of the tops is closed by a plug through which passes the relief valve of the liquid applying the pressure; the pipes and electric wires of the measuring unit are connected through the other top. It is by means of a rod screwed to this same top that the instrument is installed in the borehole and the depth and azimuth of this are determined. The relief valve is remote controlled by compressed air so that pressure can be removed after each test in order to move the dilatometer along the borehole.

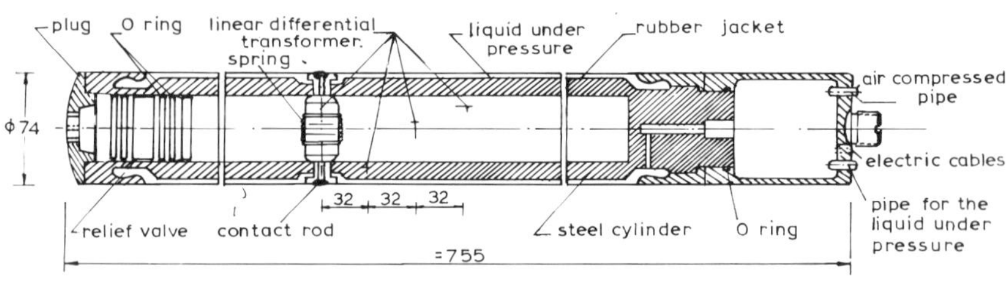

Fig. 8-60. LNEC dilatometer
(after ROCHA et al, 1966).

The differential transformers are applied on four cross-sections 32 mm (1.26 in) apart, i.e. measurements are carried out in a length of 96 mm (3.78 in) of the borehole. The stroke of the linear transformers is 5 mm (0.2 in), their accuracy about 1 μ and their sensitivity 0.1 μ. Each transformer has its metallic core and its coil assembly in contact with the rock by means of two small rods. The rods are applied against the rock by means of a spring. When the dilatometer is placed inside the borehole, the two rods of each transformer are sucked in by means of compressed air.

The instrument weighs 12 kgf (26.5 lbf) only, so that it is very easily transported and operated.

8.7.1.2. Yachiyo tube deformeter

The dilatometer developed by Yachiyo Engineering Company has a diameter of 297 mm (11.7 in) and a length of 1300 mm (51.2 in) (Fig. 8-61). The dilatometer shaft has three groups of differential transformers each of which consists of eight differential transformers mounted in radial direction in order to measure the variation in diameter in four different directions. Oil pressure is used to expand the neoprene sleeve which covers the entire dilatometer. The dilatometer requires a borehole of 300 mm (11.8 in) diameter.

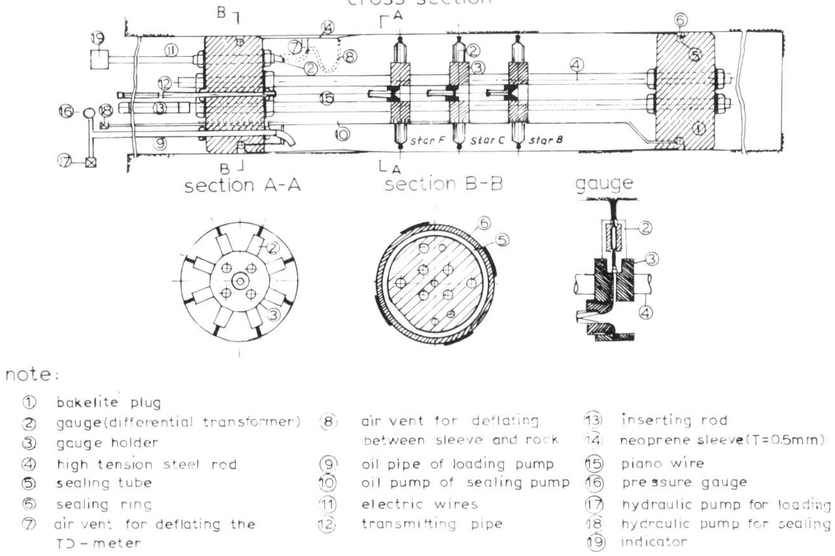

Fig. 8-61. Yachiyo tube deformeter, assembled view without neoprene sleeve.

8.7.1.3. OYO elastometer 200

Elastometer 200 (Fig. 8-62) has been developed to work at pressures up to 200 bars (20000 kPa) (2900 lbf/in^2) and is hence suitable for medium to hard rocks. The outside diameter of the dilatometer is 60 mm (2.4 in) with an effective length of 520 mm (20.5 in). The dilatometer is capable of giving a maximum displacement of 15 mm (0.6 in) (25 % of diameter) and can be used in boreholes of 62 mm (2.44 in) diameter. It uses indirect method of measurement of borehole diameter employing contact balancer system (Fig. 8-63). A known volume of liquid and air is present in a chamber with a number of sensors (usually six) which act against the innerside of the rubber sleeve of the dilatometer. The bellows system core changes in proportion to the total change of all the sensors such that

$$L = \frac{(L_1 + L_2 + \dots . L_n) A_s}{A_b} \tag{8.33}$$

where A_s = cross-sectional area of sensor and
A_b = cross-sectional area of bellows.

The displacement of the core set upon the bellows is therefore determined by the proportion of the cross-sectional area of sensor to that of the bellows. By changing the relations between the two, the sensitivity of the measuring system can be changed. The elastometer 200 is supplied with pressure and displacement recorder system with built-in error correcting circuit so that a direct record of the borehole deformation with pressure can be obtained immediately during testing.

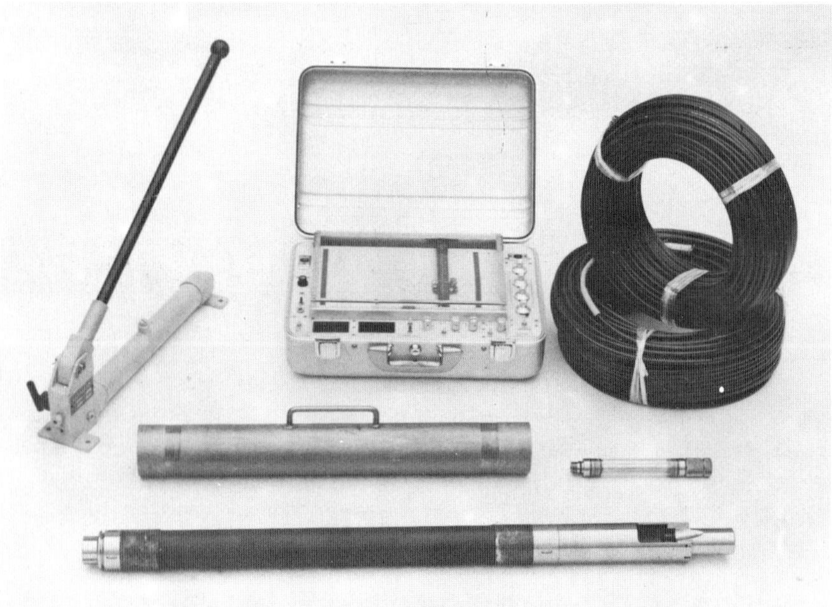

Fig. 8-62. OYO Elastometer 200
(after OYO Corporation, 1976).

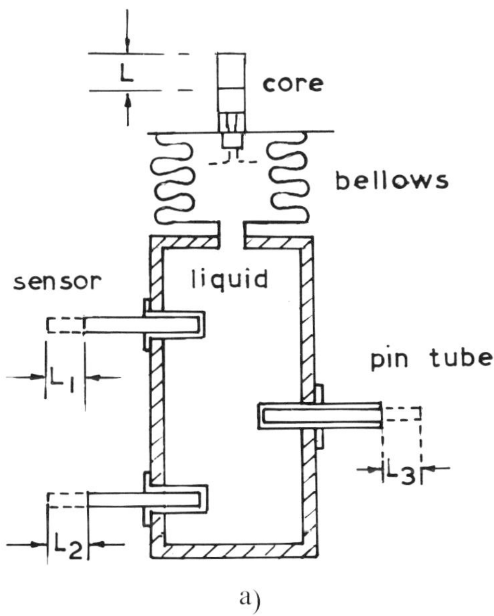

a)

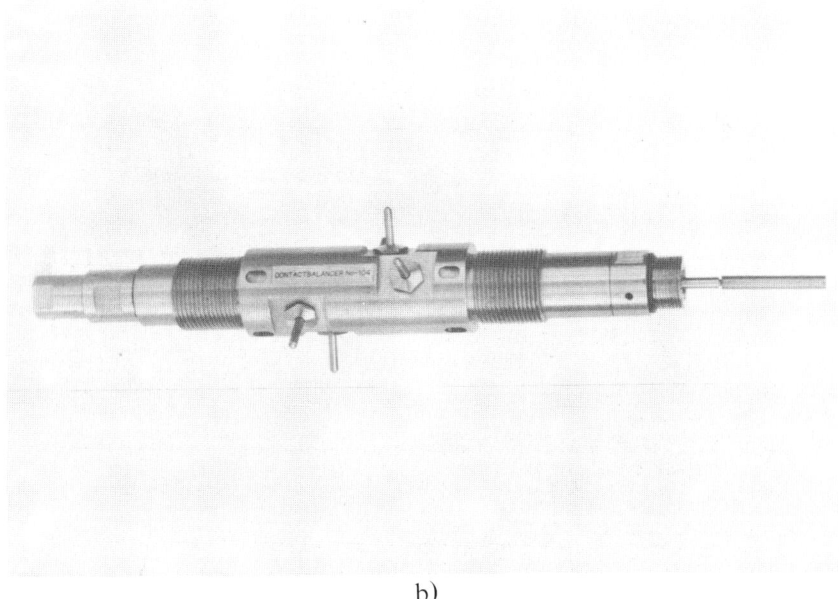

b)

Fig. 8-63. (a) Principle of contact balancer
(b) Contact balancer, OYO elastometer with sleeve removed
(after OYO Corporation, 1976).

8.7.1.4. Calculation of modulus of rock from dilatometer tests

In dilatometer tests, the diametral deformation of the hole subject to uniformly distributed radial pressure is used to define the modulus of elasticity. If a hole in an infinite elastic solid is subjected to uniform pressure such that the loaded length is large compared with the hole diameter, the stresses at any point about the hole using thick cylinder theory, are given by (Fig. 8-64)

$$\sigma_r = p\frac{a^2}{r^2} \quad \text{(compression)}$$

$$\sigma_\theta = -p\frac{a^2}{r^2} \quad \text{(tension)} \tag{8.34}$$

where a = radius of the borehole

p = pressure uniformly distributed on borehole wall and

r = distance from the centre of the borehole

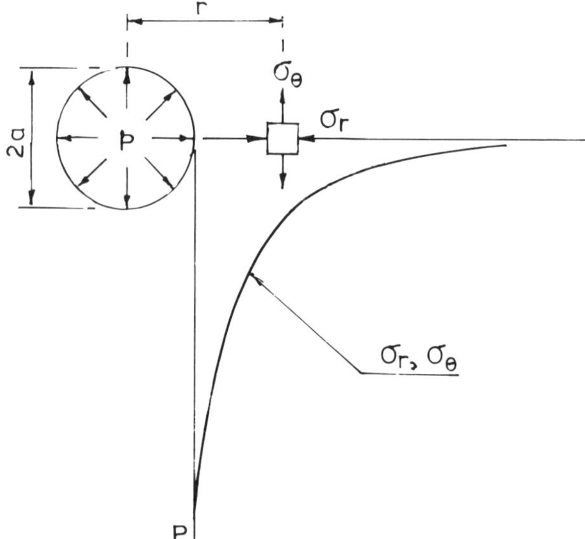

Fig. 8-64. Stress distribution around hole in an infinite elastic solid subjected to uniform pressure.

Since $\sigma_r + \sigma_\theta = 0$, stresses are zero in the elements of area normal to the axis of the borehole in plane equilibrium.

As σ_r and σ_θ decrease very rapidly with the distance to the axis of the borehole and the ratio of the length of the borehole under pressure to the diameter of the hole is large, the above equations are assumed to apply.

The deformation of the diameter of the hole is given by the expression

$$\Delta = 2\frac{1+\nu}{E}ap \tag{8.35}$$

where E and ν are the modulus of elasticity and the POISSON's ratio of the rock mass. The modulus of elasticity can therefore be obtained from the measured value of Δ by means of the expression,

$$E = 2\frac{1+\nu}{\Delta}ap \tag{8.36}$$

8.7.2. Borehole Jacks

Instead of applying a uniform pressure to the full cross-section of the borehole, high pressure can be applied to a part of the borehole surface by driving plates against the borehole walls using hydraulic pistons, wedges or flat jacks. Loading plate in contact with the borehole may be a stiff rigid plate or a flexible plate. The load can be directed in any desired direction and this offers an advantage over the dilatometers. The test can be oriented to study special geological features, e.g. it can be used to measure the force necessary to open a joint plane cutting a hole or determine deformation in any one particular direction.

A number of borehole jacks have been designed and used. Basically these can be divided into two groups; (1) using rigid jacks and (2) using flexible jacks. Rigid jacks have the disadvantage that any mismatch between the borehole and jacking surfaces (unevenness of borehole walls) results in high initial deformation and tilting of the jack. This results in lower modulus value of the rock. GOODMAN's jack, for example, has been found to give lower values and this could be one of the reasons. The jacks having flexible loading surface give a more uniform stress distribution and are hence closer to the theoretical solutions applied in interpretation of results from these tests. Details of some of these instruments are given below.

8.7.2.1. Goodman's jack

The borehole jack designed by GOODMAN, HARLOMOFF and HORNING* is useful for NX diameter boreholes. It consists of two steel plates of angular width 90° forced apart by 12 race-track shaped pistons (Fig. 8-65). Two LVDTs mounted at either end of the 20 cm (7.8 in) long plate measure the displacement of the

* U.S. Patent No. 573920; Inventors: GOODMAN, HARLOMOFF & HORNING; Licensed by Slope Indicator Co., Seattle, Washington.

plate. Two return pistons close the instrument to a thickness of 69.85 mm (2–3/4 in) to give a clearance of 6.3 mm (1/4 in). The total piston travel is 12.5 mm (1/2 in) and the LVDTs have a linear range of 5 mm (0.2 in). The LVDTs can be adjusted to start when the plates come in contact with the hole walls. The pressure applied to the jacks is 640 bars (64,000 kPa) (9282 lbf/in^2) unidirectionally and they give a total force of about 71,670 kgf (158,000 lbf). The volume of rock affected is about 0.028 m^3 (1 ft^3) (GOODMAN, VAN and HEUZE, 1968) and extends to about 114 mm (4.5 in) into rock away from borehole wall.

Modulus values can be calculated using the relationship

$$E = \frac{\Delta p}{\Delta U_d / d} K(\nu, \beta) \tag{8.37}$$

where E = YOUNG's modulus of rock
Δp = pressure increment
ΔU_d = diametral displacement increment
d = diameter of hole and
$K(\nu, \beta)$ = constant dependent upon POISSON's ratio (ν) and angle of loaded arc (β).

TABLE 16

Values of $K(\nu, \beta)$ for Use in Eq. 8.37—Analytical Solution

(after GOODMAN, VAN and HEUZE, 1968)

β	ν : 0	0.05	0.10	0.15	0.20	0.25	0.30	0.35	0.40	0.45	0.50
5.0	0.434	0.433	0.430	0.424	0.417	0.407	0.396	0.382	0.366	0.348	0.327
10.0	0.704	0.703	0.698	0.690	0.678	0.663	0.645	0.622	0.597	0.568	0.536
15.0	0.904	0.903	0.897	0.887	0.873	0.854	0.831	0.803	0.772	0.735	0.694
20.0	1.052	1.051	1.046	1.035	1.019	0.998	0.973	0.942	0.906	0.864	0.818
25.0	1.159	1.159	1.154	1.143	1.127	1.105	1.078	0.045	1.007	0.963	0.914
30.0	1.230	1.231	1.227	1.217	1.201	1.179	1.152	1.119	1.080	1.035	0.985
35.0	1.271	1.274	1.271	1.262	1.247	1.226	1.200	1.168	1.129	1.086	1.036
40.0	1.287	1.291	1.290	1.282	1.269	1.250	1.225	1.195	1.159	1.117	1.069
45.0	1.282	1.288	1.288	1.282	1.271	1.254	1.232	1.204	1.170	1.131	1.087
50.0	1.261	1.268	1.270	1.266	1.257	1.243	1.224	1.199	1.169	1.133	1.092
55.0	1.227	1.236	1.240	1.238	1.232	1.221	1.204	1.183	1.156	1.125	1.088
60.0	1.186	1.197	1.202	1.203	1.199	1.190	1.177	1.160	1.137	1.109	1.077
65.0	1.142	1.154	1.161	1.164	1.162	1.156	1.146	1.132	1.113	1.089	1.062
70.0	1.098	1.111	1.120	1.124	1.125	1.122	1.114	1.103	1.088	1.068	1.045
75.0	1.059	1.073	1.083	1.089	1.091	1.090	1.085	1.076	1.064	1.048	1.028
80.0	1.028	1.042	1.053	1.061	1.064	1.065	1.061	1.055	1.044	1.031	1.013
85.0	1.007	1.022	1.034	1.042	1.046	1.048	1.046	1.040	1.031	1.019	1.004
90.0	1.000	1.015	1.027	1.035	1.040	1.042	1.040	1.035	1.027	1.015	1.000

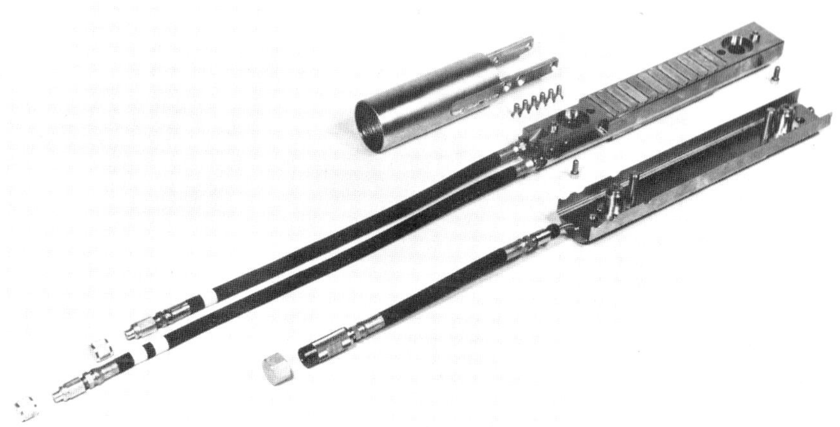

Fig. 8-65. (a) Goodman jack model 52101 hard rock jack (Courtesy Slope Indicator Co.).

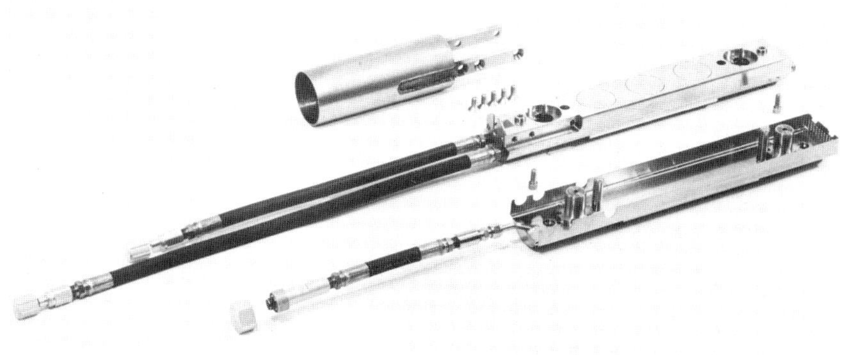

Fig. 8-65. (b) Goodman jack model 52102 soft rock jack (Courtesy Slope Indicator Co.).

The values of $K(\nu,\beta)$ are given in Table 16. Because of the finite length of the loaded area, the values of K in Table 16 have to be corrected. The E values will be about 14% higher than actual values calculated from Eq. 8.37 (for $\nu=0.25$).

8.7.2.2. *C.S.I.R.O. pressiometer*

The C.S.I.R.O. pressiometer (WOROTNICKI, ENEVER and SPATHIS, 1976) (Fig. 8-66) consists of three basic units; (a) curved quadrantal jacks, (b) the collapsible body and (c) deformation measuring system.

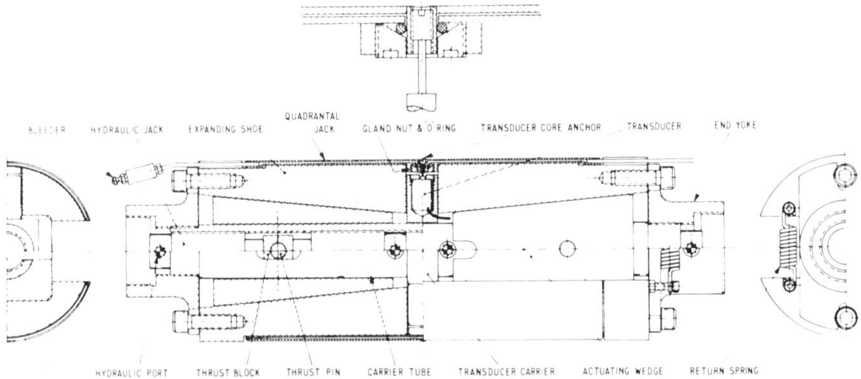

Fig. 8-66. Schematic view of C.S.I.R.O. Pressiometer.
(after WOROTNICKI et al, 1976).

(a) The curved quadrantal jacks are made of 16 gauge sheet steel, preformed to curvature and welded along the edges using standard basic coated rods. The included angle of the jacks is 120° and they have a length of 280 mm. They are fitted with centrally located glands which permit movements of the outside plate (the one in contact with the hole wall) of the jacks to be measured. Both jacks ar inflated together by a common hand pump. Hydraulic connection to the jacks is by means of small diameter steel bundy tubing. Short bleed tubes facilitate the removal of all air. In operation, the jacks fit snugly into position in the shoes, located by means of the central gland and held down firmly by the clamps over the tubes. When the device is expanded, the quadrantal jacks are positively confined by the body against the wall of the hole so that when the jacks are inflated they do not deform excessively. This ensures that the jacks can be used for several cycles of loading.

The 16 gauge jacks are found to be reusable at pressures up to 25,000 to 35,000 kPa (3,625 to 5,075 lbf/in^2) but are too stiff at low pressures, particularly on first pressurising when they give 80% of 'true' diametral displacement only at approximately 7,000 kPa (1,015 lbf/in^2).

(b) The collapsible 'body' serves to bring the jacks in contact with the rock and to support them during pressurising; after the test the 'body' is made to collapse and the jacks are withdrawn from contact with the rock so that the pressiometer can be pulled out of the hole or shifted to a new position along it.

The body consists of two solid steel 'shoes' with semi-circular outer surfaces containing recesses for the curved jacks; and with central holes for displacement measuring devices; two wedges which together with return springs serve to actuate, i.e. to move the 'shoes' out and in as required; four small cylindrical hydraulic jacks of cylindrical shape which are contained in a jack carrier tube and serve to push the wedges in or out along the axis of the pressiometer; thrust blocks and pins to transmit forces from actuating jacks to the wedges; hydraulic connections and end yokes.

The actuating jacks are commercially available single acting jacks with a 25 mm (1 in) stroke, 2 tonnef (1 97 tonf) load capacity and a maximum working pressure of 70,000 kPa (10,152 lbf/in^2). Hydraulic pressure is applied to the actuating jacks from a double acting hand pump via flexible hose and solid steel plumbing. The jacks are operating in two opposed pairs, the outer two together causing expansion and the inner two in conjunction with the return springs bringing about contraction. The actuating jack movements are transferred to the wedges, which are a sliding fit on the jack carrying tube, by means of the high strength thrust pins. The wedges move inward and outward, guided by the slots in the end plates, causing the shoes to expand and contract by approximately 6 mm (0.24 in). A low wedge angle, of 1 : 10 ensures that the wedges and shoes are 'self-locking'. Attached to the end plates are screwed bosses which facilitate connection of the device to the N.Q. drill rods used for placing the instrument into the desired position in the borehole.

(c) The deformation measuring system consists of two DCDT transducers and the free floating transducer carrier. The DCDTs are hermetically sealed, commercially available types with a range of ± 2.5 mm (± 0.1 in). The two DCDT bodies are held in fixed positions in the carrier with 30 m (98 ft) of four channel screened cable connected to each DCDT. The transducer cores are attached to the outside plates of the quadrantal jacks via glands in the jacks as shown in Fig. 8-67. The method of attachment allows for adjustment of the position of the core relative to the transducer body after the quadrantal jacks are placed in position in the expanding shoes. This allows for initial zeroing of the measuring system to ensure that the transducers are operating in their optimum range when the pressiometer is in its operating condition (i.e. expanded).

The two transducers are excited from a commom 6 V D.C. power supply. The transducers are connected together in series and the combined D.C. output is registered on a 1 microvolt sensitivity digital voltmeter (giving potential resolution of deformations of 1.2 micron). The pressiometer is suitable for borehole diameter of 150 mm (5.9 in) and the modulus of elasticity can be calculated using Eq. 8.37.

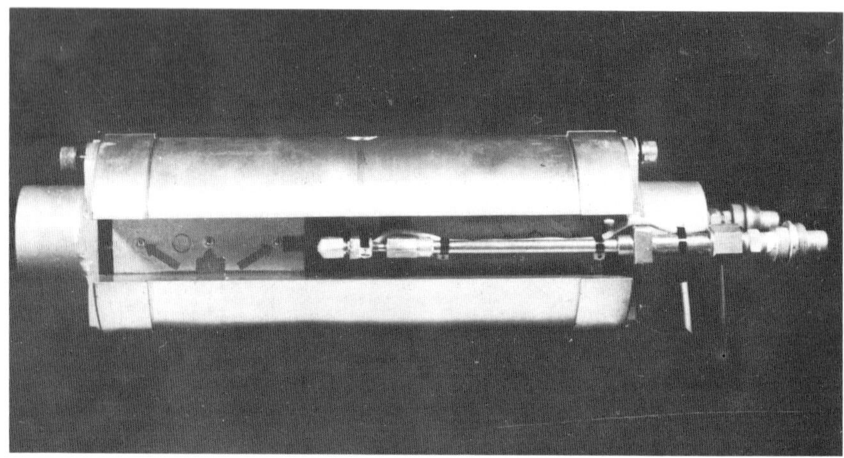

Fig. 8-67. C.S.I.R.O. Pressiometer.
(after WOROTNICKI et al, 1976).

8.7.3. Borehole Penetrometers

Borehole penetrometers are based on the concept of penetration of a small rigid die into the borehole wall. The contact pressure in these instruments is very high while the loaded volume does not exceed 100 cm^3 (6.1 in^3). The devices are used mainly to determine the response of rock for local loads such as rock bolts, e.g. U.S.B.M. borehole penetrometer (STEARS, 1965) or as stiff inclusions in stress measurements (HULT, 1963; DRYSELIUS, 1965). For modulus determination, these do not have much significance.

8.7.4. Testing Procedure in Using Borehole Deformation Instruments

Depending upon the diameter of the instrument, the size of boreholes to be drilled is about 2–3 mm (0.08–0.12 in) greater in diameter. When borehole diameter is too large the pressurising surface will use up most of its displacement before contacting the borehole wall which limits the actual borehole displacement. For this reason special care must be taken in drilling the boreholes. Diamond drilling in hard rocks gives a good smooth surface with low tolerances. Cores obtained during drilling can be used to obtain further data about the rocks at the location of the tests.

In softer rocks, protection of the boreholes can be achieved using thin plastic casing pipes which can be withdrawn immediately before the test. It is desirable to start the test immediately after drilling has reached to the depth at which the test is to be conducted.

The various methods of loading that can be used are given in Fig. 8-68.

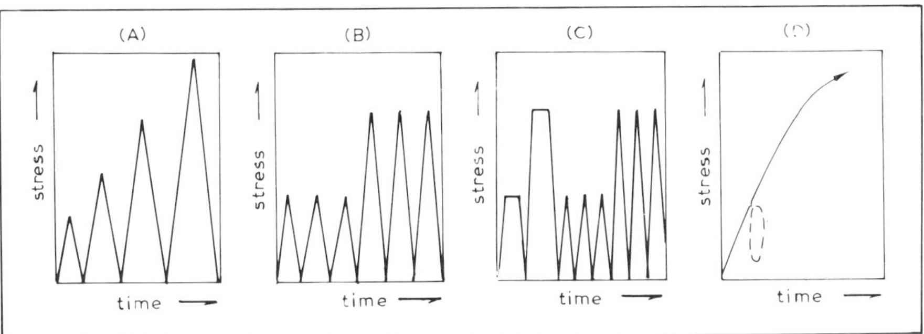

Fig. 8-68. Methods of loading using borehole deformation instruments.

(A) is the pattern of repeated loading by steps, in which loading is made by steps until the fixed load is reached. By this means it is possible to observe differences in behaviour in deformation due to various load levels, and the effect of consolidation due to repetition.

(B) is the pattern of repeated loading at the same level and repetition of the test at higher load level. The difference between (A) and (B) is that the latter puts more emphasis on observation of the consolidation effect of repetition than on difference due to load levels.

(C) is the pattern of measuring deformation of creep due to constant loading. Repeated loading after continued load is intended for observing elastic character. By this means it is possible to obtain modulus of elasticity and creep coefficient.

(D) is a pattern which puts emphasis on yielding load. It is possible to take displacement features from the curves inclination leading up to the yield, but when repetition is inserted therein as shown in the broken line, it is possible to obtain modulus of elasticity due to repetition effect.

Rate of loading should be kept as small as possible. Test should be conducted at least over an hour period and loading rate adjusted depending upon the maximum load to be achieved. In deciding the maximum pressure to be exerted on borehole walls, it is necessary to estimate in advance the deformation modulus of the material and thereby guess displacement scale for the load and decide the limit of measurement. This is also necessary for setting the gain of any displacement recording system.

8.7.5. Interpretation of Results from Borehole Jacks

JAEGER and COOK (1969) have developed the solution for stresses applied to a circular hole using complex variable method. For a uniform radial pressure p applied to the surface of a borehole over opposite arcs of $(-\alpha < \theta < \alpha$ and $\pi - \alpha < \theta < \pi + \alpha)$ (Fig. 8-69) of a circular hole in an infinite elastic isotropic body, the radial displacement of the surfaces along the line passing through the centre of the loaded arcs can be given by

$$\Delta R = \frac{Rp}{2\pi G}\left[-2\alpha - x\left(\frac{1}{2}\pi\cos\alpha + \sin\alpha\, ln\cot\frac{1}{2}\alpha - \frac{1}{2}\pi + \alpha\right)\right.$$
$$\left. - \left(\frac{\pi}{2} - \alpha - \frac{1}{2}\pi\cos\alpha + \sin\alpha\, ln\cot\frac{1}{2}\alpha\right)\right] \qquad (8.38)$$

where ΔR = radial displacement at $\theta = 0$
R = radius of the hole
p = applied radial pressure
G = shear modulus of the rock $\left(G = \frac{E}{2(1+\nu)}\right)$
E = YOUNG's modulus
ν = POISSON's ratio and
x = $(3\text{–}4\,\nu)$ for plane strain conditions

Analysis of Eq. 8.38 shows that maximum radial displacement will occur with the arc angle $2\alpha = 110°$ and that small variation about the angle would not significantly influence the radial displacement (Fig. 8-69). As long as the jacks are flexible (e.g. C.S.I.R.O. pressiometer) any variation in contact angle will not make any significant change in the E value determined. But for rigid jacks (e.g. GOODMAN's jack) the loss of contact will distribute the total force on a smaller area and hence result in increased deformation for the same internal pressure.

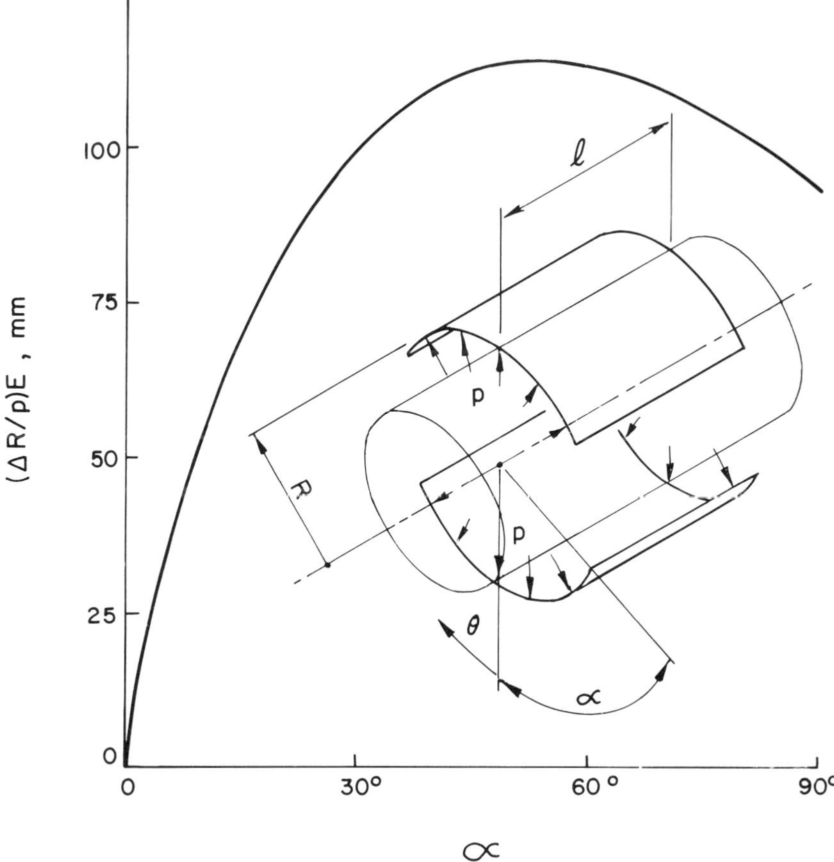

Fig. 8-69. A theoretical relationship between arc of loading and deformability of rock surrounding a borehole (after WOROTNICKI, ENEVER and SPATHIS, 1976).

The POISSON's ratio (ν) for rock must be known for the calculation of YOUNG's modulus. The influence of any error in assessing the value ν does not seriously affect the E values calculated except at high values of ν. In C.S.I.R.O. pressiometer, a variation of ν from 0.1 to 0.25 results in error in estimation of E by 1.3%. In GOODMAN's jack, taking value of ν as 0.3 instead of a true value of 0.2 results in error in the calculated value of E by 3.25% and for $\nu = 0.4$ instead of true value of 0.2 gives an error in E by 8.50%.

The influence of finite length of the loading surface is more important. Fig. 8-70 gives the variation of displacement at the border of a borehole over the width and length subjected to a uniform pressure for GOODMAN's jack. The K values

(Table 16) (for $\nu = 0.25$) for average displacement gave a value of 1.23 instead of 1.06 showing that the E values calculated using Eq. 8.37 are about 14% higher. Increase in the length of the loaded surface will decrease this error. DAVYDOVA (1968) has shown that for diameter to length ratio of about 1.2, the E values are underestimated by 15–20%. The error is insignificant when length to diameter ratio is 4–5. Table 17 gives the influence of length/diameter ratio on the displacement for rigid and flexible jacks. For flexible jack borehole displacement approaches two-dimensional value for the length/diameter ratio of 5.

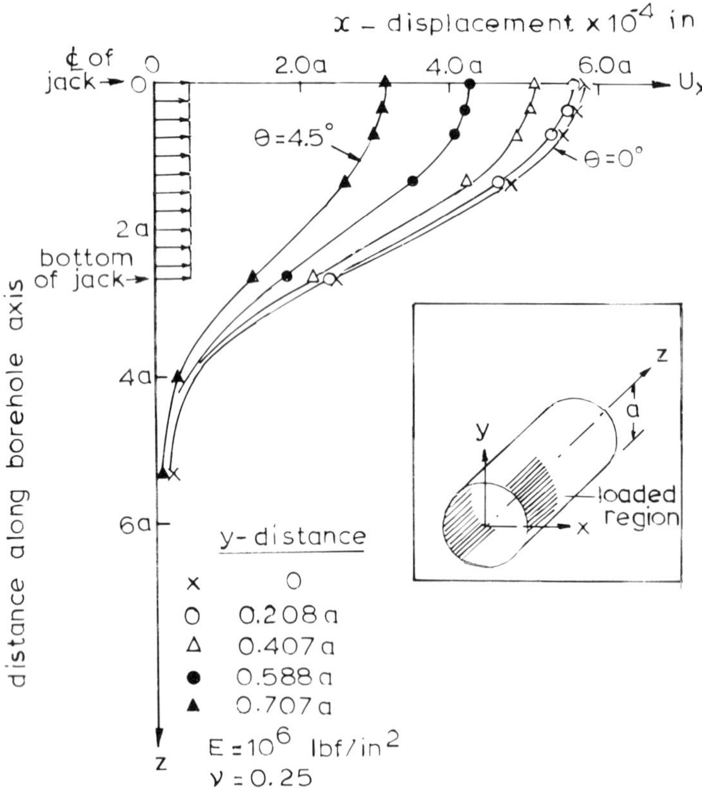

Fig. 8-70. Variation of x-displacement along the borehole (z-direction) at different points around the wall of the borehole. Prismatic space – constant pressure = 375 lbf/in^2 (after GOODMAN, VAN and HEUZE, 1968).

TABLE 17

Influence of length/diameter ratio of borehole jack on borehole deformation

(after DAVYDOVA, 1968)

Borehole displacement	0.25	0.5	Length/diameter ratio 1.0	2.0	3.0	5.0	Two-dimensional problem
U_d (Flexible jack)	0.585	0.830	1.054	1.244	1.263	1.290	1.30
U_d (Rigid jack)	0.524	0.742	0.940	1.100	1.182	1.237	1.30

TABLE 18

Influence of Crack Formation on the Measure of In Situ Modulus of Deformation

(after HEUZE, 1970)

Length of radial crack measured from borehole wall	0	$a/2$	a	5a
Apparent decrease in E %	0	13	24	29

a is the radius of the hole

As in the case of dilatometers, the borehole jacks develop tangential tensile stress on the wall of the borehole at $\theta = 90°$ (at right angle to the axis joining the opposite jacks). The value of the tensile stress is given by

$$\sigma_\theta = -4\beta \frac{p}{\pi} \tag{8.39}$$

This could result in tensile cracking at this point and this shows up on the stress-displacement curve as a point of sudden change in its slope. Appearance of a crack or the presence of a crack could result in the decrease in the value of modulus value measured depending upon the length of the crack.

Table 18 gives the influence of crack formation on the value of E in GOODMAN's jack. If fissures are present and extend to a distance b from the axis of the hole and make an angle α with one another (Fig. 8-71), the influence of these on the deformation of the borehole can be estimated as follows:

Let us assume that the state of stress in the wedge *LML'M'* (Fig. 8-72) at a point at a distance *r* from the axis of the hole is (ROCHA et al, 1966)

$$\sigma_r = \frac{ap}{r} \tag{8.40}$$

$$\sigma_\theta = \tau_{r\theta} = 0 \tag{8.41}$$

and also that the distribution of stresses beyond a cylinder of radius *b* is the same that arises in an indefinite solid with a hole of radius *b* subjected to a uniform pressure $p' = p\frac{a}{b}$. The deformation of the wedge in a radial direction is

$$\int_a^b \frac{\sigma_r}{E} dr = \frac{ap}{E} \log \frac{b}{a} \tag{8.42}$$

and the deformation of the hole considered is given by

$$\frac{bp'}{E}(1+v) = \frac{ap}{E}(1+v) \tag{8.43}$$

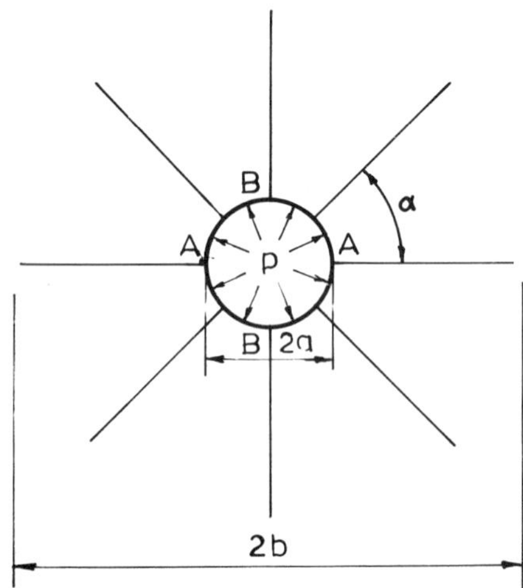

Fig. 8-71. Notation for fissures around the hole.

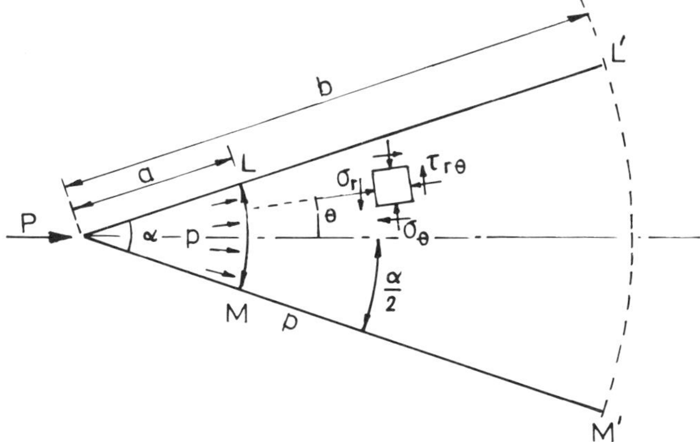

Fig. 8-72. Notation for stresses around the hole.

Therefore the deformation of the diameter of the hole in a fissured rock mass equals twice the sum of the preceding deformations

$$\Delta' = 2\,\frac{ap}{E}\left(\log\frac{b}{a} + 1 + \nu\right) \tag{8.44}$$

From the preceding assumptions on the distribution of stresses in a fissured solid, a value of Δ' independent of the angle α, i.e. of the number of fissures, and of the azimuth θ was derived. Therefore Eq. 8.44 is not correct but its error is not considerable.

The state of stress and strain in the diameter of a fissured plate with a hole are calculated by an integral numeric method recently developed by OLIVEIRA (1955). Consider the case of a hole with a diameter $2a = 7.6$ cm (3 in) under a pressure $p = 9808$ kPa (100 kgf/cm^2) (1422 lbf/in^2) and with 2,4 and 8 fissures with depths such that b/a amounts to 2,4 and 8 (Fig. 8-73). It is also assumed that $E = 9808$ MPa (100,000 kgf/cm^2) (1,422,000 lbf/in^2) and $\nu = 0.2$. The values of the deformation Δ' of the diameter, computed on the basis of Eq. 8.44 and of the deformation of the diameter at points P and Q of Fig. 8-73 designated by Δ'_P and Δ'_Q respectively, computed by that numeric method, are given in Table 19; the values of the deformation of the diameter in a solid without fissures are also indicated.

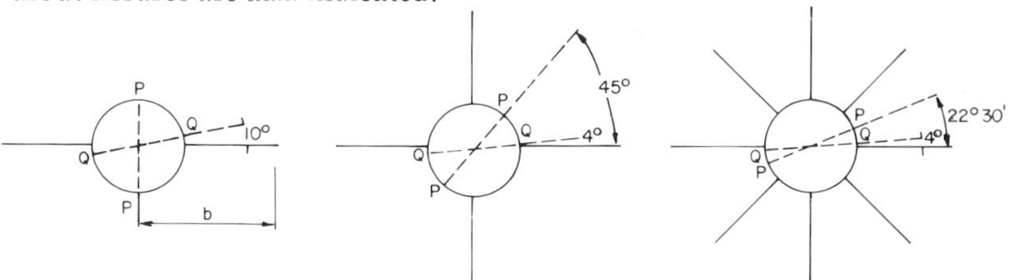

Fig. 8-73. A hole with fissures under pressure.

TABLE 19

(after ROCHA et al, 1966)

Conditions	$\frac{b}{a}$	Δ' (μ)	Δ'_P (μ)	Δ'_Q (μ)	$\frac{\Delta'_P - \Delta'_Q}{2}$ (μ)
2 fissures	2	142	130	82	106
	4	194	192	86	139
	8	246	254	94	174
4 fissures	2	142	132	101	117
	4	194	182	134	158
	8	246	246	200	223
8 fissures	2	142	142	140	141
	4	194	194	214	204
	8	246	236	242	239
No fissures	0	90	90	90	90

Table 19 shows an excellent agreement between the values of Δ' and Δ'_P, which correspond to the points farthest removed from the fissures. Therefore in the particular case in which the location of the fissures is known, very satisfactory results can be obtained from Eq. 8.44, if measurements are taken at points P. On the other hand, considerable differences are observed between Δ' and Δ'_Q, except in the case of 8 fissures. Taking the mean of Δ'_P and Δ'_Q, this value is seen to differ by less than 20% from Δ', save in the case of two fissures. If the instrument measures deformations along four diameters, it makes possible to determine a mean deformation, which, as we have just seen, fairly approaches Δ', all the more so as the case of two fissures rapidly turns into the case of four fissures. It follows that, the values of b and E being known, Eq. 8.44 yields a satisfactory value for the mean deformation of the diameter of the hole and parameters α and θ need no longer be taken into account.

As seen, in a solid with a hole stresses are zero in the surface elements normal to the axis of the hole. Yet when fissures occur, the state of stress in the wedges thus formed becomes, $\sigma_r = \frac{ap}{2}$, giving rise to a compressive stress amounting to $\sigma_z = \nu\sigma_r$ parallel to the axis of the borehole, as the equilibrium is assumed to be plane. Therefore the radial deformation at each point becomes

$$\frac{\sigma_r}{E} - \frac{\nu\sigma_z}{E} = \frac{\sigma_r}{E}(1-\nu^2).$$

For the current values of ν, ν^2 can be neglected, i.e. the influence of σ_z need not be considered in the calculation of the deformation Δ' of the diameter of the hole.

To determine the influence of the fissures on the value of the deformation of the diameter of the hole, the unit increase of the deformation of the diameter due to the occurrence of fissures is required and given by

$$\frac{\Delta'-\Delta}{\Delta}=\frac{1}{1+\nu}\log\frac{b}{a}=\frac{2.3}{1+\nu}\log_{10}\frac{b}{a} \tag{8.45}$$

The values of $\dfrac{\Delta'-\Delta}{\Delta}$ in percentage for different values of b/a and assuming $\nu=0.2$ are presented below:

b/a	1	2	4	6
$\dfrac{\Delta'-\Delta}{\Delta}$ (%)	0	58	115	150

Therefore the influence of fissuring on the deformation of the diameter proves to be very important. Notice, however, that given the limited length of the borehole subjected to pressure p, the value of b/a cannot be very large.

If σ_t is the tensile strength of the rock and fissures are assumed to extend down to a depth such that $\sigma_\theta=\sigma_t$, it follows that

$$\sigma_t=\frac{a^2p}{b^2}$$

and therefore the depth reached by the fissure is

$$b=a\sqrt{\frac{p}{\sigma_t}} \tag{8.46}$$

Consequently the deformation of the diameter is given by the expressions

$$\Delta=2\frac{ap}{E}(1+\nu) \qquad \text{for } p\leq\sigma_t \tag{8.47}$$

$$\Delta'=2\frac{ap}{E}\left(\log\sqrt{\frac{p}{\sigma_t}}+1+\nu\right) \qquad \text{for p} > \sigma_t \tag{8.48}$$

The curve representing the evolution of the deformation of the diameter in a test during which fissuring occurs in the rock mass is presented in Fig. 8-74.

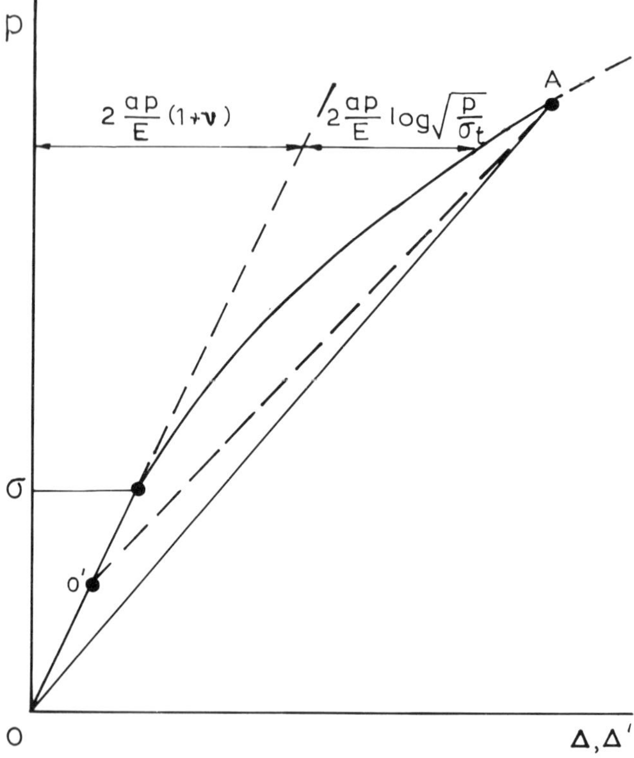

Fig. 8-74. The curve representing the evolution of the deformation of the diameter in a test during which fissuring occurs in rock mass (after ROCHA et al, 1966).

The modulus of elasticity can be calculated from the linear stretch by means of the expression

$$E = 2\frac{ap}{\Delta}(1 + \nu) \tag{8.49}$$

Since for $p > \sigma_t$, the calculation of E requires ascribing to the tensile strength of the rock a value which as a rule is not reliable, it is very advisable to begin measuring Δ at very small pressures in order to observe the linear stretch.

If the diagram obtained in a test does not begin with a well-marked linear stretch, the modulus of elasticity has to be calculated by the expression

$$E = 2\frac{ap}{\Delta'}\left(\log\sqrt{\frac{p}{\sigma_t}} + 1 + \nu\right) \tag{8.50}$$

where Δ' is the average of the values observed for the deformation of the four diameters. As for the uncertainty of the value of σ_t, notice that E changes very slowly with σ_t.

If the rock mass behaves elastically and if failure obeys the theory assumed for deriving the Eq. 8.44, the value of σ_t can even be determined from the experimentally determined curve of Δ' as function of p. In fact if we ascribe to the tensile strength in the expression of E not its true value σ_t but $n\sigma_t$, the following expression is then obtained as function of p:

$$f(p) = 2\,\frac{ap}{\Delta'}\left(\log\sqrt{\frac{p}{n\sigma_t}} + 1 + \nu\right) \tag{8.51}$$

hence

$$f(p) = \frac{2ap}{\Delta'}\left(\log\sqrt{\frac{p}{\sigma_t}} + 1 + \nu\right) + 2\,\frac{ap}{\Delta'}\log\sqrt{\frac{1}{n}} \tag{8.52}$$

$$f(p) = E + 2\,\frac{ap}{\Delta'}\log\sqrt{\frac{1}{n}} \tag{8.53}$$

This means that by ascribing successive values to σ_t and computing $f(p)$ from experimental values of p and Δ', a family of straight lines should be obtained and the true tensile strength is the value of σ_t for which $f(p)$ is constant. The existence of such a value of σ_t will confirm the reliability of the expression determined for Δ'.

Determination of deformability using borehole devices has certain advantages. These devices allow measurement of deformability at points far removed from the excavation and hence at places which are not effected by drivage of access galleries and disturbances caused due to site preparation. The method being simple requires much less site preparation, and permits a larger number of tests to be conducted in a shorter time at low costs.

The disadvantage of the method is the small volume of the rock involved in each test so that the results obtained may not be representative of the deformability of the rock mass. Besides, the borehole devices develop tensile and radial stresses in the borehole walls and the modulus values calculated should be different from those obtained using jack tests. These values are more comparable with those obtained from pressure chamber tests except for the volume effect. Figs. 8-75 and 8-76 show the results obtained at some test sites using OYO elastometer 200 and jack tests using a 30 cm (11.8 in) diameter jack. Jack test shows higher values than dilatometer test at lower dilatometer values. This is true for both repetitive and non-repetitive modulus values. Table 20 compares the results of tests using GOODMAN borehole jack and other test methods. The values obtained using the borehole jack compared to plate bearing method are higher for jointed and fractured rock but lower for the massive marble in Crestmore mine. While in jointed rocks, the volume effect seems to dominate, but in massive rock perhaps the type of stress exerts a more dominating influence.

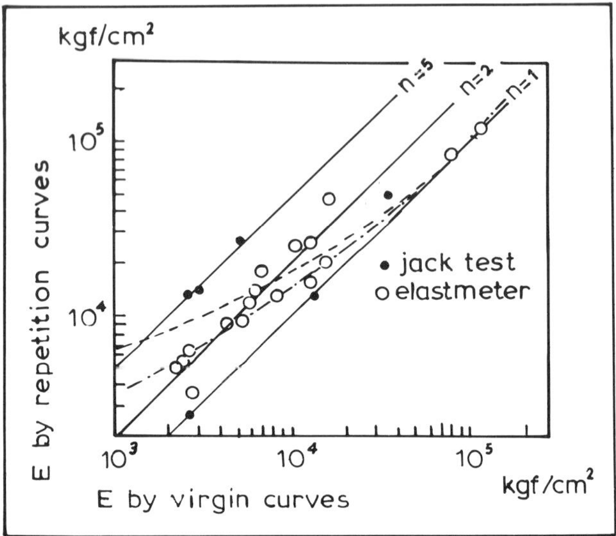

Fig. 8-75. Effect of repetition loading (after OHYA et al, 1975).

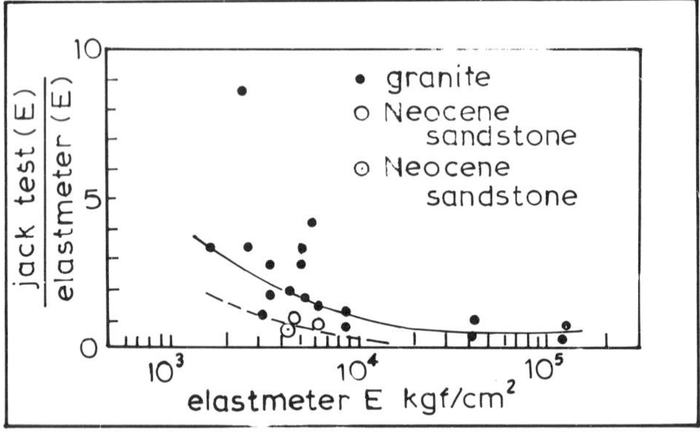

Fig. 8-76. Relation for elastmeter and jack test (after OHYA et al, 1975).

Similarly plate loading (280 mm (11 in) diameter punch) and dilatometer (160 mm (6.3 in) diameter and 1600 mm (63 in) long) tests conducted at Vouglans dam (DUFFAUT and COMES, 1966) gave deformation modulus values obtained from dilatometer consistently higher than plate loading tests. These results are given in Table 21. The ratio of the minimum deformation modulus values between the two tests is 2.00 and the elastic deformation values 1.87. The ratio drops for the corresponding maximum values to 1.84 and 1.45 respectively.

TABLE 20

Comparison of in situ, core and borehole jack tests

(after GOODMAN, VAN and HEUZE, 1968)

Site	Rock type	POISSON's ratio	YOUNG's Modulus, 10^6 lbf/in^2			
			Unconfined compression (laboratory average)	Plate bearing* (in situ)	Flat jack* (in situ)	Borehole jack* (in situ)
Tehachapi Tunnel	diorite gneiss; fractured & seamy	0.35	11.3	0.53 to 0.83		0.61 to 1.03
Dworshak Dam	granite gneiss; massive to moderately jointed	0.20	7.5	0.5 to 5.0		1.54 to 2.70
Crestmore Mine	marble; massive	0.25	6.9	1.74 to 2.72	1.79 to 2.98	1.35 to 1.70

* In the same pressure range 0–3000 lbf/in^2.

TABLE 21

Comparison of Moduli

(after DUFFAUT and COMES, 1966)

Test method		Modulus of deformation E_d, MPa	Modulus of elasticity E_e, MPa
Circular plate	Minimum	3.2	7.1
	Mean	10.0	16.0
	Maximum	18.5	26.4
Dilatometer	Minimum	6.5	13.3
	Mean	19.0	26.0
	Maximum	34.1	38.3

U.S.B.M. tests made in Idaho Springs granitic gneiss (Colorado) have shown that the jointing greatly influences the test results. The ratio of apparent value of shear modulus G_a $\left(G = \frac{E}{2(1+\nu)}\right)$ and real shear modulus G_o of rock as a function of average joint spacing $\bar{s}$ and borehole diameter d is given in Fig. 8-77. This functional relationship revealed by this figure indicates that for $\bar{s}/d < 1/4$ which seems likely for a large structure such as a tunnel, G_a or E_a of the rock mass may be no more than 1/10 or 1/4 of the value obtained from intact specimens in laboratory. U.S.B.M. dilatometer tests on weaker materials (Black shale and Gray shale, Casey Ville formation, Rock River juncture interstate bridge, compressive strength = 240 lbf/in^2 and 100 lbf/in^2 (1.65 MPa and 0.69 MPa)) showed values higher than laboratory values by a factor of 5 to 13 (HENDRON et al, 1969). It is possibly due to laboratory specimen disturbances. In another site field values using the dilatometer in sandstone and shales showed consistently lower modulus values than laboratory tests possibly because the specimens selected for laboratory tests were of better quality. Higher values of dilatometer tests approached that of laboratory tests (DIXON, 1969).

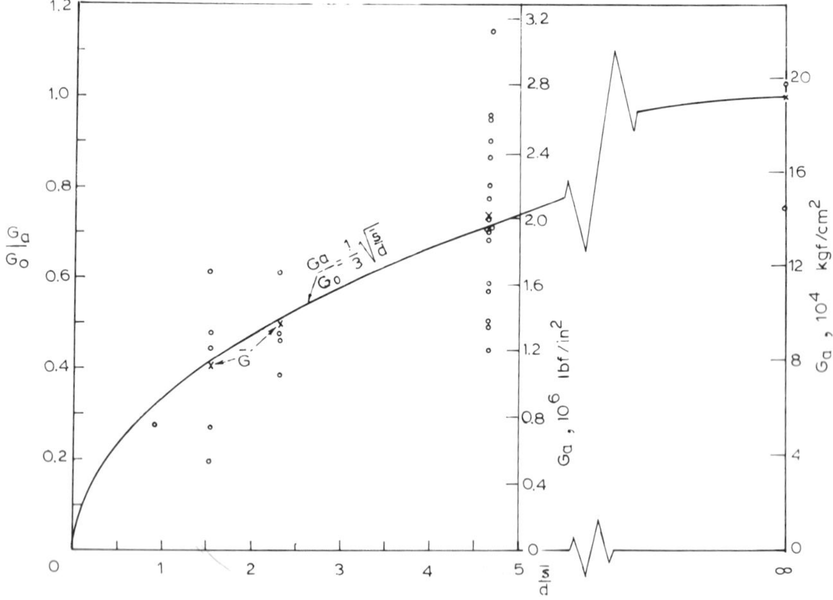

Fig. 8-77. Dilatometer test data – modulus vs $\bar{s}/d$ (after PANEK, 1970).

While comparing modulus values obtained from different tests, it is worth remembering that comparison can be valid only when the plane of rock deformation in the plate bearing test and the borehole deformation test is the same.

This is possible only when test borehole is drilled at right angle to the direction of loading in the plate bearing test. Also in case of high ratio between the principal stresses prevailing at a point, and nonlinear behaviour of the rock, the values obtained could be affected considerably.

Dilatometer or borehole jacks are very useful for first assessment of the deformability and subsequent zoning of the rock mass in dam foundations and other large structures because of the ease of access to deeper regions through galleries or surface boreholes.

8.8. Deformability of Rock Mass

The pressure-displacement curves for rock mass can be divided basically into four groups (Fig. 8-78). Curve (a) represents progressive closing of joints and increasing compaction and is the most commonly obtained type of behaviour. Curve (b) represents the behaviour of rock mass containing a large portion of weak material such as clay etc. where the solid portion is extremely small and is getting crushed as stress increases. This results in transfer of load to the weaker constituents. If a sudden change in the slope of the curve takes place without corresponding reversal of deformation, the point may conform to an internal failure. Tests on phyllite-quartzite showed a sudden change in the slope of the curve at two points—one for the phyllite phase (small crystals) and second at higher pressures for the quartzite phase (larger crystals). Curve (c) represents a general shape where the weaker material is broken or squeezed out in stages and more and more load is transferred to solid points with partial destruction of the solid—solid contact points at different stages of compaction. Curve (d) represents the behaviour of an unweathered fresh rock mass and it rarely is observed in practice except in deep excavations. The deformability of rock mass is dependent mainly upon the joints, bedding planes and softer beds in the system. When joints are filled with clay or beds contain softer materials, a certain amount of creep is observed. If clay is compact and dry, its influence is minimised. The influence of faults is more or less the same as that of joints.

In general, deformability decreases with increase in load indicating compaction of the rock mass. Strain recovery on unloading is small in the early stages and rapid recovery takes place only when a fair amount of unloading has been done (Fig. 8-79). Sliding along joints and their back sliding on destressing follows the same principles as for cracks (Section 6.7.) and occurs after a definite stress drop. This gives rise to hysteresis. Only in compact, massive rocks, the hysteresis effect is very small. The heavier the fissuration and weathering, the stronger the hysteresis effect and greater the concavity of the load-deformation curve.

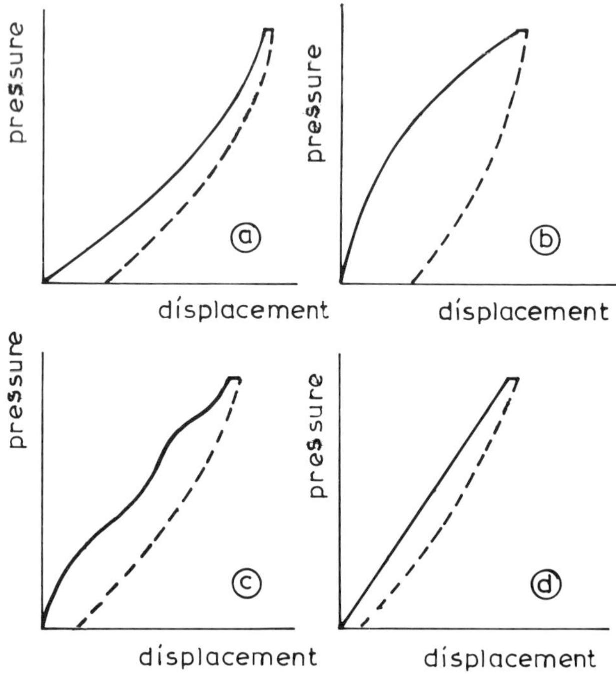

Fig. 8-78. Types of pressure – displacement curves obtained from plate loading tests.

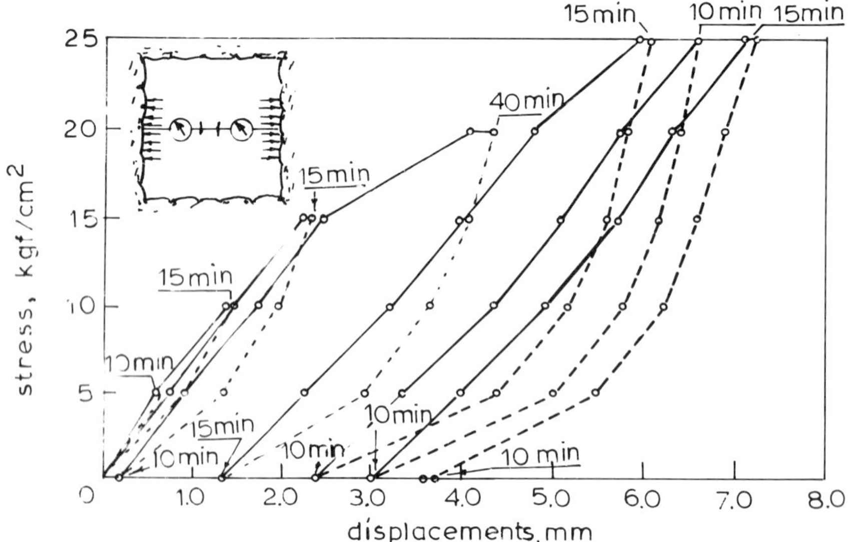

Fig. 8-79. Deformability tests in granites before grouting: Alto Rabagao Dam (after ROCHA, 1964).

The ratio of the deformation modulus during loading and unloading is considered by several investigators as a measure of the jointing and weathering. The ratio between loading and unloading modulus is high for rocks having high deformability (lower modulus) (Fig. 8-80). The unloading tangent modulus, particularly at high loads, many times lies very close to the dynamic modulus measured by geophysical methods.

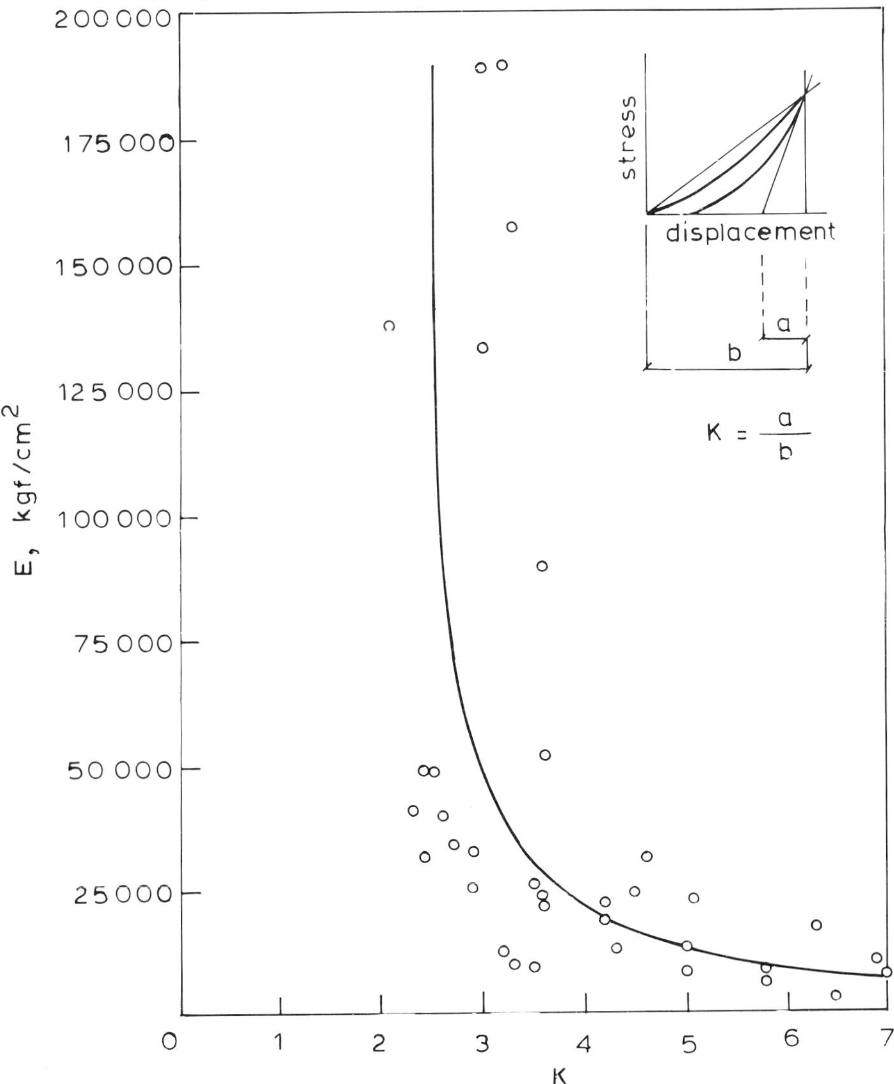

Fig. 8-80. Ratio between loading and unloading deformability: Bemposta Dam (after ROCHA, 1964).

Modulus of elasticity drops rapidly with weathering. A measure of weathering is the porosity of the rocks which increases as weathering increases. The relationship between the index *i* (ratio of water absorbed to weight of dry rock at 105°C, sometimes expressed as alteration index) and modulus of elasticity for granite is given in Figs. 8-81 and 8-82. The modulus of elasticity of granite in the two directions, horizontal and vertical, has a ratio of about $E_H/E_V = 2$. This ratio tends to be maintained even for advanced stages of alteration.

The influencing of cyclic loading is to increase the elasticity of the material. Usually in the second cycle and after, the loading and unloading curves seem more or less parallel provided the maximum load applied remains the same (Fig. 8-83).

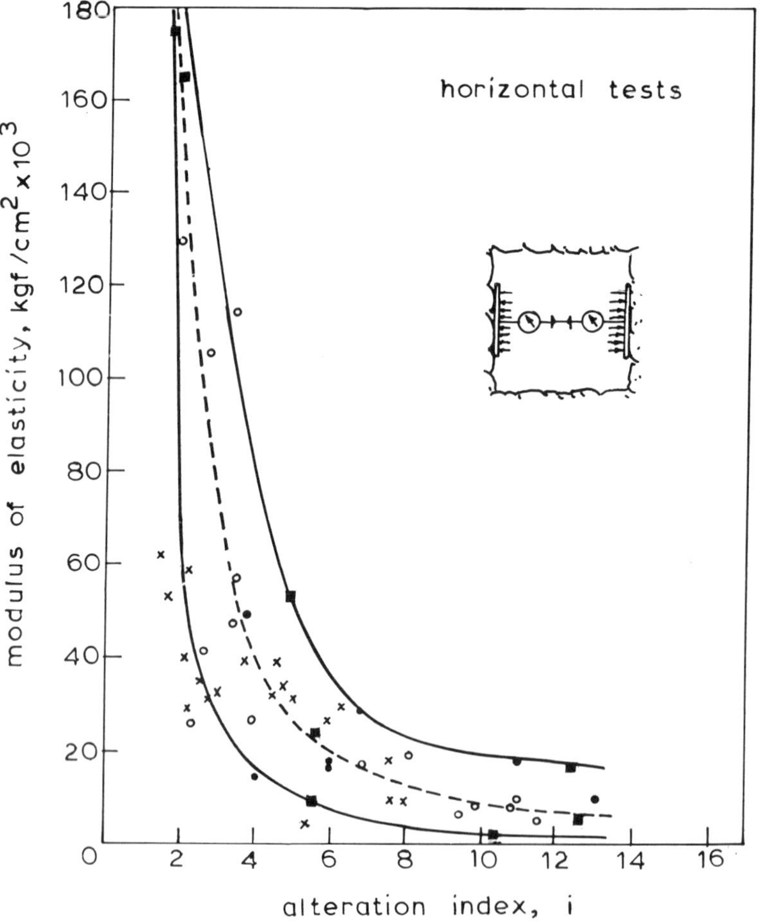

Fig. 8-81. Correlation between modulus of elasticity obtained in situ and alteration index: Alto Rabagao Dam (granite) – horizontal loading (after ROCHA, 1964).

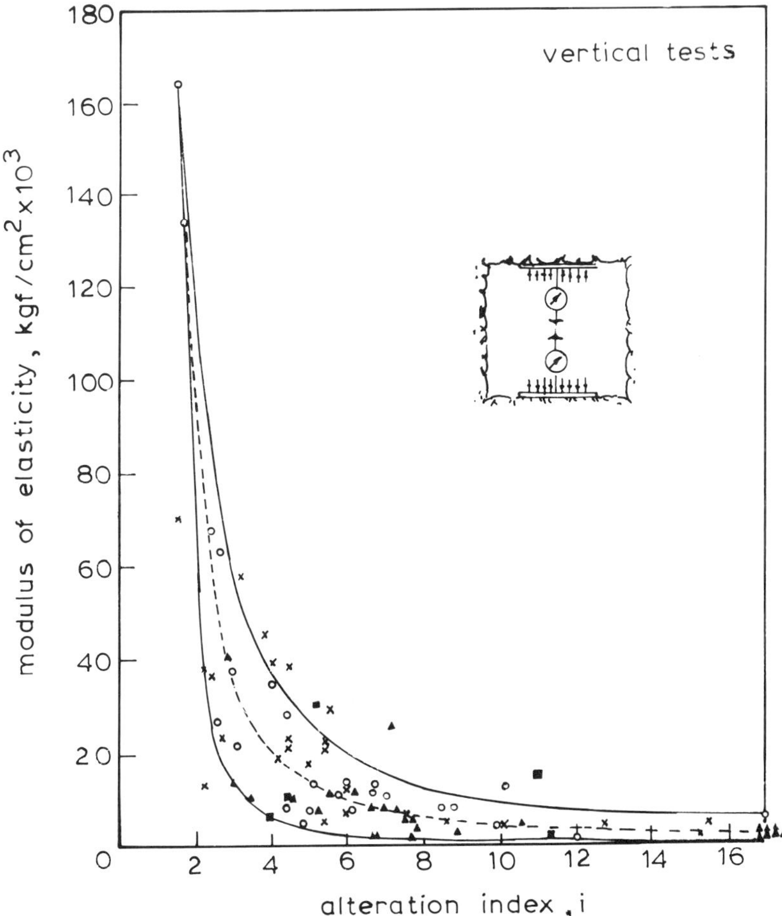

Fig. 8-82. Correlation between modulus of elasticity obtained in situ and alteration index: Alto Rabagao Dam (granite) – vertical loading (after ROCHA, 1964).

Decrease in deformability after grouting is an indication of the strengthening of the rock mass with the voids filled in with grout and exerting a binding effect. In certain cases, increase in deformability or no decrease in deformability may occur after grouting, which is an indication of damage to the rock mass due to application of higher grouting pressures, unsuccessful grouting or due to decrease in friction between the joint sets as a result of presence of grout.

As already indicated, deformability is not the same in all directions and this is evident in various tests conducted using different methods. Hydraulic pressure tunnel tests easily permit the determination of the modulus polar diagram. The deformation obtained in a marl deposit is given in Fig. 8-84. The solid line represents the total deformation and the broken line the elastic deformation. The ratio between the elastic and total deformation is not the

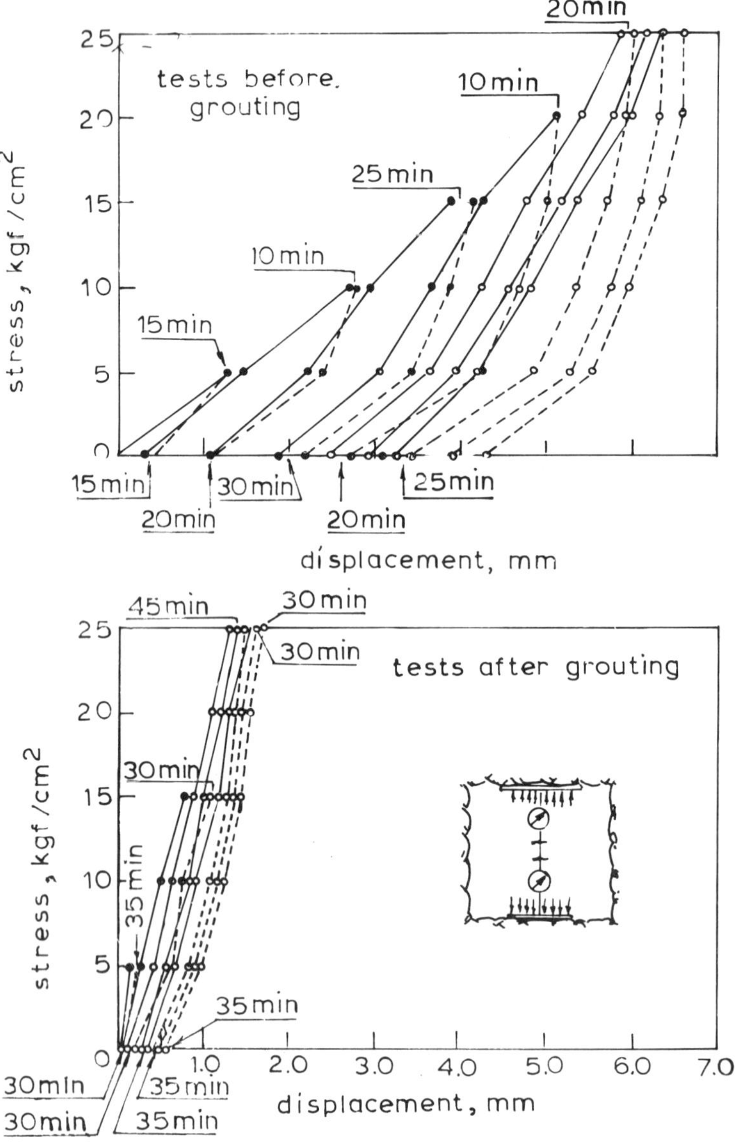

Fig. 8-83. Deformability tests in granites before and after grouting: Girabolhos Dam (after ROCHA, 1964).

same in all directions, though the general shape of the curves seems to be symmetrical. Such curves are also very useful in predicting the grouting efficiency. Fig. 8-85 shows the results of pressure chamber tests before and after grouting in schistose gneiss at Graminha Dam, Sao Paulo. The rock anisotropy is very clear in both the pre- and post-grouting tests. It is of interest to note that while largest deformation is obtained in the horizontal direction at low pressures 393 and 883 kPa (4 and 9 kgf/cm^2), (57 and 128 lbf/in^2), at higher pressures the direction of maximum deformation is in direction 2, a displacement of about 60°. This direction remains dominant in post-grouting both at low and high pressures. The anisotropy observed at low pressure is due to induced anisotropy as a result of faulting, jointing, fracturing, etc. and at high pressure is the inherent anisotropy. The cell 2 represents the direction at right angle to the schistosity and hence direction of minimum modulus. Grouting eliminates induced anisotropy but does not very much influence the inherent anisotropy.

The anisotropy ratios for some rocks are given in Tables 9 and 10 of Volume II. The range of values from in situ tests may not be close to the laboratory tests. Fig. 8-86 shows the results of in situ tests on granite and Graywacke schist. The values for modulus of deformation for stress parallel to planes of schistosity are plotted along the ordinate and for stress at right angle to the schistosity along the abscissa. Though granite shows an expected behaviour with the ratio between the moduli $\simeq$ 1:3 with the maximum value parallel to plane of schistosity and minimum at right angle to it, in schist, the directions of minimum and maximum moduli are not at right angle to each other. The

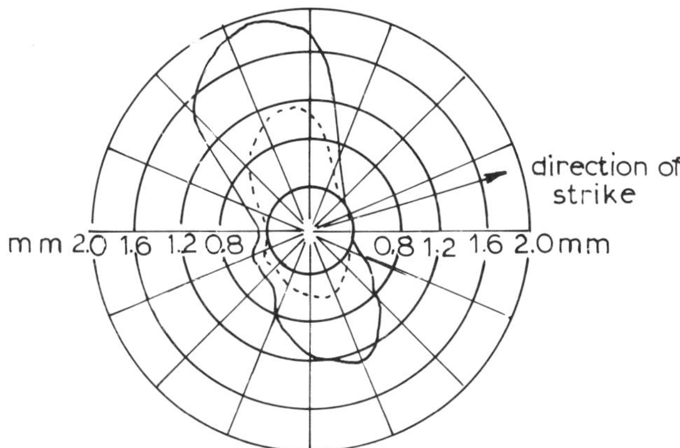

Fig. 8-84. Typical radial deformation diagram obtained in the Kaunertal pressure tunnel. Broken line, elastic deformations; solid lines, total deformations – (parallel layers of marl, 0.30 m thick, distant 2.00 m) (after SEEBER, 1964).

moduli ratio is highest for $(E_{\|}/E_{35^\circ})$. This type of behaviour could be due to the structure of the Graywacke schist which consists of highly deformable material (schist) enclosed in layers of low deformability (Graywacke).

The deformation modulus for rocks is dependent upon the stress level and usually increases with increase in stress except in certain cases in which rock disintegrates with increasing stress (Fig. 8-78b and c). In jointed rocks it increases till the joints are closed and filling material compacted. For example, for chloritic schists (UKHOV and TSYTOVICH, 1966) the following deformation modulus values were established.

For first cycle,

$$E_{\mathrm{d}} = 3.2 \times 10^3\, \sigma^{0.5} \tag{8.54}$$

For second cycle,

$$E_{\mathrm{d}} = 3.2 \times 10^3\, \sigma^{0.6} \tag{8.55}$$

Plate loading tests on horizontally bedded sandstone and mudstone in the foundation of Meadow Dam (Tasmania) up to 70 MPa (714 kgf/cm²) (10156 lbf/in²) gave the following relationship (WILKINS, 1966)

$$E_{\mathrm{d}} = K\sigma^{0.6} \tag{8.56}$$

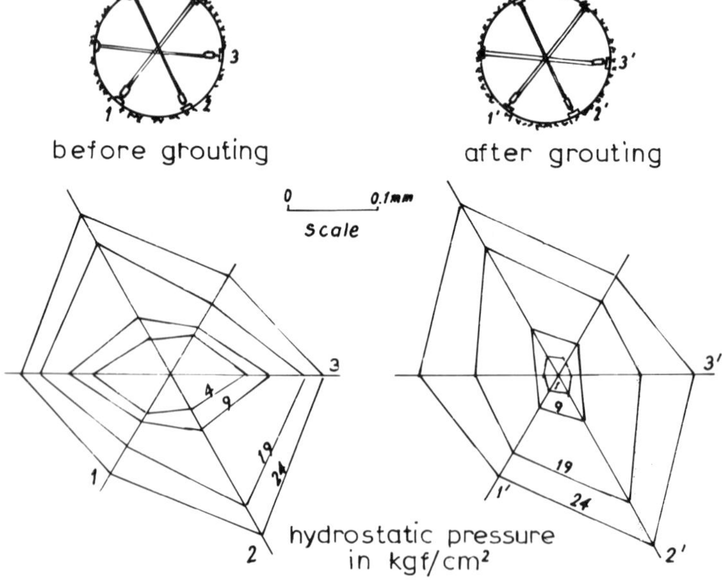

Fig. 8-85. Diametral deformation in a pressure chamber at Graminha Dam, São Pãulo (after RUIZ, 1966).

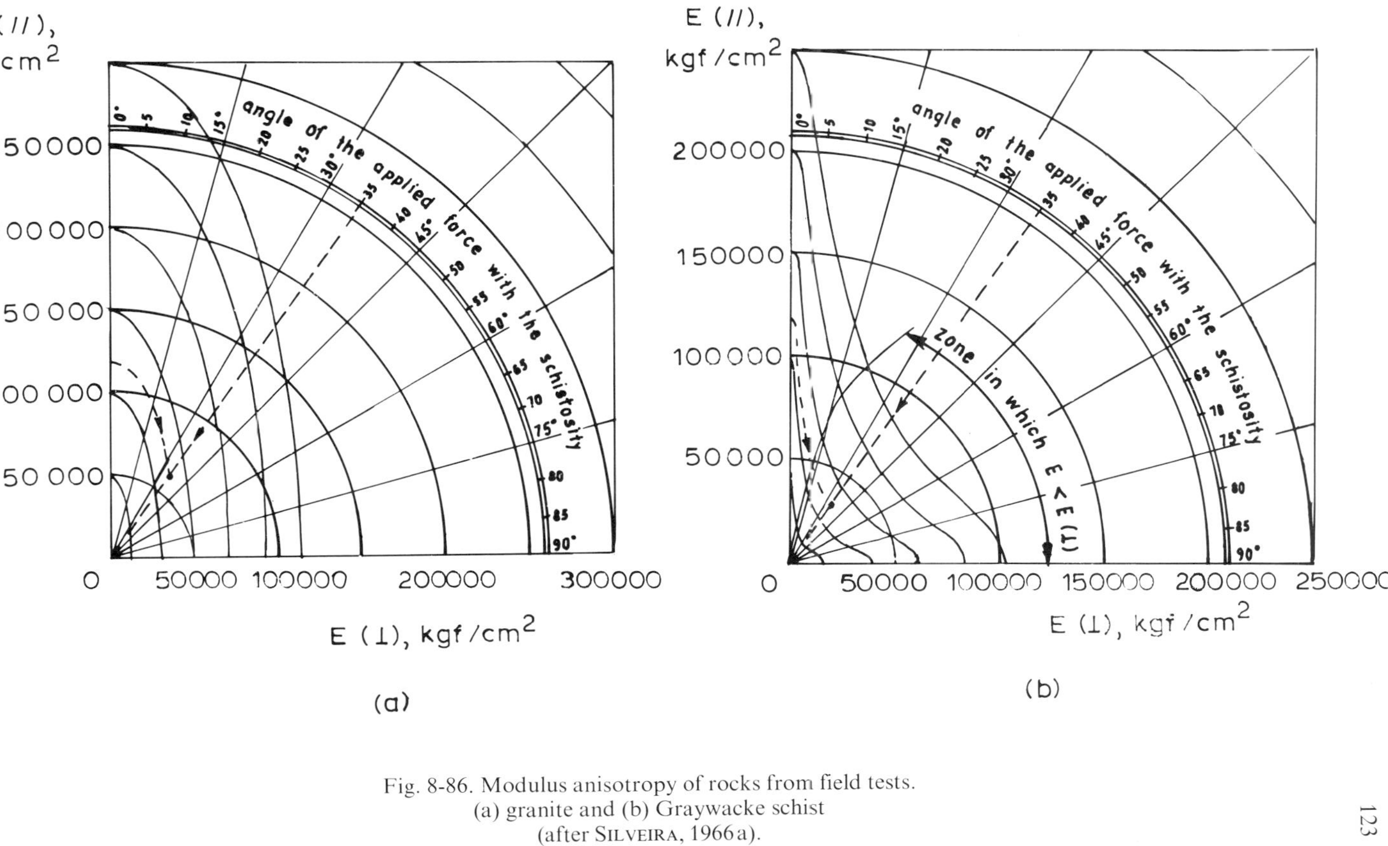

Fig. 8-86. Modulus anisotropy of rocks from field tests.
(a) granite and (b) Graywacke schist
(after SILVEIRA, 1966a).

Increase in modulus value will tend to decrease induced anisotropy, but inherent anisotropy will remain. Increase in hydrostatic pressure, however, tends to decrease anisotropy as has been observed in laboratory rock specimens (VOIGT, 1968).

Because of the large number of variables involved, no definite correlation between the modulus values determined from large field tests and laboratory tests on specimens has been observed. However, a certain trend can be seen when the ratio of field and laboratory modulus against Rock Quality Designation (*RQD*) (see Chapter 11.) is plotted. Fig. 8-87 shows the trend for Dworshak Dam where *RQD* values were determined at the test site. The reduction factor drops rapidly as the *RQD* value drops to 0.65. Fig. 8-88 represents the results of tests on 18 dam sites. Abscissa represents rock quality, i.e. square of the ratio between field and laboratory velocities $(V_F/V_L)^2$ and ordinate represents the reduction factor, ratio between rock mass modulus

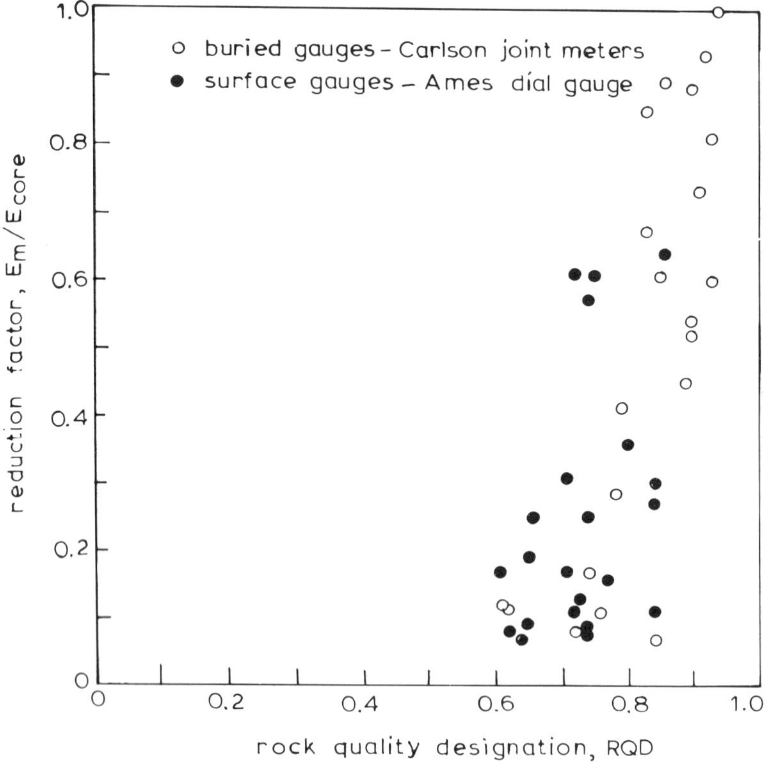

Fig. 8-87. Variation of reduction factor with rock quality from plate jacking tests, Dworshak Dam (after DEERE et al, 1966).

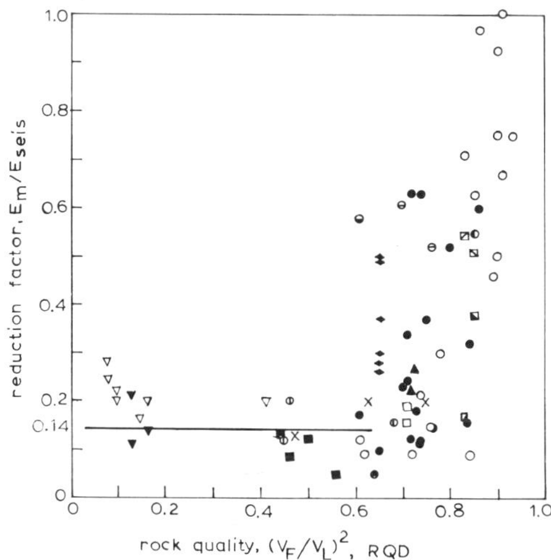

◪ Dworshak dam – pressure chamber test (F) – buried gauges, Shannon & Wilson (1964)

◪ Dworshak dam – pressure chamber test (F) – surface gauges, Shannon & Wilson (1964)

▨ Dworshak dam – pressure chamber test (E) – buried gauges, Shannon & Wilson (1964)

◩ Dworshak dam – pressure chamber test (E) – surface gauges, Shannon & Wilson (1964)

● Dworshak dam – jacking tests – surface gauges, Shannon & Wilson (1964)

○ Dworshak dam – jacking tests – buried gauges, Shannon & Wilson (1964)

■ Latiyan dam – Iran, Lane (1964)

▲ Kariba dam – slightly weathered gneiss, Lane (1964)

▼ Kariba dam – heavily jointed quartzite, Lane (1964)

x Nevada test site – Dacite Porphyry, Judd (1965)

✦ Morrow point dam, Bureau of Reclamation (1965), Rice (1964)

□ Ananaigawa dam, Kawabuchi (1964)

⦶ Agri river, Italy, Lotti and Beamonte (1964)

⊖ Koshibu dam – jacking tests

◒ Koshibu dam – pressure chamber tests

◐ El Noville – Mexico, Deere (1963)

▽ Onodera (1962)

⬓ Vajont dam, Italy, upper slope, pressure chamber test, Link (1964), Jaeger (1964)

Fig. 8-88. Variation of reduction factor with rock quality (after DEERE et al, 1966).

and seismic modulus (E_m/E_{seis}). The trend is fairly consistent with the reduction factor falling rapidly as *RQD* or $(V_F/V_L)^2$ drops to 0.65. At this point the reduction factor has a value of about 0.14. A further decrease in the value of rock quality does not seem to have any influence on reduction factor. A rough idea could be obtained about the reduction factor by entering the *RQD* value from exploration data into these figures. While correlating *RQD* at a particular test area with the results of any modulus, it is essential to take the weighted *RQD* of the rocks under the test area. The weighting factor is determined by the stress distribution at different depths away from the loaded surface and the contribution of the modulus values for each layer. For example, for a plate jacking test, the *RQD* index is weighted against the BOUSSINESQ stress distribution. Fig. 8-89 shows the variation in the *RQD* values at a plate jacking test site and the weighting factors.

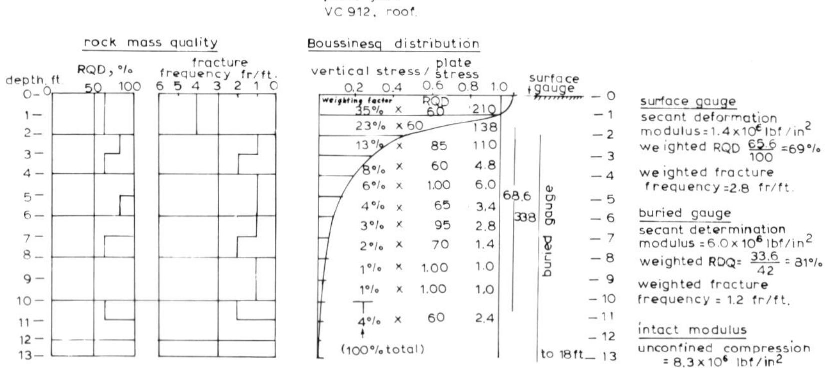

Fig. 8-89. Illustration of a method for obtaining a weighted rock quality beneath a plate jacking test (after DEERE et al, 1966).

Not only the static but also the dynamic elastic modulus values of rocks fall rapidly as the fracture frequency increases and the ratio between the two moduli drops rapidly. Tables 22 and 23 indicate this influence on 3 types of rocks. The ratio between the two moduli remains much less affected if stress acting is high (Fig. 8-90) which tends to eliminate the influence of fractures.

A correction between jointing, rock substance cohesion (c_1) and rock mass cohesion (c_m) has been obtained by MANEV and AVRAMOVA-TACHEVA (1970) (Fig. 8-91). The coefficient j is obtained using the relationship

$$j = \frac{n_1 + n_2 + n_3 + \ldots . n_n}{n_1 k_1 + n_2 k_2 + n_3 k_3 + \ldots . n_n k_n} \tag{8.57}$$

where $n_1, n_2, n_3 \ldots$ = number of pieces in the core of the rock bed corresponding to a class and

$k_1, k_2, k_3 \ldots$ = mean length of the core piece in the class.

TABLE 22

Test Data on the Moduli of Static Deformation and Dynamic Elasticity for Specimens with Varying Jointing (Tight Joints), $\sigma = \text{kgf/cm}^2$

(after MOGILEVSKAYA, 1970)

Joint frequency	Veined limestone		Limestone and calcite breccia	
	$E_{st} \times 10^{-5}$ kgf/cm^2	$E_{dyn} \times 10^{-5}$ kgf/cm^2	$E_{st} \times 10^{-5}$ kgf/cm^2	$E_{dyn} \times 10^{-5}$ kgf/cm^2
10	8.2	8.4	9.8	5.8
30	5.2	7.0	9.4	5.4
60	9.2	6.0	2.8	5.0
80	2.6	5.6	2.6	4.9
80	2.9	5.0	2.3	4.8

TABLE 23

Test Data on Deformability on Monolithic and Jointed (Throughout Joints) Specimens, 100 kgf/cm^2

(after MOGILEVSKAYA, 1970)

Rock	Joint frequency	$E_{st} \times 10^{-5}$ kgf/cm^2	$E_{dyn} \times 10^{-5}$ kgf/cm^2
Veined Limestone	0.8	1.3	4.4
	5.2	5.3	5.7
	4.2	5.1	6.1
Limestone Breccia	0.9	1.4	3.5
	0.8	1.2	3.4
	3.3	3.3	4.4
	1.4	—	—
	1.6	—	—
Calcite	0.8	1.4	3.8
	1.7	—	—
	1.2	—	3.4

The relationship represented in Fig. 8-91 can be expressed as

$$\frac{c_m}{c_1} = Ae^{-b(j-2)} + B \tag{8.58}$$

where $A = 0.114$
$B = 0.02$ and
$b = 0.48$.

This has correlation coefficient of 0.88.

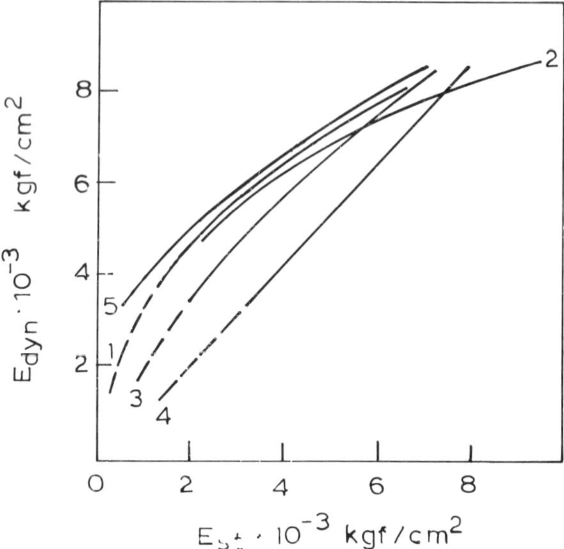

Fig. 8-90. Diagram of correlation between the static modulus of deformation, E_{st}, and the dynamic modulus of elasticity E_{dyn} according to laboratory (1–4) and field test (5) data. 1 – E_{st} and E_{dyn} measured with $\sigma = 40$ kgf/cm^2, 2 – E_{st} and E_{dyn} measured with $\sigma = 300$ kgf/cm^2, 3 – E_{st} measured with $\sigma = 40$ kgf/cm^2, E_{dyn} $\sigma = 0$. 4 – E_{st} measured with $\sigma = 300$ kgf/cm^2, E_{dyn} measured with $\sigma = 0$; 5 – E_{st} measured on plates with $\sigma = 100$ kgf/cm^2, the dead weight of the rock mass being taken into account (after MOGILEVSKAYA, 1970).

The function j usually decreases with increase in the depth at which rocks lie and has been found to be related as

$$j = B + \frac{A}{H + 0.3} \tag{8.59}$$

where $A = 166$
$B = 2.5$ and
$H =$ depth from surface.

The function j is similar to RQD and corresponds to RQD values for a highly jointed rock. Knowing the value of j and c_1, the value of c_m can be calculated. It may however be kept in view that the intensity of jointing increases near the excavations due to the creation of an opening and damage caused due to excavation processes. As such the value of j will have to be modified accordingly. The effect of excavation usually extends to a distance about 2–3 times the radius of the excavation and the increase in the value of j may be 100–150% (MANEV and AVRAMOVA-TACHEVA, 1970).

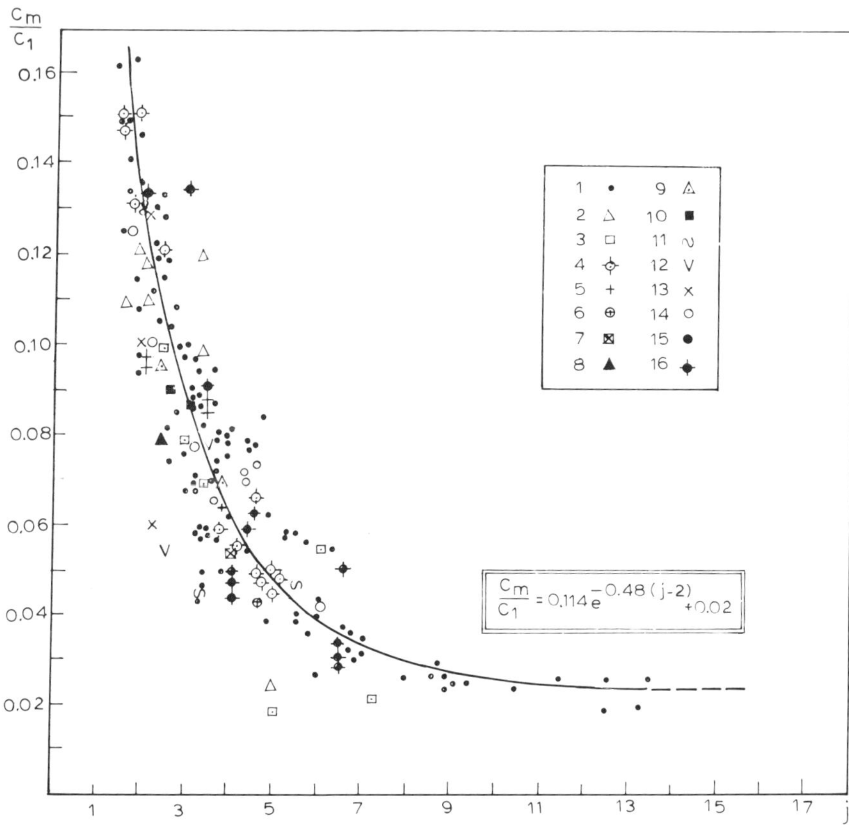

Fig. 8-91. Dependence between intensity of jointing and coefficient of structural looseness.
1 – gneiss, 2 and 5 – marble, 3 – amphibolite, 4 – gneiss in situ, 6 – calc-schist, 7 – siderite, 8 – gabbro-diorite, 9 – monzonite-porphyry, 10 – diorite-porphyry, 11 – hornfelsite, 12 – monzonite-gabbro-porphyry, 13 – andesite, 14 – andes.tuff, 15 – andes.tuff breccia, 16 – theoretical
(after MANEV and AVRAMOVA-TACHEVA, 1970).

A detailed discussion on classification of rock is given in Chapter 11. Based upon field tests alone, a simple classification system used by KUJUNDZIC, COLIC and RADOSAVLJEVIC (1970) for Derdap hydroelectric power and navigation system is useful in classifying the rocks at any site into different groups. The types of rocks at this dam site are mostly schistose but have a wide range in their properties. The classification is given in Table 24.

TABLE 24

Classification of rock mass and their mechanical properties

(after KUJUNDZIC, COLIC and RADOSAVLJEVIC, 1970)

Property	Classification Category					
	I	II	III	IV	V	VI
1. Deformation modulus E_d, kgf/cm$^2 \times 10^3$	< 4	4–8	8–12	12–20	20–30	> 30
2. Modulus of elasticity E_e, kgf/cm$^2 \times 10^3$	< 10	10–20	20–30	30–45	45–75	> 75
3. Seismic wave velocity km/s	< 3.2	3.2–3.5	3.5–3.8	3.8–4.1	4.1–4.5	> 4.5
4. Joint cohesion, c, kgf/cm^2	< 0.5–1.0	1–1.5	1.5–2.0	2.5–3.0	3.5–5	> 5
5. Peak joint friction angle	< 25°	25°–30°	30°–40°	40°–50°	50°–60°	> 60°
6. Residual joint friction angle	< 25°	25°–30°	~ 30°	~ 35°	35°–40°	40°–45°

Category—I : Tectonically highly fractured with clay bands and altered schist with clay filled joints.

Category—II : Strongly laminated gneiss mainly biotite with intercalations of altered and soft schist with clay.

Category—III : Tectonically fractured biotite—laminated gneiss with intercalations of amphibolite, granite and chlorite schist.

Category—IV : Mainly biotite-gneiss with intercalations of biotite quartzite gneiss with small amount of chlorite gneiss.

Category—V : Mainly biotite-gneiss with little tectonic damage.

Category—VI : Mainly biotite-quartzite-gneiss with least tectonic damage.

TABLE 25

Correlation of results of rock deformation moduli ($\times 10^3$ kgf/cm^2)

(after GALLICO, 1974)

Dam	Foundation rock	E_d^*	E_{re}^*	E_p^*	E_{ss}^*	E_{ors}^*	Remarks
Kurobe	weathered granite	50–250	40–80	10–30	20–50	10–50	variable between banks & from top to bottom
Dez	conglomerate	–	50	50–80	50–110	60	homogeneous
Soledad	volcanic tuffs and lava	40–200	100–200	–	10–60	30–120	variable from top to bottom
Santa Rosa	sound rhyolite	280	200–400	–	80–130	150–250	scattered joints
Novillo	sound rhyolite	300	300–370	–	150	100–150	variable from top to bottom
Ocumarito	weathered basalt	160	150–400	500–800	–	150	variable from surface to inside bank
Tachien	stratified slate and quartzite	–	200–700	120–200	50	120	parallel to bedding
				50–150	40	60	normal to bedding
Inguri	weathered & tectonised limestone	40–130	100–200	20–130	–	30–90	variable from top to bottom
Oroville	amphibolitic schist	333–1054	734–1034	95–1116 (flat jack) 82–112		41–510	sp.gr. = 2.9

* E_d = in situ dynamic test; E_{re} = rock element, laboratory; E_p = in situ plate-load test; E_{ss} = in situ pressure-chamber and radial jack test; and E_{ors} = overall rock system.

As already indicated, the modulus values determined from different tests are influenced by the test configuration and the volume effect. Volume effect which is a manifestation of joints and cracks is a major factor reducing the modulus values as one moves across the spectrum from sonic tests to laboratory tests, plate bearing tests, pressure tunnel tests and overall rock system behaviour on actual construction of the structure. Tables 25, 26 and 27 give results of tests at different dam sites using different methods. More results are given in Appendix III.

TABLE 26

Rock properties at different dam sites obtained by using different methods

(after CORDING, HENDRON and DEERE, 1971)

Location, Size, Depth of Cavern	Rock Properties	Measured Field and Lab Moduli, 10^6 psi		$\frac{E_{Field}}{E_{Lab}}$
Nevada Test Site, Cavities I and II Hemisphere on end: Height $H = 140'$ Width $B = 80'$ Length $L = 120'$ Depth $D = 1300'$ (Vertical stress, σ_v = 1000 psi, horizontal stress, σ_h = 500 psi.)	Massive bedded tuff, *RQD*: Excellent, Unconfined compressive strength, σ_c = 1500 psi, water content, $w = 21\,\%$	E_{lab} $E_{lab\ sonic}$ $E_{seismic}$	: 0.5 : 1.5 : 1.0	1.0 est.
Nevada Test Site, Cavity III, hemisphere on end: H = 80', B = 50' L = 75', D = 350' ($\sigma_v = \sigma_h$ = 400 psi. estimated).	Granite (quartz monzonite), iron-stained joints. Blocky. *RQD* Fair to Good. (75 % average) Major joint set parallel to wall, occasional shear zones	E_{lab}	: 10	0.4 est
Tumut 1, H = 110', $B = 77'$ $L = 300'$, $D = 1100'$ (σ_v = 1500 psi, σ_h = $0 \rightarrow 1800$ psi)	Granite, Granite Gneiss, σ_c = 20,000 psi, *RQD*: Fair-Good (estimate). High angle fault intersects one wall. Joint spacing approx. 1' to 5'.	E_{lab} $E_{lab\ sonic}$ $E_{flat\ jack}$ $E_{pressure\ chamber}$	: 8 : 8 : 5 : 2	0.25–0.6
Tumut 2, $H = 110'$, $L = 300'$, $B = 60'$, $D = 1000'$	*RQD*: Good to Excellent (est.)	E_{lab} $E_{lab\ sonic}$ $E_{flat\ jack}$ $E_{pressure\ chamber}$ $E_{plate\ bearing}$	: 8 : 8 : 6 : 3 : 1	0.13–0.75
Morrow Point Powerplant, Colorado $H = 100 \rightarrow 138'$, $B = 57'$, $D = 400'$ $L = 207'$ (σ_v = $400 \rightarrow 2000$ psi)	Micaceous Quartzite, Mica Schist, $\sigma_c = 6000 \rightarrow$ 16,000 psi, *RQD* Good to Excellent (est.)	E_{lab} (micaschist) E_{lab} (micaceous quartzite) $E_{plate\ bearing}$	: 1.3 : 4 : 1.3	0.33–1.0

Observed Cavern Displacements, δ, inches	Modulus Determined from δ, inches	$\frac{E_\delta}{E_{Lab}}$	$\frac{E_\delta}{E_{Field}}$	Comments
Elastic, crown and sidewalls: 0.15″ to 0.4″	0.5	1.0	1.0	
Shallow, crown: 1″ to 2″	(0.05–0.15)	(0.1–0.3)	(0.1–0.3)	Displacements occurred at 3′ to 5′ depth behind shallow slabs
Deep-seated, cavity II wall: 1″ to 2″	(0.05–0.15)	(0.1–0.3)	(0.1–0.3)	Displacements occurred at 10′ to 30′ depth over 80′ × 100′ area of wall. Stabilized with additional bolts
Crown: 0.015″–0.035″	4	0.4	1.0	
Wall, upper: 0.05″	3	0.3	0.75	Larger displacements due to opening of joints 3 to 5′ behind and parallel to wall
Wall, middle: 0.125″	2	0.2	0.50	
Crown:	2*	0.25	0.50	
Wall, at springline: 0.15″	2	0.25	0.50	* From strains in machine hall roof
Wall, near fault, at springline: 0.4″	(0.7)	(0.9)	(0.18)	
Crown:	5*	0.6	1.2	* From strains in machine hall roof
Walls: 0.2″–0.6″	2	0.25	0.50	By precise survey
Walls:	0.8–7.0	0.1–0.9	0.2–1.7	From extensometers
Crown: 0.1″ to 0.4″	0.5–2.1	0.17–0.7	0.5–1.0	* Shear zone on *a*- line wall formed wedge which moved into cavity
*b*line Wall: 0.3″	~ 3.0	~ 1.0	~ 1.0	
a-line Wall: 2.3″	(0.4)	(0.13)	(0.2)	

TABLE 26 (continued)

Location, Size, Depth of Cavern	Rock Properties	Measured Field and Lab Moduli, 10^6 psi		$\frac{E_{\text{Field}}}{E_{\text{Lab}}}$
Oroville Power-plant, California, $H = 120'$, $B = 69'$, $L = 550'$, ($\sigma_v = \sigma_h = 500$ psi)	Amphibolite *RQD*: Fair to Good (est.)	E_{lab}	: 13	
		$E_{\text{lab sonic}}$	: 16	
		E_{seismic}	: 5 to 15	
		(10 average)		
		$E_{\text{flat jack}}$	: 1.5 to 16	0.12–1.0
		$E_{\text{plate bearing}}$	: 1.2 to 1.8	0.12
Oroville Tunnel $B = 35'$		E_{lab}	: 13	
Poatina Power Station, Tas. $H = 85'$, $B = 45'$, $L = 300'$, $D = 500'$, ($\sigma_v = 1200$ psi, $\sigma_h = 1800$–2400 psi)	Thin to massive bedded mudstone, $\sigma_c = 5{,}000$ psi, $w = 1.5\%$	Loaded perpendicular to bedding:		
		E_{lab}	: 4.5	
		$E_{\text{flat jack}}$	: 2.4	0.53
		Loaded parallel to bedding:		
		E_{lab}	: 6.3	
		E_{flatjack}	: 3.2	0.51
Kariba Powerplant, Rhodesia, $H = 132'$, $B = 75'$, $L = 468'$, $D = 200'$	Biotite gneiss *RQD*: Fair to Good, (est.)	E_{lab}	: 9	
		E_{seismic}	: 11	0.1
		$E_{\text{plate bearing}}$	: 0.9	
Lago Delio $H = 195'$, $B = 68'$, $L = 620'$, $D = 520'$, ($\sigma = 800$ to 1600 psi)	Fine grained paleozoic gneiss, near vertical foliation perpendicular to axis of cavern. Local weathered zones	E_{seismic}	: 7.6 to 8.6	
		$E_{\text{pressure chamber}}$	: 1.4 to 3.8	0.3
		$E_{\text{plate bearing}}$	: 1 to 3	0.25
Kisenyama, Japan $H = 165'$, $B = 83'$, $L = 200'$, $D = 810'$	Chert, sandy slate	E_{seismic}	: 8.5	
		$E_{\text{plate bearing}}$	: 0.7	0.08
Niagara Tunnel $B = 50'$, $D = 300'$	Bedded limestone, sandstone, shale, Roof in hard limestone, wall in shale, sand-stone, and dolomite.	E_{lab}	: 1–10	

Observed Cavern Displacements, δ, inches	Modulus Determined from δ, inches	$\frac{E_\delta}{E_{Lab}}$	$\frac{E_\delta}{E_{Field}}$	Comments
Crown: *0.002″ to 0.059″ (.012 average)	8.2 average (1.6 low value)	0.63	~ 1.0	* measured with 20′ extensometers ** measured with 40′ extensometers
Wall**: 0.054″ to 0.247 (0.114 average)	2	0.15	0.2–1.0	** Vertical shear intersected wall at shallow angle
Left wall: ** 0.250″			(~ 0.2)	
Crown: Extensometers	0.6–7.5	0.05–0.58		
Crown: 0.10″	1.6	0.36	0.67	* After horizontal shear developed at intersection of crown and haunch
0.18″*	(0.9)	(0.2)	(0.37)	
Wall: 0.4″	2.5	0.4	0.78	
Wall Less than 0.5″	0.5–1.0	0.1	0.1	Arch concreted. Rock bolts used as required by miners
Wall* 1.5″	(0.2)	(0.02)	(0.20)	* Fault zone at NW corner of powerhouse reinforced with tendons
Crown: 0.5″*	0.7	0.09	0.36	* Occurred within 25′ of surface ** End wall is parallel to foliation
Wall: 0.7″	1.2	0.15	0.60	
Endwall**: 1.0	0.5	0.06	0.24	
Crown: Excellent Rock 0.08″	6	0.70		* Chert layer separated from slate causing large displacement. ** Struts installed across cavern for fear of collapse when displacement occurred. Struts were later removed and wall was concreted
Poor Rock 0.28″	0.8	0.10		
0.60″*	(0.4)	(0.05)		
Walls: left 0.4″	5	0.59		
right 1.5″**	(1)	0.12		
Crown: 0.05″	2	~ 0.5		Steel rib support, heading and bench.
Wall: (1.8″)	(0.1)	(0.05)		Horizontal offset on bedding planes observed.

TABLE 27
Modulus of Deformability of Various Rocks
(after ROCHA, 1964)

Rock Type	Site	Modulus of deformability (10^3 kgf/cm^2) Lab. E_r	In situ E_m	$\frac{E_m}{E_r}$
Granite	Alvarenga (P)	520	490	1/1.1
Granite	Alto Rabagao (P)	26	9	1/2.9
Granite	Alto Lindoso (P)	320	60	1/5.3
Granite	Vilarinho (P)	430	15	1/29
Gneiss	Cabora-Bassa (P)	800	650	1/1.2
Schist	Cedillo* (S)	900	400	1/2.2
Schist	Cedillo** (S)	650	120	1/5.4
Schist	Alcantara* (S)	1400	50	1/28
Conglomerate	Avlaki (G)	600	60	1/10
Sandstone	Cambambe (P)	650	86	1/7.6
Siltstone	Avlaki (G)	150	15	1/10
Claystone	Karum (I)	115	70	1/1.6
Marl	Karum (I)	470	430	1/1.1
Limestone	Karum (I)	700	600	1/1.2
Limestone	Karum (I)	500	75	1/67
Quartzite	Alvito (P)	430	4	1/108
Quartzite	Alvito (P)	330	70	1/4.7

* Parallel to schistosity ** Perpendicular to schistosity
P – Portugal S – Spain
G – Greece I – Iran

8.9. Bearing Capacity of Rock

The bearing capacity of rock is defined as the maximum load exerted through a rigid punch per unit area that the rock is capable of withstanding before failure. The bearing capacity is of importance in the design of strata control systems in coal mines and in rock drilling. The method of determining the bearing capacity is based upon forcing an indenter into rock as in the case of a plate bearing test. The load and penetration of the indenter is continuously monitored. The test procedure usually adopted is as follows:

(a) Exert an even increasing load on the surface of given dimensions and measure the penetration placed close to the end of the loading surface. This maximum value gives an initial estimate of the failure load.

(b) Apply a seating load and adjust the penetration measuring equipment to read zero.

(c) Increase the load by approximately one quarter of the estimated failure load.

(d) Record the deflection (penetration) by maintaining the load constant until the deflection rate drops to 4×10^{-7} m/s (0.001 in/min).

(e) Repeat (c) and (d) for loads approximately half and three-quarters of the estimated failure load.

(f) Unload the seating load after the third increment and record the rebound deflection.

(g) Reload to the third incremental level and record the deflection.

(h) Load to failure and record the deflection.

The type of the load vs penetration curve is dependent upon rock type. Fig. 8-92 shows a typical load deflection curve for an elastic-plastic rock. After a very limited linear behaviour (I), a range of nonlinear behaviour (II) is observed leading to a perfectly plastic behaviour (III).

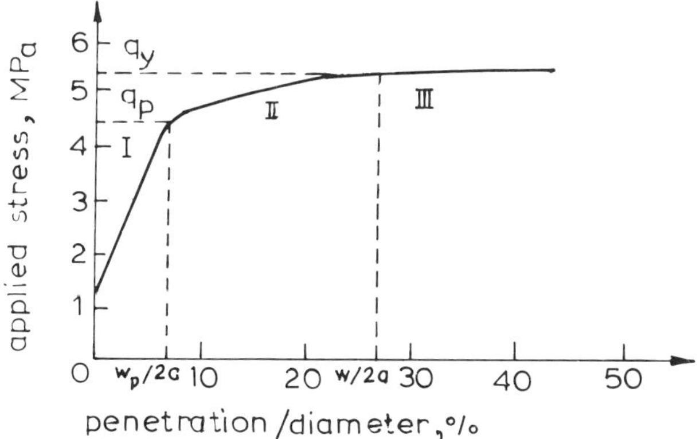

Fig. 8-92. Yielding failure of an "elastic-plastic" rock (after COUETDIC and BARRON, 1975).

Fig. 8-93 shows the load vs penetration curve for an elastic brittle-plastic rock. A steep linear behaviour (I) is followed by a short nonlinear portion up to the peak (II) and then a post-peak drop (up to 20%) (III) followed by typical residual (plastic) behaviour (IV).

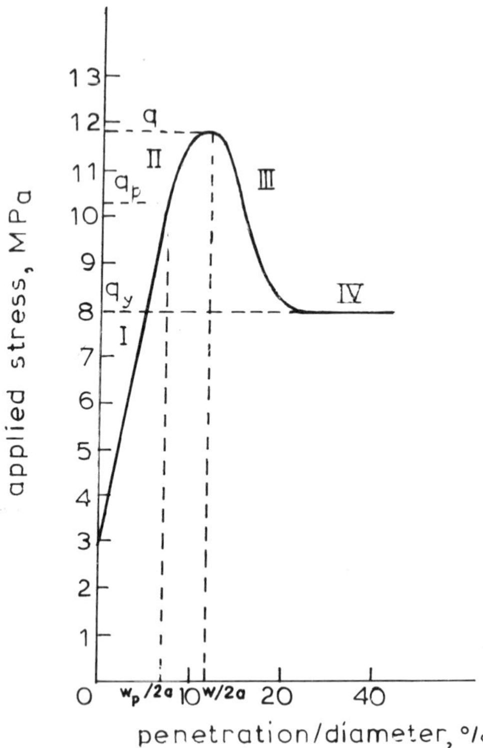

Fig. 8-93. Behaviour of an "elastic brittle-plastic" rock (after COUETDIC and BARRON, 1975).

In general, the behaviour is dependent upon a large number of factors such as presence or absence of fluids, temperature, ambient pressure, indenter shape, and rock porosity. The mode of failure of solid dry rock under low pressures and room temperature and under a flat ended punch is dependent upon its porosity. Schematically this can be outlined as in Fig. 8-94. In hard dense rocks (e.g. quartzites) Hertzian tensile cracks will initiate at the rim of the indenter forming a truncated cone beneath the indenter (ROESLER, 1956) with stable growth with increasing load. With increased loads, the cone may eventually crush with unstable crack growth and with fractures reaching the surface. In comparatively more deformable rocks initiation of Hertzian cracks is less important and a bulb of rock beneath the indenter may attain a failure stage before the formation of lateral cratering and appearance of cracks on the surface. Radial tensile cracks may form around the bulb as load on the indenter is further increased.

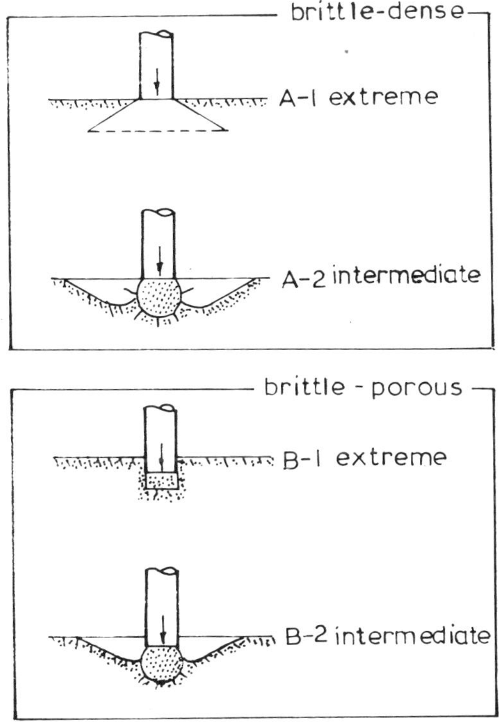

Fig. 8-94. Assumed indentation failure modes for extreme and intermediate cases of brittle-dense and brittle-porous rocks (after LADANYI, 1968).

In very porous rocks, the mode of failure may be merely a structural collapse beneath the indenter with practically no or limited damage to the surrounding material (CHERCASOV, 1967). The bearing capacity in such materials is almost equal to the uniaxial compressive strength of rock. This type of behaviour has been observed for coal (POMEROY, 1958). For less porous materials (porosity 10–25%) failure may start with local structural collapse and compaction (strain hardening) and may approach that of intermediate dense rocks.

For extremely soft rocks and particularly in the presence of pore fluid, plastic failure will occur (Fig. 8-95) where PRANDTL (1920) curvilinear lines of failure may appear and theories of soil mechanics may be applicable (TERZAGHI, 1943; MEYERHOFF, 1951). In such cases, material under a punch can be treated as an ideal MOHR-COULOMB material.

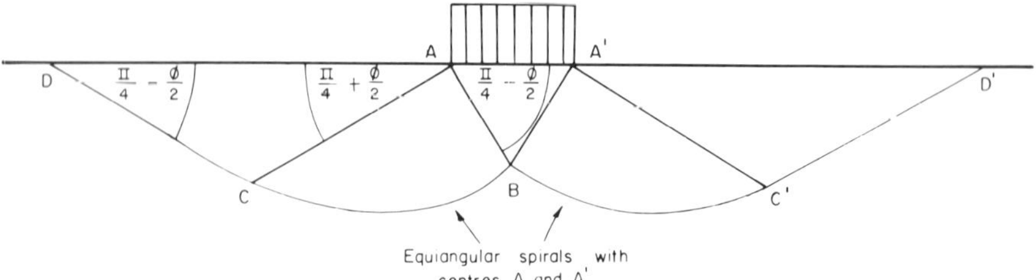

Fig. 8-95. Rupture surface below a smooth foundation on a weightless cohesionless soil, ϕ-angle of shear resistance, AA′-loaded surface, ACD-zone of passive RANKINE failure, ABC-zone of radial shear, ABA′-zone of active RANKINE failure.

Depending upon the mode of failure assumed, a number of theories have been put forward to calculate the bearing capacity of materials. Table 28 summarises the various failure criteria. The relationships have been expressed in terms of $\bar{\sigma}_f$ (ratio of the bearing capacity with the uniaxial compressive strength). The various approaches are dependent upon the shape of the failure surface. Failure surfaces have been assumed to be either broken planes in a wedge type failure or curvilinear surfaces or combinations. The variation in the results is very much evident from the relationships. Fig. 8-96 gives the variations in $\bar{\sigma}_f$ as a function of friction angle ϕ. Indications are that the expansion of a spherical bulb approach suggested by LADANYI (1967) is the most satisfactory approach for most rocks. Fig. 8-97 gives the geometry of the testing area under a circular footing of diameter $2a$ after the completion of the test in coal seams (COUETDIC and BARRON, 1975). A hemispherical ball exists underneath the loading plate consisting of a highly compacted coal (region 1) surrounded by a less compacted coal (region 2), and a disturbed elliptical shape zone (region 3) with radial cracks surrounding this bulb. Major axis and direction of this elliptical zone are governed by local bedding and fissuring.

Bearing capacity determined from penetration of an indenter in brittle rocks is sensitive to size and shape of the indenter. JENKINS (1960) found that for circular footings, the bearing capacity of coal mine floors is proportional to a power law of the form

$$\sigma_f = Kd^{\beta} \tag{8.60}$$

where d = diameter of the footing
β = 1 for cohesionless floors,
0.5 for brittle floors, and
0 for perfectly plastic and soft floors and
K = constant.

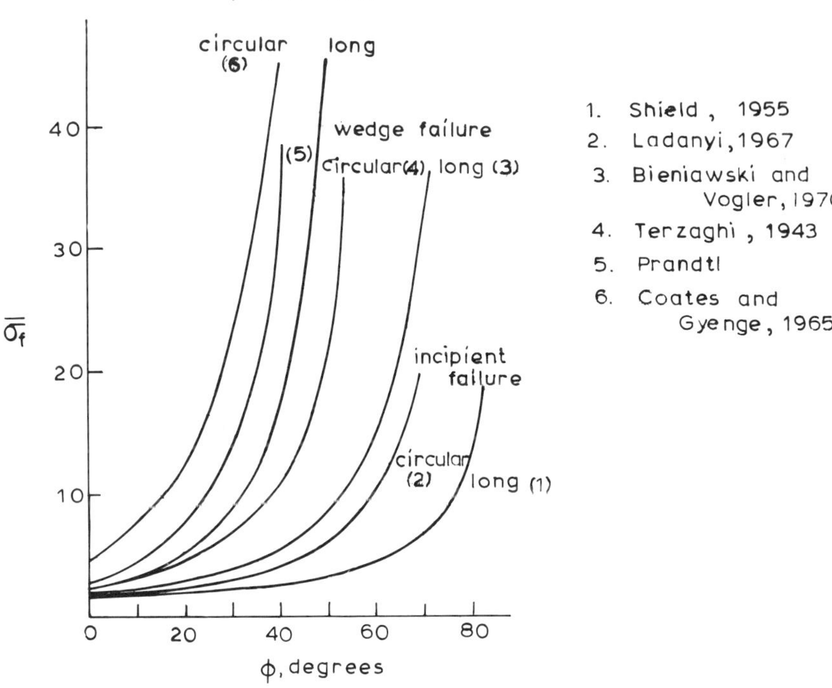

Fig. 8-96. Bearing capacity function ($\bar{\sigma}_f$) as a function of ϕ (after COUETDIC and BARRON, 1975).

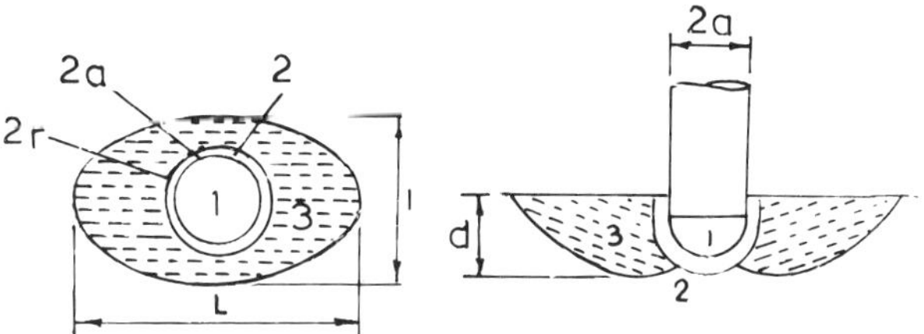

Fig. 8-97. Hemispherical bulb under the indenter.

The type of load-penetration curves obtained for there different materials are given in Fig. 8-98. The value of β is determined to a large extent by the mode of failure, being more severe, the more brittle the material (GONANO and BROWN, 1975).

TABLE 28

Bearing Capacity Theories

(after COUETDIC and BARRON, 1975)

		Failure type	Failure criterion	Footing	No.	$\bar{\sigma}_f$ = Bearing Capacity/Uniaxial Compressive Strength $= \sigma_f/\sigma_c$	Sketch
Yielding Ground	Lower Bound	**Wedge Analysis**	COULOMB or Modified GRIFFITH	long	1.	$\bar{\sigma}_f = K_p + 1$	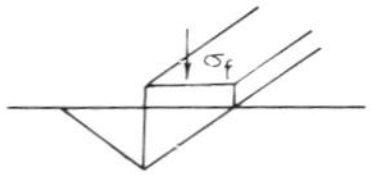
				circular	2.	$\bar{\sigma}_f = (K_p + 1)(1 + 0.2\,K_p)$	
			GRIFFITH	long		$\bar{\sigma}_f = 3$	
			FAIRHURST	long		$\bar{\sigma}_f = 4(n+1)^{\frac{1}{2}}/[1 + (n+1)^{\frac{1}{2}}]$	
		Shield & limit analysis theorem	COULOMB	circular	3.	$\bar{\sigma}_f = \dfrac{\dfrac{c}{2}\tan^3\left(\dfrac{\pi}{4}+\dfrac{\phi}{2}\right)[4 + \sin\phi + \sin^2\phi + (1+\sin\phi)\sqrt{(4+\sin^2\phi)}] + 2c\tan\left(\dfrac{\pi}{4}+\dfrac{\phi}{2}\right)}{c\cot\phi\left[\tan^2\left(\dfrac{\pi}{4}+\dfrac{\phi}{2}\right)-1\right]}$	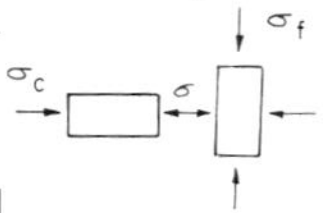
	Upper Bound	Plastic Equilibrium	COULOMB	long	4.	$\bar{\sigma}_f = [K_p\exp(\pi\tan\phi) - 1]\,(K_p - 1)$	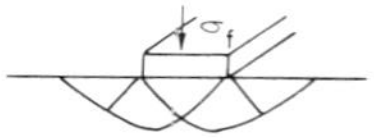
				circular	5.	$\bar{\sigma}_f = [K_p\exp(\pi\tan\phi) - 1]\,(1 + 0.2\,K_p)/(K_p - 1)$	
		Approximation	COULOMB	circular	6.	$\bar{\sigma}_f = 9.1\,K_p^2/\cot\phi\,(K_p - 1)$	

TABLE 28
(Continued)

	Failure type	Failure criterion	Footing	No.	$\bar{\sigma}_f$ = Bearing Capacity/Uniaxial Compressive Strength $= \sigma_f/\sigma_c$	Sketch
Elastic-Brittle Plastic Ground	Expansion of a Spherical Hole	Intact rock (E, υ, n) Crushed rock (c_c; ϕ_c)	circular	7.	$P_{ult} = (\sigma_c + c_c \cot\phi_c) \left[3\left(1 - \frac{1-\upsilon}{\sqrt{(2n)}}\right)\frac{\sigma_c}{E} + e^*_{av}\right]^{-x} - c_c \cot\phi_c$ $X = 4\sin\phi_c/3(1-\sin\phi_c)$ $\sigma_f = P_{ult}(1+\tan\phi_c) + c_c$	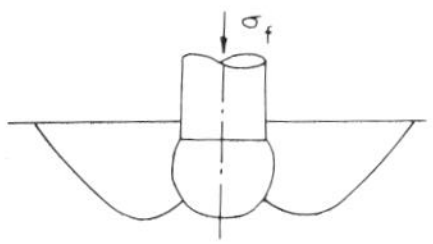
Brittle Ground	Incipient failure	GRIFFITH	long circular		$\bar{\sigma}_f = 2.17$ $\bar{\sigma}_f = 2.71$	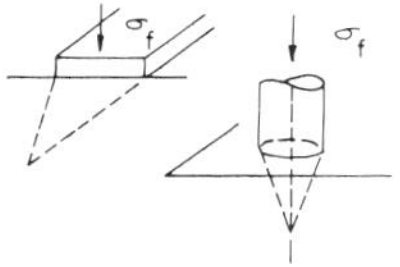
		Modified GRIFFITH	long	8.	$\bar{\sigma}_f = \frac{\pi}{2}(1-\sin\phi) \Big/ \left[\cos\phi - \left(\frac{\pi}{2} - \phi\right)\sin\phi\right]$	
			circular	9.	$\bar{\sigma}_f = [(1-\sin\phi)\sqrt{(3-\sin\phi)}]/[\sqrt{(1+\sin\phi)^3} - 2\sin\phi\sqrt{(3-\sin\phi)}]$	

e^*_{av} = Average volume strain (positive for volume decrease) undergone by the rock when passing from intact to crushed state
n = ratio between compressive and tensile strength
E = YOUNG's modulus
$K_p = \tan^2(\pi/4 + \phi/2)$
ϕ = angle of internal friction
υ = POISSON's ratio
c = cohesion

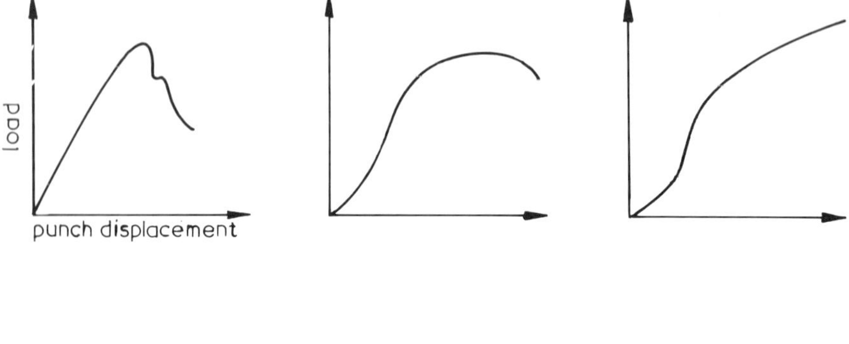

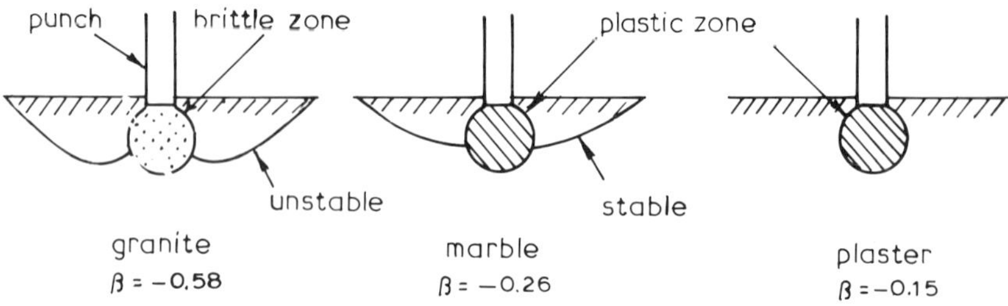

Fig. 8-98. Bearing capacity of materials under a rigid punch (after LAMA and GONANO, 1976).

In situ tests in mines have shown that the bearing capacity values have a generally negative skew distribution and as the dimension of the punch increases so does the scatter of the results (JENKINS, 1958). The presence of water and rate of loading are the two most important factors influencing the bearing capacity of mine floors in U.K. mines where the floors generally consist of shales and mudstones.

8.10. Shear Strength of Rock In Situ

The shear strength of rock is one of the most important parameters used in design. Its determination involves the measurement of two related parameters, cohesion and angle of friction. The most common method of determining these two parameters is the direct shear test (Vol. I, Chapter 4). A number of variations have been used in applying it to in situ conditions. Some of the commonly used tests are:

(1) Inclined load test
(2) Parallel load test
(3) Torsion test.

In all these tests, it is important to recognise the plane of interest and align it with the shear plane of the test. At least 5 tests should be performed at each place or on each type of plane.

8.10.1. Inclined Load Test

This is the most commonly used direct shear method. The method has been recommended by the International Society for Rock Mechanics (1974) as one of the standards for determining the direct shear strength of rock in situ. Fig. 8-99 represents the general arrangements. The rock blocks are of square base of suitable dimensions and height. They are enclosed in rigid metallic frame bonded to them with mortar and concrete. Their displacements are measured with reference to a point fixed at a sufficient distance from the position of the blocks and regarded as fixed.

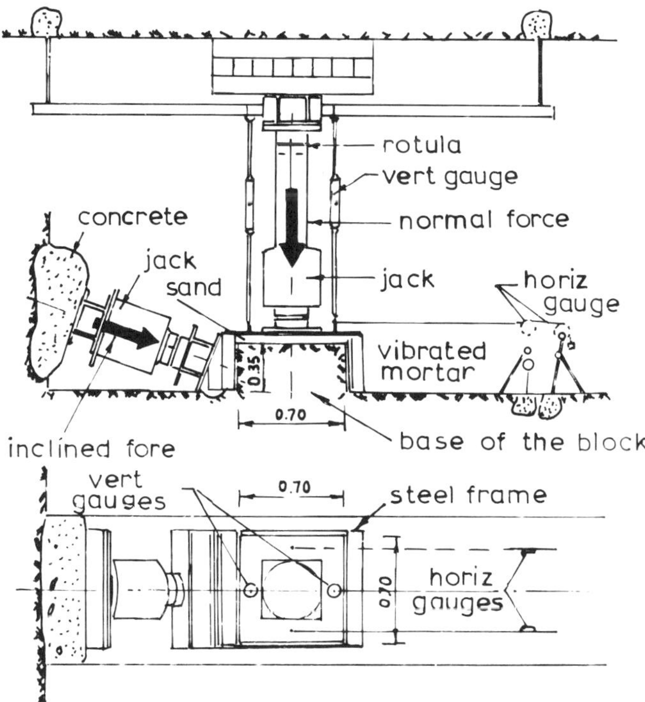

Fig. 8-99. Loading system for in situ shear test (after SERAFIM and LOPES, 1962).

The blocks are stressed with jacks acting vertically (normal to the base) till stabilisation of deformation of the blocks is reached. The lateral force, i.e. the shearing force, is applied in steps, waiting for the stabilisation of the displacement for each step until failure is reached. The shearing strength of the blocks is taken to be the tangential stress for which the vertical movement of the side of the block is inverted (point *A* of Fig. 8-100 b) or the maximum tangential stress or the residual stress after large displacement. The tests are carried out for

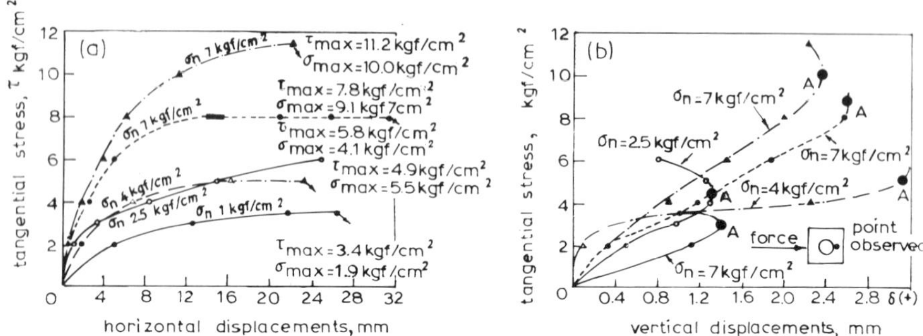

Fig. 8-100. Displacements measured in field shear tests (after SERAFIM and LOPES, 1962).

different normal loads. The test procedure recommended by I.S.R.M. is given below.

1. Test equipment: The following equipment is required for conducting a shear test.

(a) Equipment for cutting and encapsulating the test block, rock saws, drills, hammer and chisels, formwork of appropriate dimensions and rigidity, expanded polystyrene sheeting or weak filler, and materials for reinforced concrete encapsulation.

(b) Equipment for applying the normal load including:

(i) Flat jacks, hydraulic rams or dead load of sufficient capacity to apply the required normal loads.

(ii) A hydraulic pump if used should be capable of maintaining normal load to within 2% of a selected value throughout the test.

(iii) A reaction system to transfer normal loads uniformly to the test block, including rollers or a similar low friction device to ensure that at any given normal load, the resistance to shear displacement is less than 1% of the maximum shear force applied in the test. Rock anchors, wire ties and turnbuckles are usually required to install and secure the equipment.

(c) Equipment for applying the shear force including:

(i) One or more hydraulic rams or flat jacks of adequate total capacity with at least 70 mm (2.75 in) travel.

(ii) A hydraulic pump to pressurise the shear force system.

(iii) A reaction system to transmit the shear force to the test block. The shear force should be distributed uniformly along one face of the specimen.

The resultant line of applied shear forces should pass through the centre of the base of the shear plane with an angular tolerance of $\pm 5^\circ$.

(d) Equipment for measuring the applied forces including:

(i) One system for measuring normal force and another for measuring applied shearing force with an accuracy better than $\pm 2\%$ of the maximum forces reached in the test. Load cells (dynamometers) or flat jack pressure measurements may be used. The gauges should be calibrated both before and after testing.

(e) Equipment for measuring shear, normal and lateral displacements.

(i) Displacements should be measured (for example using micrometer dial gauges) at eight locations on the specimen block or encapsulating material.

(ii) The shear displacement measuring system should have a travel of at least 70 mm (2.75 in) and an accuracy better than 0.1 mm (0.004 in). The normal and lateral displacement measuring systems should have a travel of at least 20 mm (0.8 in) and an accuracy better than 0.05 mm (0.002 in). The measuring reference system (beams, anchors and clamps) should, when assembled, be sufficiently rigid to meet these requirements. Resetting of gauges during the test should if possible be avoided.

2. *Preparation of test block*: Before starting preparation of the test block it is important to decide upon the size of the test block.

The shearing resistance of the rock masses in a test is dependent upon the size of the test specimen. Results of tests by LNEC (Portugal) at four dam sites are given in Fig. 8-101. The figure shows that the results of both cohesion and coefficient of friction converge as the size of the test specimen increases. There seems to be no significant improvement by increasing the size of the specimen for determining shear parameters beyond a certain limit. This limit seems to be about 4000 cm^2 (620 in^2). Tests conducted by URIEL in Spain on specimens with 3 m^2 (32 ft^2) area gave results which fit on to lines defined from smaller specimens. It is therefore clear that results from laboratory and in situ tests on smaller specimens may quite significantly represent the results of a large rock mass.

The reason of this observation lies in the mechanism of failure in a shear test. Firstly, strain distribution in a direct shear test is not uniform and the crack propagation that starts near the loading end does not propagate along the imposed surface, but follows some form of least energy consumption path not necessarily following existing plane along its whole length. Experience has shown that 700 mm × 700 mm (28 in × 28 in) specimens are quite sufficient for in situ shear tests (SERAFIM and GUERREIRO, 1966). I.S.R.M. recommends test specimen size of 700 mm × 700 mm × 350 mm (28 in × 28 in × 14 in). The size and shape of the test block may be adjusted to coincide with the natural joints or fissures. This reduces disturbances due to preparation. The preparation of the blocks for shear test should be carried out with minimum disturbance. Any blasting work should be limited to 3–4 m (10–13 ft) away from the block except in certain circumstances where clean blasting can be carried. In such cases the distance may be decreased to 2–3 m (8–10 ft). The amount of explosive should

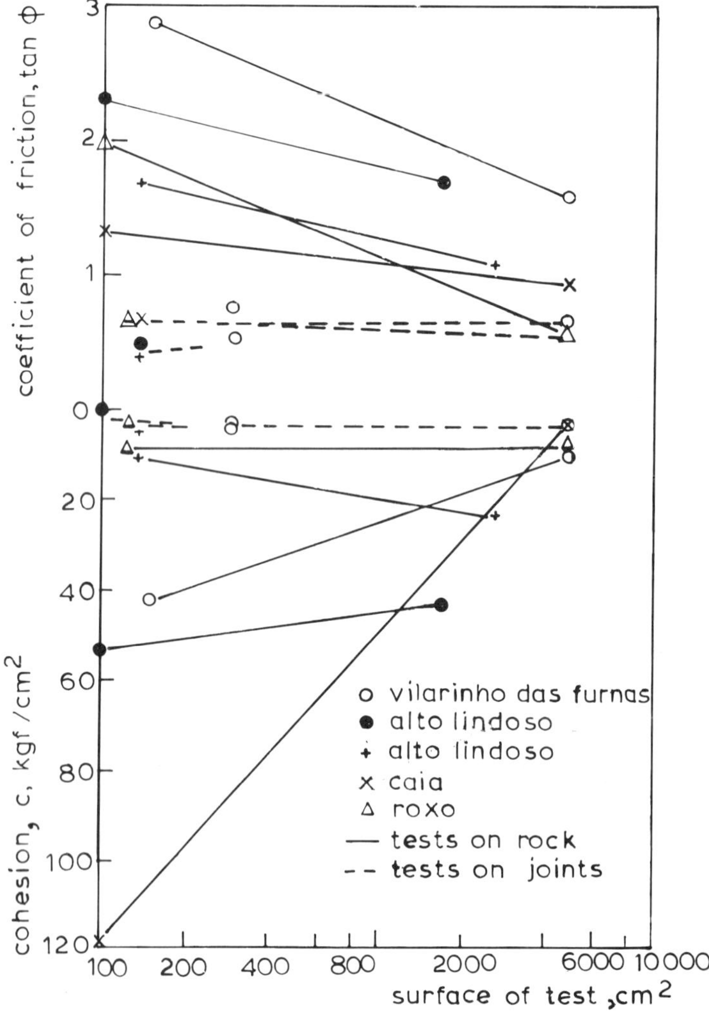

Fig. 8-101. Influence of surface area on the frictional behaviour of rocks in in situ tests (after BARROSO, 1966).

be reduced as the test site is approached. Actual block preparation should be done manually using mechanical and pneumatic disintegrators.

The amount of excavation required depends upon the size of the block to be tested, the size of the loading equipment, the method of test (direct or inclined shear test). Usually a working space of 2–3 m (7–10 ft) is required around the test block for manoeuvring of equipment during preparation of the specimens and installation of loading jacks.

The location of the plane (shearing plane), the properties of which are to be determined, is important. Before deciding, the various planes may be identified into different groups such as bedding planes; clean joints; filled joints > 1 mm (0.04 in) thick; 1–10 mm (0.04–0.4 in) thick and 10–30 mm (0.4–1.2 in) thick. This classification should be related to the minimum and maximum size of the asperities. All joints with thickness of fillings greater than maximum asperity height, should be grouped under one group and all clean joints under another group.

Where the specimen contains cracks or other planes of discontinuities, the blocks should be enclosed from all sides using steel straps and anchors leaving only the plane to be tested free. The top block should be encased first before excavating the lower portions of the blocks.

It is important that the base of the test block should coincide with the plane to be sheared and the direction of shearing should correspond if possible to the direction of anticipated shearing in the full scale structure to be analysed using the test results. The block and particularly the shear plane should, unless otherwise specified, be retained as close as possible to its natural in situ water content during preparation and testing, for example by covering with saturated cloth. A channel approximately 20 mm (0.8 in) deep by 80 mm (3.1 in) wide should be cut around the base of the block to allow freedom of shear and lateral displacements.

A layer at least 20 mm (0.8 in) thick of weak material (e.g. foamed polystyrene) is applied around the base of the test block, and the remainder of the block is then encapsulated in reinforced concrete or similar material of sufficient strength and rigidity to prevent collapse or significant distortion of the block during testing. The encapsulation formwork should be designed to ensure that the load bearing faces of the encapsulated block are flat (tolerance ± 1 mm (0.04 in)) and at the correct inclination to the shear plane (tolerance $\pm 2°$).

Reaction pads, anchors, etc. if required to carry the thrust from normal and shear load systems to adjacent sound rock, must be carefully positioned and aligned. All concrete must be allowed time to gain adequate strength prior to testing.

It is advisable particularly if the test horizon is inclined at more than 10–20° to the horizontal to apply a small normal load to the upper face of the test specimen while the sides are cut, to prevent premature sliding and also to inhibit relaxation and swelling. The load, approximately 5–10% of that to be applied in the test, may for example be provided by screw props or a system of rock bolts and cross-beams, and should be maintained until the test equipment is in position.

3. *Test procedure*: The test procedure consists of two parts, consolidation and shearing.

Consolidation: The consolidation stage of testing is to allow pore water pressures in the rock and filling material adjacent to the shear plane to dissipate under full normal stress before shearing. Behaviour of the specimen during consolidation may also impose a limit on permissible rate of shearing. All displacement gauges should be checked for rigidity, adequate travel and freedom of movement, and a preliminary set of load and displacement readings is recorded. Normal load is then raised to the full value specified for the test, recording the consequent normal displacements (consolidation) of the test block as a function of time and applied loads. Fig. 8-102 gives the example of layout of test data sheet. The consolidation stage may be considered complete when the rate of change of normal displacement recorded at each of the four gauges is less than 0.05 mm (0.002 in) in 10 minutes. Consolidation curves for a typical test are given in Fig. 8-103.

Shearing: Shear force is applied only after consolidation has been completed. The shear force is applied either in increments or continuously in such a way as to control the rate of shear displacement. Approximately 10 sets of readings should be taken before reaching peak strength. The rate of shear displacement should be less than 0.1 mm/min (0.004 in/min) in the 10 minute period before taking a set of readings. This rate may be increased to not more than 0.5 mm/min (0.002 in/min) between sets of readings provided that the peak strength itself is adequately recorded. For a 'drained' test particularly when testing clay filled discontinuities, the total time to reach peak strength should exceed $6.t_{100}$ as determined from the consolidation curve (Fig. 8-103). If necessary the rate of shear should be reduced or the application of later shear force increments delayed to meet this requirement. The requirement that total time to reach peak strength should exceed $6.t_{100}$ is derived from conventional soil mechanics consolidation theory (GIBSON and HENKEL, 1954) assuming a requirement of 90% pore water pressure dissipation. This requirement is most important when testing a clay filled discontinuity. In other cases it may be difficult to define t_{100} with any precision because a significant proportion of the observed consolidation may be due to rock creep and other mechanisms unrelated to pore pressure dissipation. Provided the rates of shear specified are followed, the shear strength parameters may be regarded as having been measured under conditions of effective stress (drained conditions).

DIRECT SHEAR TEST DATA SHEET

Client: Project: Concrete Dam Location: Alcântara Loc. No.: Block No.:

TEST BLOCK SPECIFICATION See drawings & photographs Nos.:

General rock description, index properties and water conditions
Phyllite, sound to moderately weathered Normal conditions

Description and index properties of surface to be sheared Type;
Dip; Dip direction; Roughness; Filling & alteration;
Persistance; Spacing of set; Surface dimensions; 0.70 x 0.70 Initial area A; 0.490 m^2

FORCES

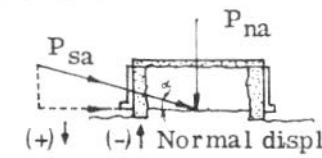

$P_n = P_{na} + P_{sa} \sin\alpha$
$P_s = P_{sa} \cos\alpha$
$\sigma_n = P_n/A$
$\tau = P_s/A$

P_{sa} P_{na} (+)↓ (−)↑ Normal displacement
$\alpha = 15^o$ σ_n (nominal)

1	2		3					4		5			6	7	8	9	10
Time elapsed	Applied normal force P_n		Normal displacement Δ_n					Applied shear force P_a		Shear displacement Δ_s			Contact area A (corrected)	P_{na}	σ_n	P_{sa}	τ
(min)	Reading	Force (kN)	Reading 1	2	3	4	Average (mm)	Reading	Force (kN)	Reading 1	2	Average (mm)	m^2	(kN)	(MP_a)	(kN)	(MP_a)
10		196	0.100	0.070	0.130	0.070			0	0	0	-	0.490	196		0	
35		233	0.130	0.085	0.140	0.090			137	0.05	0.05	0.05		"		142	
48		270	0.050	0.065	0.285	0.290			275	0.55	0.35	0.45		"		284	
64		306	-0.200	0.010	0.435	0.495			412	1.35	1.10	1.22		"		426	
87		343	-0.710	-0.205	0.600	0.720			549	2.55	2.30	2.42		"		568	
109		380	-1.165	-0.445	0.680	0.850			686	3.90	3.50	3.70		"		710	
131		417	-1.675	-0.615	0.710	0.970			824	5.15	4.60	4.88		"		853	
154		453	-1.965	-0.745	0.720	1.050			961	6.10	5.50	5.80		"		995	
172		490	-2.245	-0.880	0.720	1.105			1098	7.20	6.50	6.85		"		1137	
189		527	-2.480	-1.055	0.695	1.165			1235	8.20	7.40	7.80		"		1279	
206		504	-2.750	-1.205	0.640	1.165			1373	9.45	8.45	8.95		"		1421	
234		601	-3.075	-1.505	0.465	1.100			1510	11.00	10.00	10.50		"		1563	
252		637	-3.350	-1.830	0.280	0.910			1647	12.45	11.40	11.92		"		1705	
264		674	-3.675	-2.185	0.050	0.720			1784	14.00	12.80	13.40		"		1847	
276		711	-4.005	-2.665	-0.290	0.360			1922	15.55	14.40	14.98		"		1989	
289		748	-4.585	-3.125	-0.890	-0.020			2059	17.60	16.45	17.02		"		2132	
293	Rupture	784	-4.975	-3.375	-1.250	-0.290			2196	20.00	19.55	19.78		"		2274	

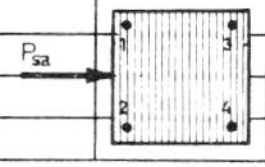

P_{sa} 1 2 3 4

Calibration data

Remarks

Tested by; Dates;
Checked by; Date;

Fig. 8-102. Example layout of direct shear test data sheet (after ISRM, 1974).

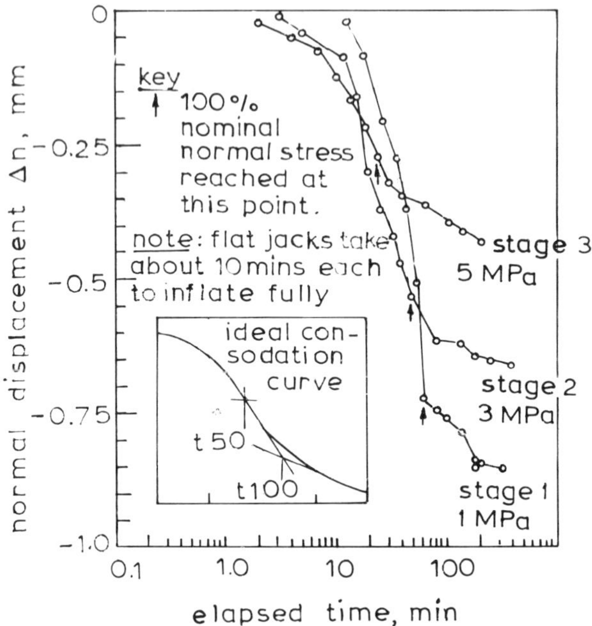

Fig. 8-103. Consolidation curves for a three stage direct shear test, showing the construction used to estimate t_{100} (after ISRM, 1974).

The displacement of the block should be measured at several points along and at an angle to the shearing plane (Fig. 8-104). The normal displacement gives dilation of the block and its different values at the front and rear (away and near to the jack applying shearing force) permit calculation of tilt of the block.

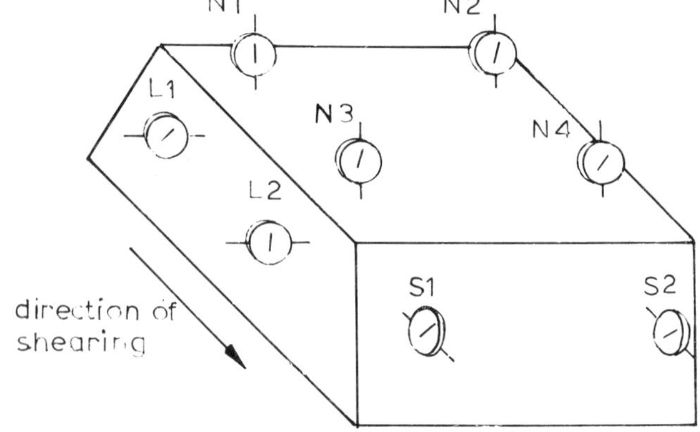

Fig. 8-104. Arragement of displacement gauges S1 & S2 for shear displacement, L1 & L2 for lateral displacement, N1–N4 for normal displacement. (after ISRM, 1974).

It has been found that the horizontal displacement of the block is not uniform along the whole length of the shearing surface. The displacements are highest at points near to the shear loading jack and lowest at points away from it (Fig. 8-105). As a result the strain energy distribution around the blocks is obviously not constant—being highest near the loading end and least on the far end (Fig. 8-106). This when coupled with nonuniform dilation of the block could result in tensile fracture at the base of the block rather than shear failure. Obviously, therefore, the interpretation of results obtained from such tests is very difficult.

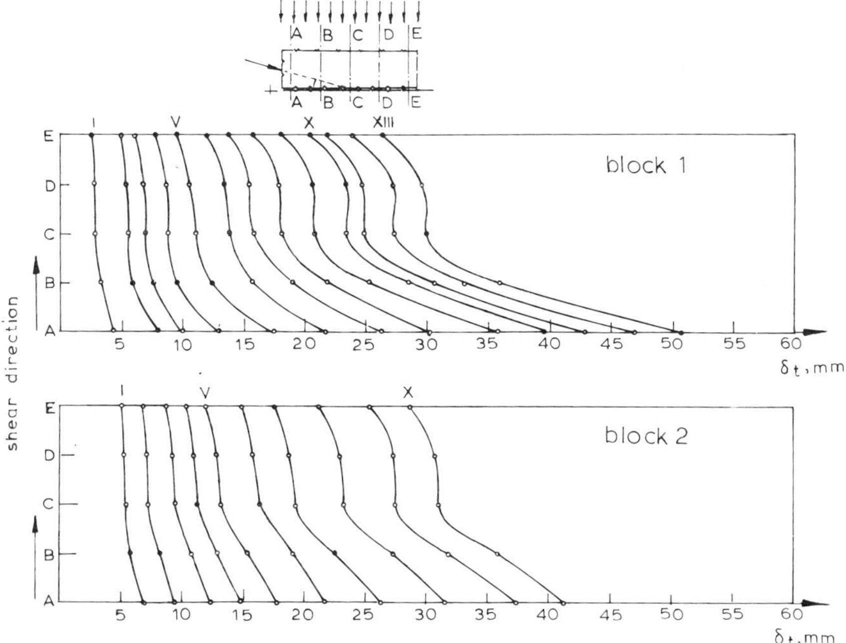

Fig. 8-105. Measured deformations of the individual profiles A to E on the blocks No. 1 and 2 in the direction of shearing, and for the various cycles of loading (after KRSMANOVIC and POPOVIC, 1966).

Displacement measurement can be done using LVDTs or mechanical dial gauges mounted on a fixed frame the position of which is not affected by the application of normal or shearing force.

After reaching peak strength, readings should be taken at increments of from 0.5–5 mm (0.02–0.2 in) shear displacement as required to adequately define the force-displacement curves. The rate of shear displacement should be 0.02–0.2 mm/min (0.0008–0.008 in/min) in the 10 minute period before a set of readings is taken, and may be increased to not more than 1 mm/min (0.04 in/min) between sets of readings.

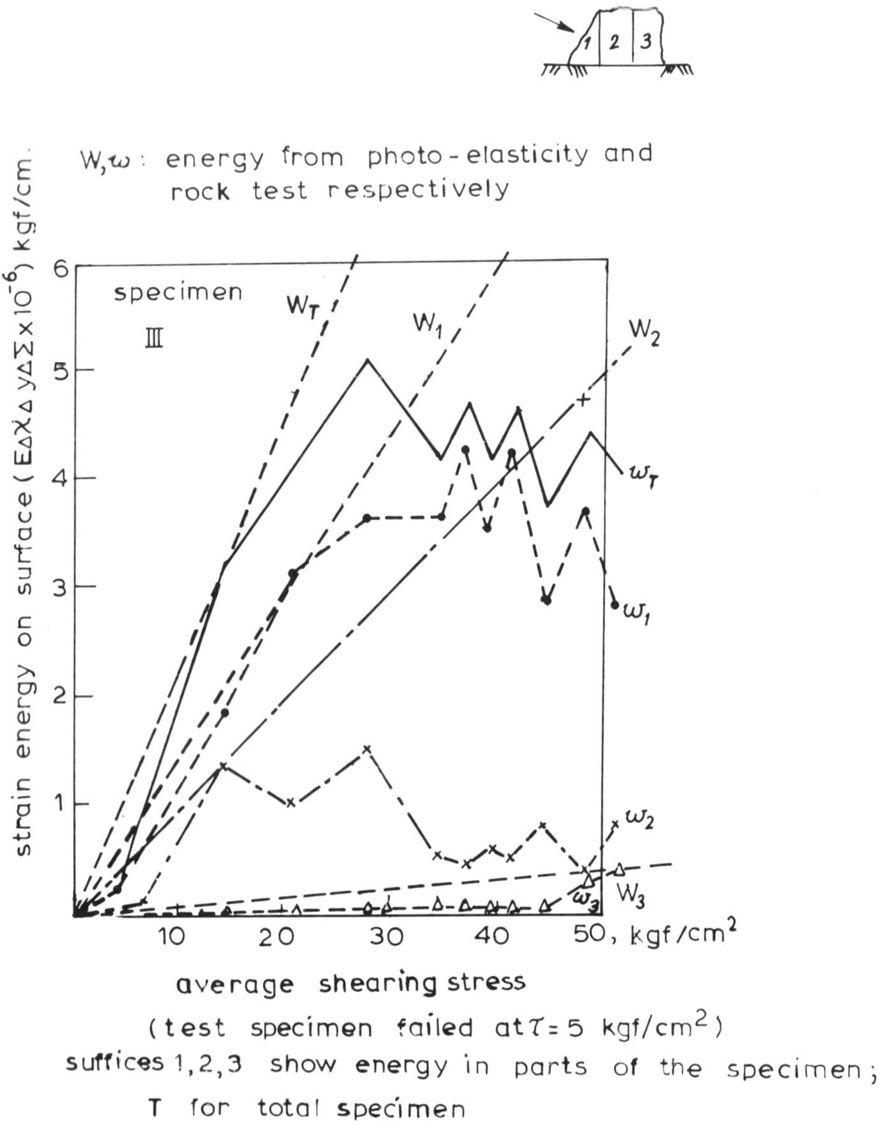

Fig. 8-106. Accumulated strain energy and given shearing stress (after KIMISHIMA, 1970).

It may be possible to establish a residual strength value when the specimen is sheared at constant normal stress and at least four consecutive sets of readings are obtained which show not more than 5% variation in shear stress over a shear displacement of 1 cm (0.4 in). An independent check on the residual friction angle should be made by testing in the laboratory two prepared flat surfaces of the representative rock. The prepared surfaces should be saw-cut and then ground flat with No. 80 silicon carbide grit. Having established a residual strength the normal stress may be increased or reduced and shearing continued to obtain additional residual strength values. The specimen should be reconsolidated under each new normal stress and shearing continued. After the test, the block should be inverted, photographed in colour and fully described. Measurements of the area, roughness, dip and dip direction of the sheared surface are required, and samples of rock, in filling and shear debris should be taken for index testing.

A more or less complete data about the behaviour of a joint can be obtained from a single test by applying load in stages and measuring displacements as the block just begins to slide. Fig. 8-107 represents the results obtained from such a test. Normal load is applied in stages and its course is shown in quadrant III. Quadrant II gives the displacements at a normal load as the shear stress is increased. Immediately when the block has displaced a little and curve tends to become horizontal, normal load is increased followed by an increase in the shear load (2b). The test is continued step by step till the maximum normal stress to which the test surface is to be subjected is reached.

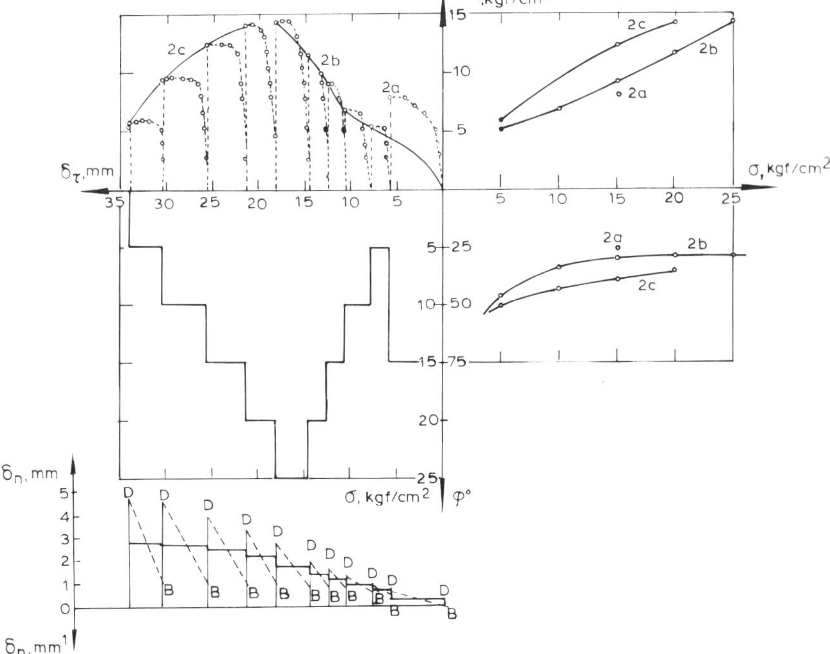

Fig. 8-107. Diagram of the obtained shear strengths; on the right – the shear strengths, on the left – the normal pressures and the deformations, σ_n and δ_τ (after KRSMANOVIC and POPOVIC, 1966).

Load is then reduced in steps and test continued further (2c). Quadrant I represents the corresponding peak shear values for the normal loads. Two curves shown in this quadrant represent the $\tau-\sigma_n$ values for the loading (2b) and unloading (2c) phases of the test. Point (2a) represents the initial shear stress value at the initial normal load. The tangent values of the (2b) and (2c) curves in quadrant I are represented in quadrant IV and these represent the coefficient of friction at these loads both in the loading and unloading phases.

Load measurement on the specimen is best done using pressure transducers measuring the pressure in the hydraulic jacks. The hydraulic jacks can be calibrated against the pressure transducer outfit.

4. Calculations and reporting of results: A consolidation curve (Fig. 8-103) is plotted during the consolidation stage of testing. The time t_{100} for completion of "primary consolidation" is determined by constructing tangents to the curve as shown.

Displacement readings are averaged to obtain values of mean shear and normal displacements Δ_s and Δ_n. Lateral displacements are recorded only to evaluate specimen behaviour during the test, although if appreciable they should be taken into account when computing corrected contact area.

Shear and normal stresses are computed as follows:

$$\text{Shear stress} \quad \tau = \frac{P_s}{A} = \frac{P_{sa}\cos\alpha}{A} \tag{8.61}$$

$$\text{Normal stress} \quad \sigma_n = \frac{P_n}{A} = \frac{P_{na} + P_{sa}\sin\alpha}{A} \tag{8.62}$$

where P_s = total shear force
P_n = total normal force
P_{sa} = applied shear force
P_{na} = applied normal force
α = inclination of the applied shear force to the shear plane and
A = area of shear surface overlap
(corrected to account for shear displacement).

If α is greater than zero the applied normal force should be reduced after each increase in shear force by an amount $P_{sa}\sin\alpha$ in order to maintain the normal stress approximately constant. The applied normal force may be further reduced during the test by an amount

$$\frac{\Delta_s(\text{mm}) \times P_n}{700}$$

to compensate for area changes.

For each test specimen graphs of shear stress (or shear force) and normal displacement vs shear displacement are plotted to show the nominal normal stress and any changes in normal stress during shearing. Values of peak and residual shear strength and the normal stresses, shear and normal displacements at which these occur are abstracted from these graphs.

Graphs of peak and residual shear strength vs normal stress are plotted from the combined results for all test specimens. Shear strength parameters ϕ_a, ϕ_b, ϕ_r, c' and c are abstracted from these graphs as shown in Fig. 8-108

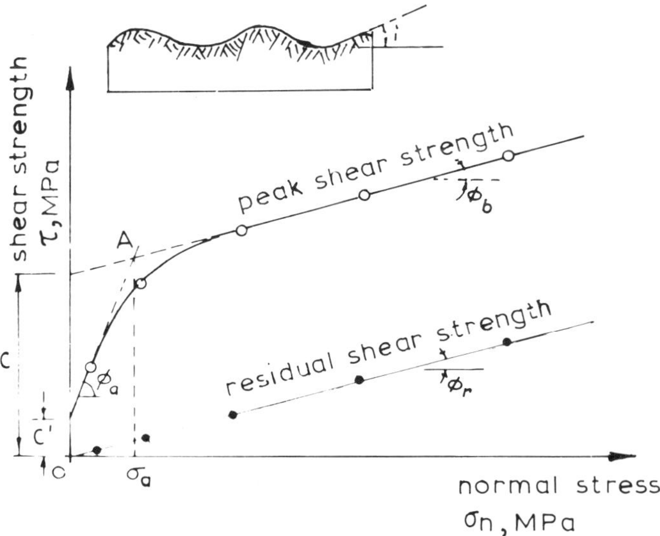

Fig. 8-108. Shear strength – normal stress graph.

where ϕ_r = residual friction angle.

ϕ_a = apparent friction angle below stress σ_a; point A is a break in the peak shear strength curve resulting from the shearing off of major irregularities on the shear surface. Between points 0 and A, ϕ_a will vary somewhat; measure at stress level of interest. Note also $\phi_a = \phi_u + i$, where ϕ_u is the friction angle obtained for smooth surfaces of rock on rock and angle i is the inclination of surface asperities.

ϕ_b = apparent friction angle above stress level σ_a (point A); note that ϕ_b will usually be equal to or slightly greater than ϕ_r and will vary somewhat with stress level; measure at the stress level of interest.

c' = cohesion intercept of peak shear strength curve; it may be zero.

c = apparent cohesion at a stress level corresponding to ϕ_b.

The report should include the following:

(i) A diagram, photograph and detailed description of test equipment and a description of methods used for specimen preparation and testing.
(ii) For each specimen a full geological description of the intact rock, sheared surface, filling and debris preferably accompanied by relevant index test data (e.g. roughness profiles; ATTERBERG limits, water content and grain distribution of filling materials).
(iii) Photographs of each sheared surface together with diagrams giving the location, dimensions, area, dip and dip direction and showing the directions of shearing and any peculiarities of the blocks.
(iv) For each test block a set of data tables, a consolidation graph and graphs of shear stress and normal displacement vs shear displacement. Abstracted values of peak and residual shear strength should be tabulated with the corresponding values of normal stress, shear and normal displacement.
(v) For the shear strength determination as a whole, graphs and tabulated values of peak and residual shear strength vs normal stress, together with derived values for the shear strength parameters.

8.10.2. Variations of Inclined Load Test

Variations of inclined load test have been used by some investigators.

PROTODYAKONOV (1961) described the method of testing in situ shear strength at an underground face (Fig. 8-109). In this method a ledge is left on the face on an inclined plane and loaded with a hydraulic jack. Shearing takes place along the inclined plane (A–B). Knowing the coefficient of internal friction of rock, the shear strength can be calculated from the relationship

$$\tau_t = (F \sin 2\theta / 2a) - (\mu F \cos^2 \theta / a) \tag{8.63}$$

where τ_t = shear strength
F = force at failure
θ = angle of inclination of the ledge with the horizontal
a = area of cross-section of the loaded surface and
μ = coefficient of internal friction of rock.

TAKANO and FURUJO (1966) used a single jack inclined method. Here the angle of inclination of the jack θ can be varied (Fig. 8-110a) which gives different values of shear stress and normal stress along a predetermined plane. The method is similar to the one discussed in Chapter 4, Section 4.4.4. (Vol. I). The results of shear tests on black schist on 1 m × 1 m (3.3 ft × 3.3 ft) surface are given in Fig. 8-110b. The method is very simple and less time consuming. The direction of loading can be so chosen that the normal and shear stresses acting on the testing surface corresponds approximately to the normal stress acting along the actual foundation. It must, however, be borne in mind that it cannot be used to determine the shear strength of material. Rather it gives the shear behaviour of a predetermined surface as in the case with other tests.

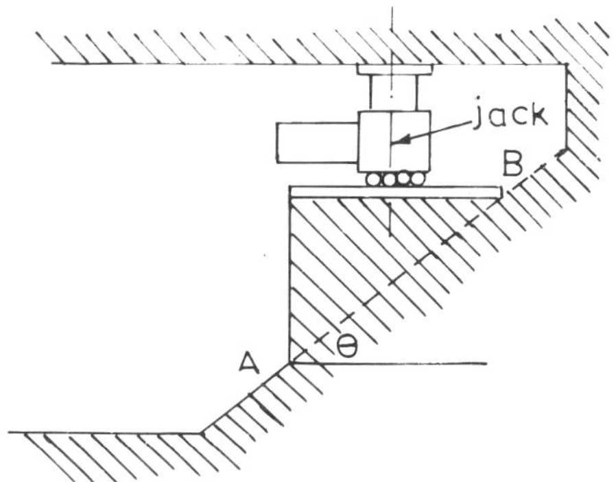

Fig. 8-109. Testing of rock for shearing with compression at an underground face (after PROTODYAKONOV, 1961).

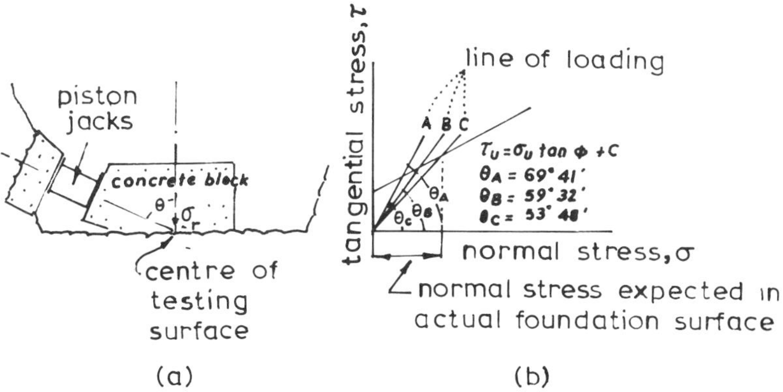

Fig. 8-110. Loading system and loading directions (after TAKANO and FURUJO, 1966).

8.10.3. Shear Test with Water Saturation

Protection of the moisture conditions during the preparation of shear test is very important for moisture sensitive materials. Moisture changes occur on exposure of the material and during the time elapsed for the actual conduct of test may run to several days. Test samples should be taken to ascertain moisture contents over the time period. During the preparation of direct shear test, the top surface, once ready, should be protected by loading plate before the side surfaces are excavated and prepared. This helps to protect the moisture and also protects against dilation (swelling), a part of which is definitely time-dependent.

If the joint or rock is satured or has to be submerged in water, tests ought to be carried out under simulated conditions. The test requires saturating the rock for a sufficiently long time before the conduct of the test. The time required to saturate depends upon the permeability of the rock and the joint-fill and may take from a few days to several weeks. Test arrangement used by ROMERO (1966) is given in Fig. 8-111. A sample is trimmed with adequate dimensions in the direction of the shear to a depth of about 15 or 20 cm (6 or 8 in) beyond the desired plane of failure. A metallic frame with top in level with the test surface is arranged and fastened with cement to the sample.

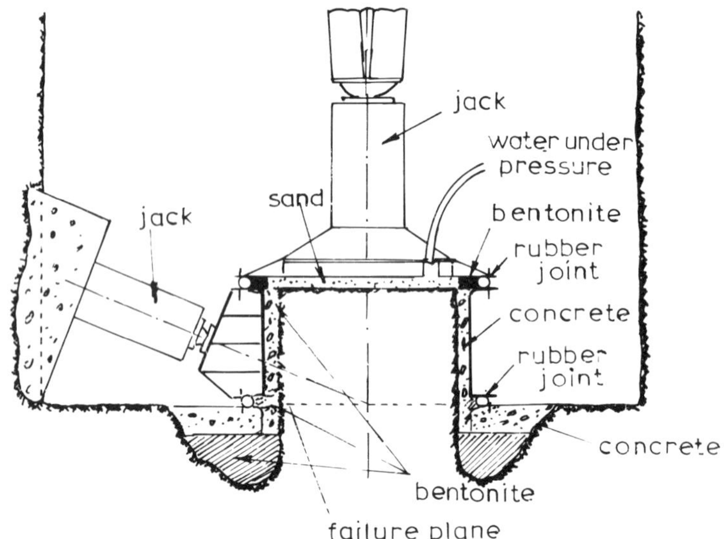

Fig. 8-111. Layout of the in situ shear test with saturation under pressure (after ROMERO, 1966).

Another frame of height greater than the sample height is placed on the lower frame with their contact plane flush with the test shear plane, and finally the sample is enclosed with a metallic plate covering the sample and the upper frame using neoprene joints. To assure the tightness of the whole set bentonite is placed in the inside. These three elements are fastened with screws compressing the joints. A layer of fine sand is put between the cover plate and the upper part of the rock. Water is injected under pressure through a hole in the cover plate. Vertical force equal to the water pressure is applied before injecting water for saturating the sample and slowly raised to increase the effective stress under which the shear has to take place. This vertical stress is maintained during the whole period of saturation which may oscillate, according to the type of rock, between several days and two months. After this period, the water pressure is dropped and simultaneously the vertical pressure of the jack is lowered by the same value. In this way, the effective stress on the shear plane remains constant. The screws fastening the two frames and the rubber joint are taken off. Afterwards, the tangential stress is increased by stages with one horizontal or inclined jack until the failure occurs.

The results of tests at some Spanish dam sites using this method are given in Table 29. A considerable drop in cohesion values is noticed in argillaceous marl compared to the other two. The drop in angle of friction, however, is not so marked.

TABLE 29

Results of tests at some Spanish dam sites

(after ROMERO, 1966)

Dam	Type of rock	Size of the sample cm (in)	Test type	Water pressure of saturation MPa (lbf/in²)	Parameter of shear strength	
					Cohesion MPa (lbf/in²)	Angle of friction
Renegado	calcareous marl	50 × 50	with saturation	0.39	0.18 to 0.43	36° to 49°
		(19.7 × 19.7)	natural moisture	(56.8)	(25.6 to 62.5)	
Ribarroja	gypsum marl contact	50 × 50	with saturation	0.49	0.23 (32.9)	27.6°
		(19.7 × 19.7)	natural moisture	(71.0)	0.14 (20.6)	28.8°
Santomera	argillaceous marl	50 × 50	with saturation	0.24	0.07 (10.1)	38.5°
		(19.7 × 19.7)	natural moisture	(35.5)	0.35 (51.1)	41.5°

8.10.4. Parallel Load Test

The difference between an inclined load test and parallel load test is that in parallel load test, shear force is applied parallel to the shear plane. This test is therefore a direct field application of the laboratory test. The general arrangement remains the same as for the inclined shear test. The method has the disadvantage that the amount of excavation work required around the test specimen for placement of jacks is more. The method has the advantage that the results of the test are comparable with laboratory tests and that it gives more uniform stress distribution along the shear plane.

An arrangement of parallel load test used at Kurobe Dam is given in Fig. 8-112.

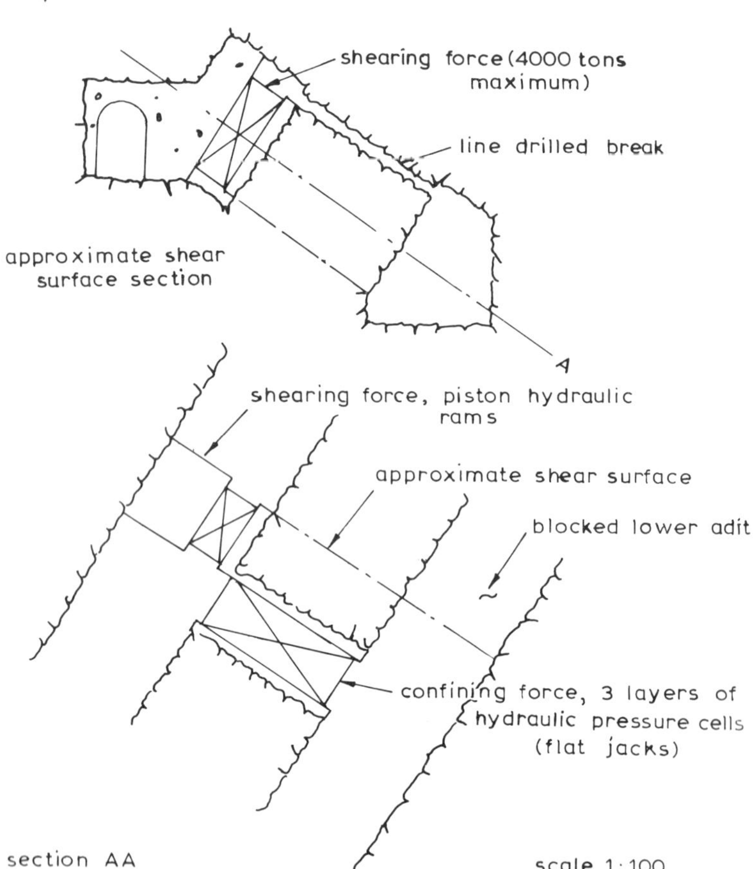

Fig. 8-112. Large-scale biaxial shear test, Kurobe Dam, Japan (after JOHN, 1961).

The test was one of the large scale tests ever attempted to determine the shear strength of the rock mass. The approximate area of the tested surface was 8.8 m^2 (94 ft^2). The equipment used could apply 40 MN (4000 tonf) of shearing force (4.6 MPa) (663 lbf/in^2) and measure deformation to an accuracy of 0.005 mm (0.0002 in). To prepare for the test:

(i) The rock was first pressure grouted.

(ii) The test adits were driven to 10 m (33.8 ft) from the test block, using regular excavation techniques (1.2 m (3.9 ft)) round and 1.5 kg/m^3 (2.5 lb/yd^3) powder factor).

(iii) From 10 m to 0.5 m (32.8 to 1.6 ft) the adit excavation was done using 0.6 to 0.8 m (2.0 to 2.6 ft) rounds and 1 kg/m^3 (1.7 lb/yd^3) powder factor.

(iv) The remaining adit excavation and the test block were prepared by cutting and chipping of the rock as no explosives were permitted.

(v) After the test block was prepared, it was carefully mapped geologically with special attention to all rock fabric.

(vi) The next step was to pour the concrete reaction blocks, part of which were reinforced.

(vii) To make the test, the confining pressure was raised to the desired amount and then the shearing force applied in increments. Each loading increment was held until no further deformation was recorded. A special bearing arrangement permitted tangential movement of the confining pressure jacks.

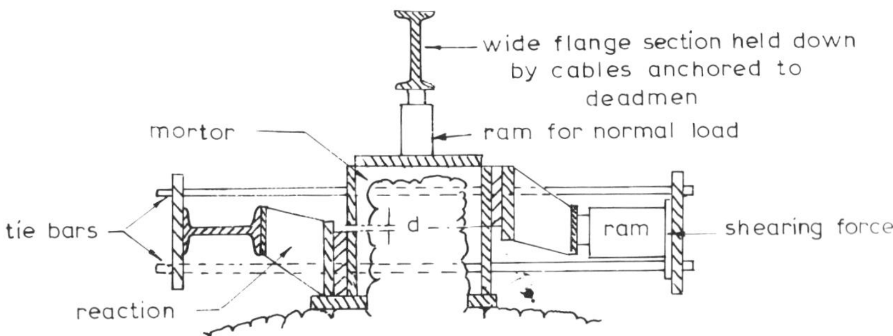

Fig. 8-113. Typical set up for shear test when abutments are not available for reactions (after DODDS, 1970).

A typical set-up for shear test when abutments are not available for reactions is given in Fig. 8-113 (DODDS, 1970). The test block was 25.4 cm (10 in) high and 30.5 cm (12 in) square. The test was completed in the following steps:

(i) The test block was enclosed in two steel shear boxes and the spaces between the box and the rock were filled with a quick-setting, high-strength mortar. Greased 0.64 cm (1/4 in) thick metal spacers were placed between the top and bottom shear box.

(ii) The top of the specimen was then capped with mortar and a steel plate.

(iii) The vertical load assembly consisting of a 300 kN (30 tonf) ram rode on greased rollers on the top steel plate to allow for tangential movement.

(iv) The shearing force was applied using a 300 kN (30 tonf) ram and frame as shown in Fig. 8-113.

(v) The tests were made using normal loads between 20 and 120 kN (2 and 12 tonf). The shearing loads were applied in increments of 2224 N (500 lbf) and deformations measured to an accuracy of 0.0025 cm (0.001 in). The shear load at each increment was held constant until movement stopped or was negligible.

The distance d, i.e. the width of the possible shearing plane (Fig. 8-113) is very important. Its increase makes the test more representative of heterogeneity and fracturing in the rock, but simultaneously its increase results in greater bending effect This disturbs the uniform distribution of normal stress. If tension due to bending is to be equal to compression, then it can be proved that

$$T = \frac{aN}{3d} \qquad (8.64)$$

where T = shear force
N = normal force and
a = width or length of the specimen inclined with the shearing force.

For value of T above this limit, tensional cracks will occur which will affect the influence of cohesion in the test result. The value of d is usually taken between $a/4$ and $a/8$ provided N is not very small.

The nonzero value of d results in nonuniform normal stress distribution but it is not a serious factor to influence the results to a considerable extent (ROCHA, 1964).

Shear strength results are sensitive to the value of d and shear strength decreases with increase in d.

When tests are conducted to determine cohesion and friction angle between rock and concrete, it is advisable to keep $d = 0$ so that shear takes place through the contact surface. Stress distribution on the shear surface tends to be more uniform when a soft layer is present at the shear surface and hence simulates the effect of d. The tensile stresses present in the close vicinity of the shear plane (not at the shear plane) at the end regions of the samples are insignificant for normal to shear force greater than one (KUTTER, 1971).

8.10.5. Torsion Test

The torsion test first suggested by RINAGL (EVERLING, 1963) is used to measure the shear strength on specimens with least disturbance and in inaccessible areas in boreholes. The test is suited for the measurement of shear strength of joints but its application to determine the shear strength of material is doubtful (Vol. I, Section 4.2). The method is based upon exerting a torsional stress on the surface to be tested. The method has been recommended by the I.S.R.M. (1974) for shear testing of planes of weaknesses.

The test consists of drilling a hole (BX size) perpendicular to the plane to be tested (Fig. 8-114) to a depth about 1 m (3.3 ft) deeper than the test plane. A rock bolt is grouted using a wedge anchor or grout. A 300 mm (12 in) diameter concentric hole is drilled using appropriate drilling fluid (water or air) so as not to wash away any soft material present in the plane to be tested. The annular space surrounding the test core is cleaned and filled with weak material (e.g. sponge rubber gasket) to a depth of about 5–10 mm (0.2–0.4 in) above the

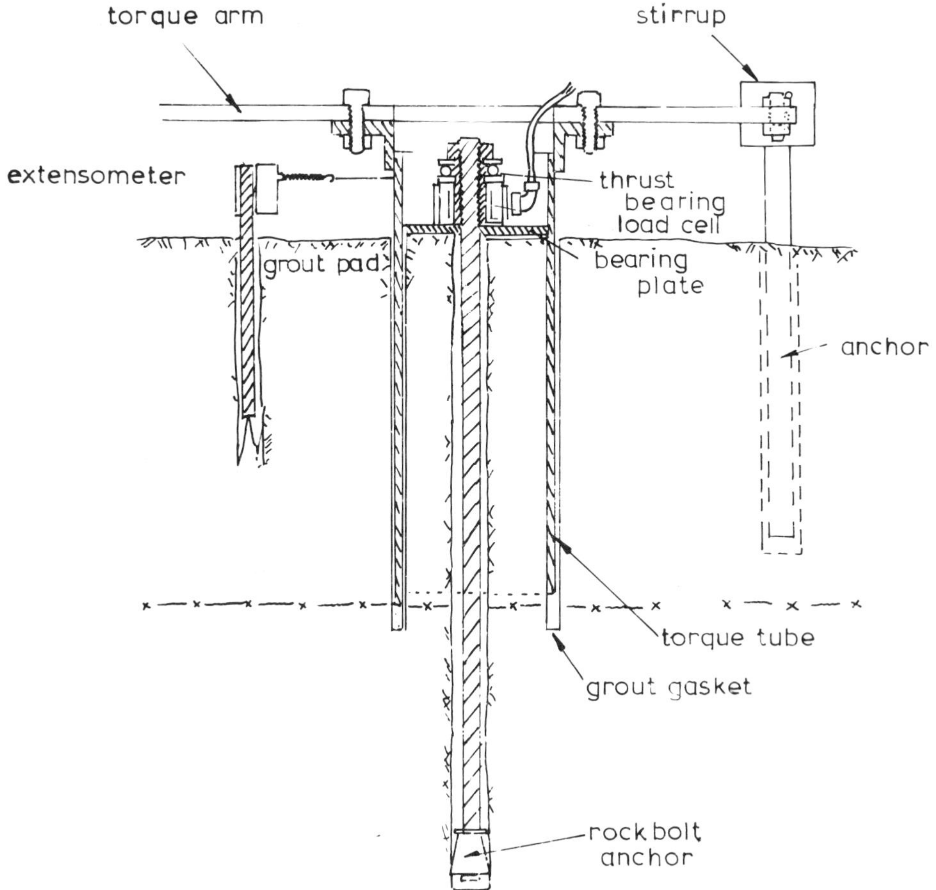

Fig. 8-114: Torsional shear apparatus: Section (after SELLERS, 1974).

test surface. A torque tube of appropriate length with greased outside surface is inserted to rest on the weaker filler surface and grouted in position. Two holes are drilled at 2–3 m (7–10 ft) from the centre hole in which 35–56 mm (1-3/8–2-1/4 in) anchors are installed. These three holes must be carefully aligned so that the applied torque is as pure a couple as possible. A rock bolt load cell is attached to the end of the central bolt and using a thrust bearing between the load cell and a nut, load of approximately 1–2 bar (14.5–29 lbf/in^2) is brought on the test surface. This load can be varied during the test.

A torque arm of about 1 m (3.3 ft) length is fitted to the top of the torque tube to which hydraulic jacks are attached using shackles, torque load cells and chains etc. (Fig. 8-115). Extensometers are installed to measure the twist.

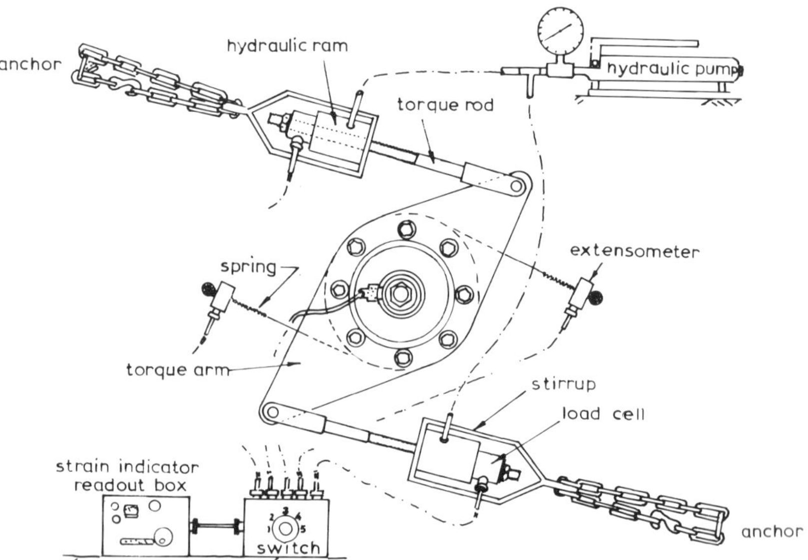

Fig. 8-115. Torsional shear apparatus: Plan view (after SELLERS, 1974).

While the normal load is held constant, a gradually increasing torque is applied to the test cylinder. At the point of failure, the extensometer reading increases rapidly (Fig. 8-116). Continued slow application of hydraulic pressure can produce continued twisting at constant torque.

The test may be continued either to give residual shear strength at a constant rate under constant normal load or the normal load may be varied to determine the influence of different normal loads and obtain possible MOHR envelope and friction angle. The results can be evaluated as follows.

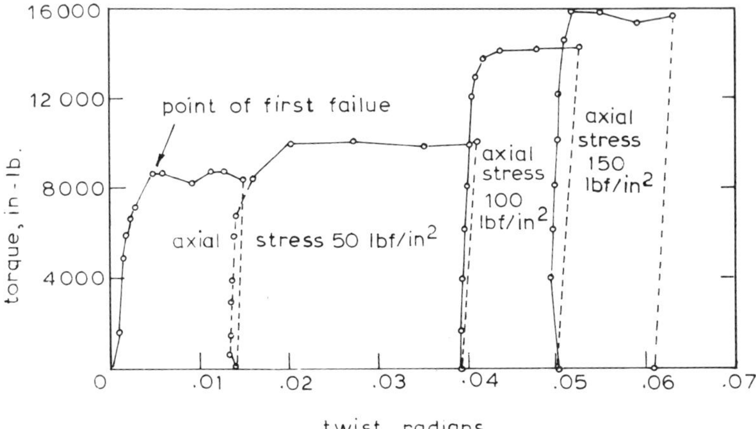

Fig. 8-116. Torsional shear test: torque/twist curves (after SELLERS, 1974).

Torque values are calculated using the load cell calibration curves to obtain the loads (kN) then multiplying by the torque radius (mm) to obtain the torque (m-N). The total torque T is the sum of the two applied torques.

Normal loads are calculated from the load cell calibration curve, and converted into normal stresses σ_n using the equation:

$$\sigma_n = \frac{4P_n}{\pi(D^2 - d^2)} \tag{8.65}$$

where P_n = applied normal load
D = outer diameter of test core
d = inner diameter of test core.

Shear strengths τ corresponding to the peak and residual portions of the torque-displacement graph may be calculated using one of two assumptions:

In strong, elastic rock assuming maximum stress at the outer circumference of the test core

$$\tau = \frac{0.5\,TD}{J} \tag{8.66}$$

In softer rock, assuming constant tangential stress at every radius across the failure plane

$$\tau = \frac{0.365\,TD}{J} \tag{8.67}$$

where $J = \dfrac{\pi(D^4 - d^4)}{32}$

The latter assumption gives a more conservative cstimate and should always be used for calculating residual shear strength; usually also for peak strength calculation.

Graphs of peak and residual shear strength vs normal stress are plotted from the combined results for all test specimens. Shear strength parameters may be abstracted from these graphs as shown in Fig. 8-108.

8.11. Factors Influencing Shear Strength of Rock In Situ

The results of in situ shear tests are given in Tablc 30. The table indicates that in general the angle of friction for rocks usually does not exceed 55°. The angle of friction decreases slowly with the degree of alteration while cohesion drops much more rapidly. Cohesion values for rock mass are low, usually under 1.0 MPa (10 kgf/cm^2) (145 lbf/in^2).

Shear strength of rock in situ is invariably lower than laboratory strength. A number of factors influence the in situ values. Among these the most important are: moisture and size effect. The percentage of moisture plays an important role particularly for moisture sensitive rocks such as claystones, shales, etc. Shear strength values calculated from plate loading tests (punch tests) and direct shear tests also differ considerably, because of the sensitiveness of shear results to the imposed displacement surface in a direct shear test and a larger number of surfaces available in a punch test. The results, therefore, are lower in a punch test than the direct sheart test (BUKOVANSKY, 1962; 1966).

The presence of cracks in in situ specimens influences the shear strength and cohesion values because the material does not need shearing along these planes and which ultimately appears in calculations. If care is taken to account for the discontinuities in the test, the shear strength determined from laboratory may be much closer to the in situ values. KIMISHIMA (1970) calculated the actual sheared area in in situ tests and worked out the values obtained to compare with laboratory tests. For four different sites, the in situ values were about 63–84% of the laboratory values. The directly determined values were only about 3–11% of the laboratory values.

The in situ shear strength of rocks decreases with increase in weathering. The influence of degree of weathering (denoted by density (ρ) or alteration index (i)) on peak shear strength values obtained at Alto Rabagoa Dam are given in Fig. 8-117. It is seen that shear strengths improve very rapidly for i values lower than 6. For i values greater than 15, specific gravity ($\rho = 1.8$), material is no longer rock but soil that disintegrates in water. Fig. 8-118 gives the values of cohesion and angle of friction for different values of alteration index i. Even the residual soil from this altered granite has an angle of friction of 35°, far higher than clay.

TABLE 30(a) In situ shear strength parameters of rocks

Rock	Specimens		Number of tests	Cohesion MPa	Angle of friction	Reference
	Cross-section (m)	Height (m)				
Granite (Portugal	0.7 × 0.7	0.7	44	0.5–1.3	41°–62°	Rocha, 1964
Shale (Portugal)	0.7 × 0.7	0.7	9	0.2	69°	Rocha, 1964
Granite (Japan)	3.5 × 2.5	1.8	9	1.0–2.0	45°–50°	Nose, 1964
Fissured granite (USSR)	12.0 × 8.0	7.0	2	0.3	42°	Evdokimov & Sapegin, 1970
Limestone (Switzerland)	1.5 × 1.5	1.5	2	0.0	45°	Locher, 1968
Sandstone/Basalt (Brazil)	6.1 × 5.5	4.6	1	0	78°	Ruiz & De Camargo, 1966
Basalt (Brazil)	2.0 × 2.0	2.0	5	0.2–0.7	53°–73°	Ruiz et al, 1968
Breccia (Brazil)	0.7 × 0.7	0.3	9	0.4–1.0	57°–61°	Ruiz & De Camargo, 1966
Marl (Spain)	1.0 × 1.0	1.5	4	0.1–0.6	28°–49°	Romero, 1968
Marl (Spain)	0.5 × 0.5	0.5	3	0.1–0.4	29°–49°	Romero, 1966
Marl (France)	0.6 × 0.6	0.6	6	0.9–2.1	18°–28°	Comes & Fournier, 1968
Limestone (Yugoslavia)	2.8 × 1.8	1.5	6	0.1	30°–45°	Krsmanovic & Popovic, 1966
Marl (Morocco)	0.7 × 0.6	0.3	9	0.6–1.2	26°–36°	Morocco, 1968
Quartzite (Spain)	0.7 × 0.7	0.7	8	0.4–0.7	51°–56°	Serafim & Guerreiro, 1968
Quartzite/Schist (Spain)	0.7 × 0.7	0.7	8	0.1–0.3	32°–40°	Serafim & Guerreiro, 1968
Limestone/Marl (Spain)	0.7 × 0.7	0.7	12	0.2–0.7	44°–62°	Serafim & Guerreiro, 1968

TABLE 30(b)

(after ROCHA, 1964)

Rock, Dam	Type of test	Quality index of the rock i %	Number of tests	Cohesion c $kg cm^{-2}$	Angle of friction ϕ^0	Coefficient of friction $\tan\phi$
Granites Alto Rabago	Rock	3	44	13	62	1.9
		5		5	57	1.5
		7		3	52	1.3
		10		2	46	1.0
		15		1	41	0.8
	Concrete-rock	6.2 to 7.3	8	2	56	1.5
Shales	Rock Normal to schistosity	0.8 to 1.7	9	2	69	2.6
Bemposta	Concrete-rock Parallel to schistosity	1.0 to 1.4	5	2	60	1.7
	Concrete-rock Parallel to joints	1.3	3	2	63	1.9
Shales Valdecañas (completed)	Rock Normal to schistosity	0.9 to 1.0	4	29	55	1.4
		1.3 to 2.0	3	7	64	2.0
	Concrete-rock Parallel to schistosity	1.0	3	4	62	1.9
Shales	Rock Parallel to schistosity	Little altered	4	4	59	1.7
Miranda	Rock Normal to schistosity	Little altered	10	6	64	2.0
	Concrete-rock Parallel to schistosity	Little altered	8	4	62	1.9
	Concrete-rock Normal to schistosity	Little altered	8	7	60	1.7
Shales	Rock Normal to schistosity	Variable	16	1	70	2.7
Alcantara	Concrete-rock	Variable	28	1	56	1.5

TABLE 30 (b) (continued)

Rock, Dam	Type of test	Quality index of the rock i %	Number of tests	Cohesion c $kg\,cm^{-2}$	Angle of friction ϕ^0	Coefficient of friction $\tan\phi$
Shales Cambambe	Rock Parallel to schistosity	Very altered	4	2	50	1.2
Sandstone (Cambambe)	Rock Parallel to schistosity	Little altered	4	1	60	1.7
	Concrete-rock Parallel to schistosity	Little altered	4	2	53	1.3

TABLE 30 (c)

Comparisons between field and laboratory shear strength tests on rock joints

Rock	Cohesion, MPa		Angle of friction		Source
	field	laboratory	field	laboratory	
Granite (Japan)	1–2	2–4	45°–50°	65°–70°	NOSE, 1964
Marl (Morrocco)	0.6–1.2	2–5	26°–36°	27°–36°	MOROCCO, 1968

Influence of degree of alteration and the agreement between in situ and laboratory shear strength values are quite remarkable. Fig. 8-119 represents the results of in situ shear strength tests and laboratory triaxial strength tests. At lower normal pressures and for highly weathered rocks, this agreement is more pronounced. The results deviate at higher normal pressures with laboratory triaxial test results higher than in situ direct shear test results.

As in laboratory tests, anisotropy of rock is clearly seen in in situ shear tests. Typical results are shown in Fig. 8-120. The cohesion values differ widely, and difference in frictional coefficients is comparatively small. Cohesion values perpendicular to the cleavage are higher than parallel to the cleavage.

While the shear strength of rock is size dependent, the shear behaviour of joints (already indicated) may give results quite comparable to those obtained from small laboratory tests.

TABLE 30 (d) Shear characteristics of foundation rocks from results of field shear tests of footings and blocks (after FISHMAN, 1976)

Dam	Foundation rock	Shearing-resistance parameters: experimental (limiting) c, kg/cm²	tg φ	tg ψ	computed (with correction factors of 0.8 for tan φ and 0.5 for c) c, kg/cm²	tg φ	tg ψ	computed (actually used in design) c, kg/cm²	tg φ	tg ψ	Compressive strength (axial) σ_c, kg/cm²: rock mass	concrete footings
		Concrete footings				Along concrete-roc contact						
Shirokov (USSR)	Sandstones and argillites	5,0	1,6	1,85	2,5	1,28	1,40				35	
Bratsk (USSR)	Diabases: Sound	14,0	1.82	2,52	7,0	1,46	1,31	3,0	0,7	0,85	109	240
	Diabases: Heavily cracked	1,4	0,64	0,71	0,7	0,52	0,56	Removed from foundation of dam			5	320
Krasnoyarsk (USSR)	Granites: Sound	14,0	1,8	2,5	7,0	1,44	1.79	3,0	0,7	0,85	103	180
	Granites: Crushing zone	4,0	1,26	1,46	2,0	1,01	1.11	Removed from foundation of dam			23	350
Chirkey (USSR)	Limestones	8,0	1,1	1,5	4,0	0,88	1,08				42	70
Charvak (USSR)	Limestones	5,0	1,0	1,25	2,5	0,8	0,92				25	
Nurek (USSR)	Sandstones (sound)	16,5	1,96	2,78	8,25	1,57	1,98				154	360
	Siltstones (heavily cracked)	6,0	1,1	1,4	3,0	0,88	1,03				31	360
Mogilev-Podol'sk (USSR)	Granites	8,0	1,4	1,8	4,0	1,12	,32				50	190
Zeya (USSR)	Diorites	14,5	1,65	2,37	7,2	1,32	1,68	3,0	0,7	0,85	102	220
Sayany (USSR)	Paraschists	25,0	1,2	2,45	12,5	0,96	1,59		8,0	1,0	140	250
	Orthoschists	26,5	1,42	2,75	13.2	1,14	1,80				167	320
Inguri (USSR)	Limestones	10,0	1,35	1,85	5.0	1,08	1,33	6,0	1,0	1,3	61	100
Andizhan (USSR)	Chloritic schists											
	Sound	22,0	1,0	2,1	11,0	0,8	1,35	2,5	0,65	0,78	107	
	Heavily cracked	6,0	1,3	1,6	3,0	1,04	1,19	Removed from foundation of dam			35	
Toktogul (USSR)	Limestones	26,0	1,6	2,8	13,0	1,2	1,85	{1,0 / 2,0}	{0,7 / 0,85}	{0,75 / 0,95}	172	380
Kirov (USSR)	Sandstones	20,0	1,7	2,7	10,0	1,36	1,86				146	
	Metamorphic schists	15,6	1,1	1,8	7,8	0,88	1,27				40	
Ust'-Ilim (USSR)	Diabases	8,1	1,0	1,4	4,0	0,8	1,0	3,0	0,7	0,85	40	
Kurpsai (USSR)	Sandstones and argillites	5,0	1,54	1,79	2,5	1,23	1,36	3,0	0,9	1,05	34	70
		10,0	1,35	1,85	5,0	1,08	1,33				60	330
Konstantinovka (USSR)	Granites: Charnockites	5,0	1,25	1,5	2,5	1,0	1,13	2,0	0,65	0,75	20	100
	Granites: Magmatites	3,4	0,9	1,08	1,7	0,72	0,81				6	
		[illegible]	[illegible]	[illegible]	[illegible]	[illegible]	[illegible]	1,5	0,85	0,92	103	

Naglu (Afghanistan)	Gneisses	Cracked	8,5	0,7	1,12	4,25	0,56	0,77	2,0	0,8	0,9	33	200
		Chrushing zone	0,6	0,75	1,05	0,3	0,6	0,75	Removed from foundation of dam			1,2	
Chaira (Bulgaria)			9,0	1,2	1,65	4,5	0,96	1,18	3,5	0,85	1,02	50	
Kurobe-IV (Japan)	Granites		10,0	1,6	2,1	5,0	1,28	1,53				70	
Aradaze (Japan)	Sandstones and schists		17,6	1,37	2,25	8,8	1,1	1,54				108	
			15,2	1,25	2,01	7,6	1,0	1,38				86	
Shiyushida (Japan)	Diabases		5,0	1,35	1,60	2,5	1,08	1,20				31	
Shimuke (Japan)	Andesites		1,0	1,3	1,35	0,5	1,04	1,29				7	
Amagaze (Japan)	Schists		17,6	1,73	2,61	8,8	1,38	1,82				131	
			16,6	1,15	1,93	8,4	0,92	1,34				89	
Shimouke (Japan)	Green schists		20,0	1,0	2,0	10,0	0,8	1,30				97	
Buttress dam (Japan)	Black schists		7,0	1,11	1,46	3,5	0,89	1,06				37	
Grande Dixence (Switzerland)	Gneisses								0	0,7	0,7		
Shasta (USA)	Grandiorites								0	0,76	0,75		
Grimzel (Switzerland)	Granites								0	0,96	0,96		
										0,7	0,7		
Gabril (Portugal)	Dolomites								0	0,7	0,7		
Schräh (Switzerland)	Granites								0	0,7	0,7		
Rigovayo (Spain)	Granites								0	0,7	0,7		
Suiana (Italy)	Sandstones								0	0,7	0,7		
Burgillo (Italy)	Limestones								0	0,7	0,7		
Francis (USA)	Limestones								0	0,55	0,55		
			Rock blocks				Throughout rock mass as a whole						
Inguri (USSR)	Limestones		7,0	1,20	1,55	3,5	0,96	1,14	4,0	0,8	0,1	39	
									2,0	0,8	0,9	—	
Chirkey (USSR)	Limestones		14	1,1	1,8	7,0	0,88	1,23	2,0	0,7	0,8	72	
Namakiwan (USSR)	Breccia and sandstones (tuffaceous)		9,5	1,65	2,12	4,75	1,32	1,56				71	
Sayany (USSR)	Paraschists		18,0	1,30	2,20	9,0	1,04	1,49	5,0	1,0	1,25	106	
Toktogul (USSR)	Limestones		17,5	1,34	2,21	8,75	1,07	1,51	3,3	0,92	1,08	105*	
									3,5	1,00	1,17		
Miatla (USSR)	Sandstones, limestones, argillites		5,0	1,05	1,3	2,5	0,84	0,97	1,0	0,55	0,6	25	
									0,5	0,40	0,42		
Kurobe-IV (Japan)	Granites		10,0	1,6	2,1	5,0	1,28	1,53				70	
Tawa (India)	Sandstones and schists		5,5	1,73	2,0	2,75	1,38	1,52				41	
Alto-Rabagao (Portugal)	Granites	Sound	14,0	1,9	2,6	7,0	1,52	1,97				114	
		Medium-sound	7,0	1,7	2,05	3,5	1,36	1,54				51	
		Weathered and fractured	5,0	1,6	1,85	2,5	1,28	1,40	Removed from foundation of dam			35	
			4,0	1,3	1,5	2,0	1,04	1,14				24	
			2,0	1,2	1,3	1,0	0,8	0,85				11	

Note: 1. The shear coefficient tan ψ for all structures was computed for the same value $\sigma = 20\ kg/cm^2$.
2. The strength of rock mass in uniaxial compression, with the exception of the blocks of the Toktogul hydroproject was obtained by indirect means from the limiting shear parameters using the Mohr-Coulomb relationship:
$\sigma_c = 2c/\tan(45 - \psi/2)$.
3. Shearing along exposed cracks not included in tha table.
4. Tan φ = shear/normal force.
*–results of direct fracture test of rock block insitu.

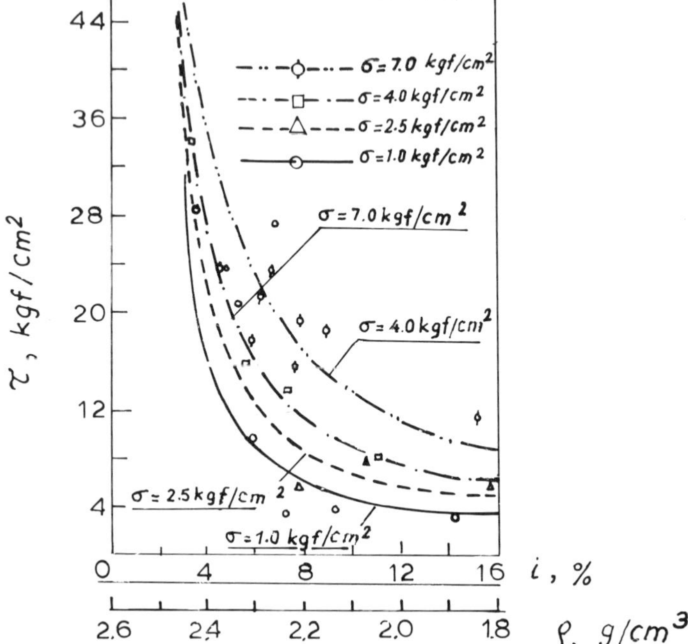

Fig. 8-117. Shear strenght τ as a function of alteration index i (after ROCHA, 1964).

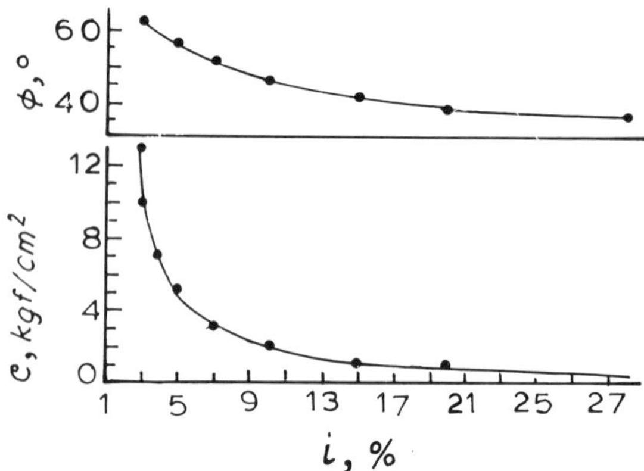

Fig. 8-118. Cohesion c and friction angle ϕ as a function of alteration index i (after ROCHA, 1964).

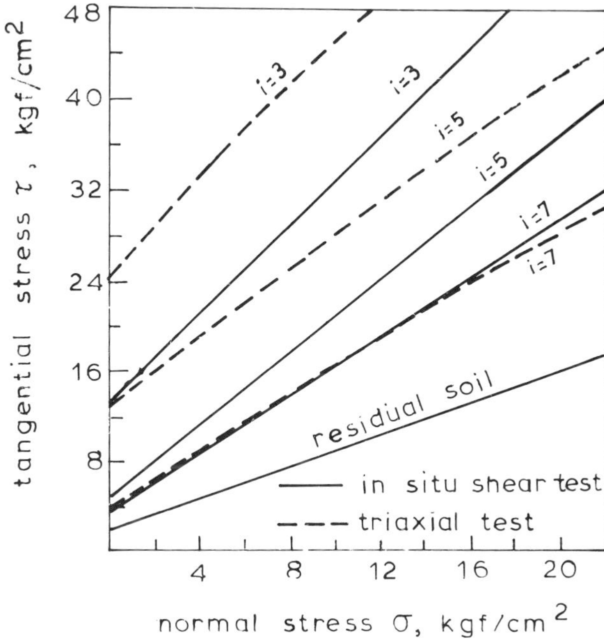

Fig. 8-119. Comparison of triaxial envelopes with results of field shear tests (after SERAFIM and LOPES, 1962).

KRSMANOVIC and POPOVIC (1966) conducted large-scale in situ tests for the shear strength of limestone along the surfaces of discontinuity of the various categories (clean fissures, thin fissures filled with clayey material and fissures with layers of clayey materials). The tests were performed with shear surfaces of about 5.00 m^2 (53.8 ft^2) and the normal pressures of various intensities up to the maximum of 2.45 MPa (355 lbf/in^2) were applied. Later on, on the same places, the shear strengths of the concrete-limestone contact surfaces were tested as well. With certain exceptions, in situ test results and the results of large scale laboratory tests, have shown a very good congruency. The differences appeared only there where the test conditions were not identical, or the materials of the specimens were different from one another in a greater degree.

Similar tests were made on the Morrow Point Project of the U.S. Bureau of Reclamation to provide correlation with laboratory specimens of the same size (DODDS, 1970). The tests were made in mica schist. Essentially the same equipment was used for both laboratory and field tests. From their results the Bureau of Reclamation came to a conclusion that the results of field and laboratory tests are sufficiently close to permit the use of laboratory tests when it is not feasible to make in situ studies.

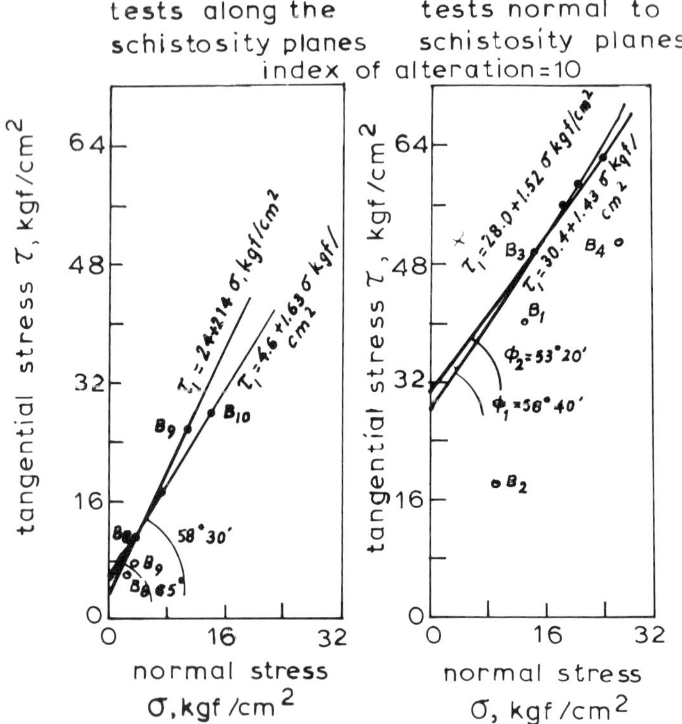

Fig. 8-120. In situ shear tests on schists at Valdecanas dam site, showing effect of anisotropy on shear strength (after Anonymous, 1963).

The shear strength of joints defined by the peak of stress-displacement curve increases with increasing normal stress suggesting that peak strength is dependent upon normal stress. This peak strength consists of two components, cohesion and friction.

The shear displacement is also accompanied by dilatancy normal to displacement plane and the influence of dilatancy has to be evaluated in attempting to find the true shear strength of the material. The method of calculation is shown in Fig. 8-121. Fig. 8-122a gives the shear behaviour of gneiss sheared along the heads of the outcrops of nearly vertical planes of foliations (curve 1) and when the direction of shear stress is perpendicular to the heads (curve 2). Curve 3 corresponds to line 2 for a partially weathered gneiss and curve 4 represents the behaviour of stratified marly sandstone with the direction of displacement as in 2 and 3. If the dilatancy component is substracted from these values a new set of curves is obtained (Fig. 8-122b). These curves then represent truely the shear behaviour. The component of friction increases with normal stress linearly. The component of cohesion occurs with small normal stress and decreases with increase in normal stress (MENCL, 1962).

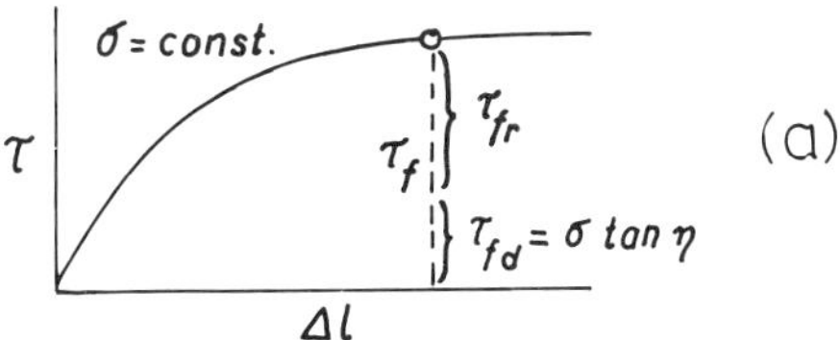

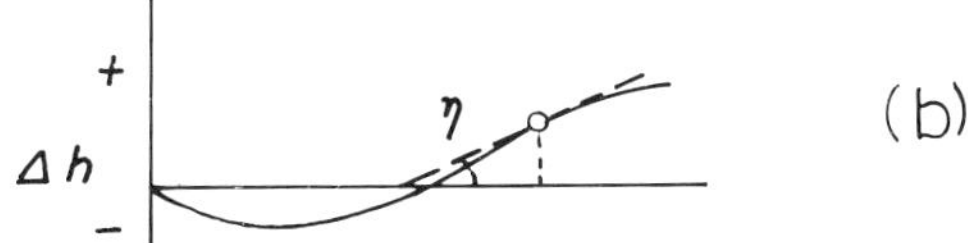

Fig. 8-121. (a) Stress-displacement curve of a field shear test. (b) Change in volume in the course of the test (after MENCL, 1966).

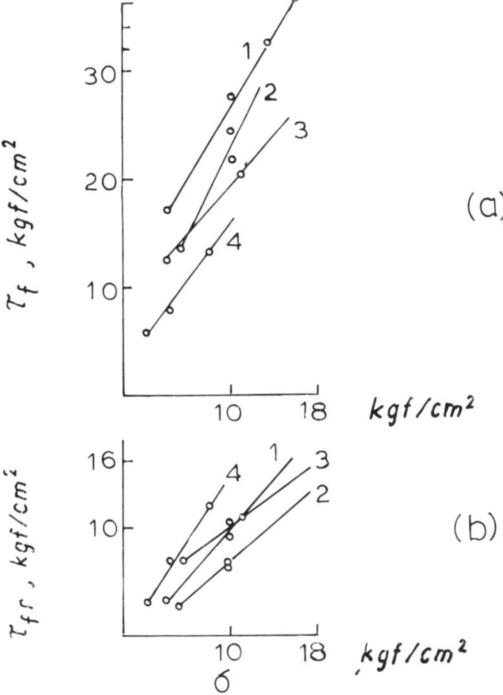

Fig. 8-122. (a) MOHR lines of shear strength of three rock masses: 1,2 – gneiss, 3-gneiss attacked by weathering, 4 – tiny-stratified sandstone
(b) The lines after the portion due to dilatancy is substracted (after MENCL, 1966).

In situ shear tests on discontinuities in limestone both clean and filled joints have shown that at normal stress of 2.5 MPa (25 kgf/cm^2 (356 lbf/in^2), the ratio $\tau_{max}/\sigma_n = 1$ while at a smaller normal stress (1.23 MPa) (12.5 kgf/cm^2) (178 lbf/in^2), this ratio increased to 2. The unevenness of the surface resulted in increase in this ratio to 3 at normal stress of 1.23 MPa (12.5 kgf/cm^2) (178 lbf/in^2) (KRSMANOVIC and POPOVIC, 1966). The cohesion at higher normal stresses does not play any important role.

The decrease in strength with displacement is dependent upon the surface roughness and normal stress. Though laboratory tests show clearly that rough mating surfaces show very quick drop in shear strength after peak is reached and smooth surfaces almost maintain their strength, or even show increased strength as displacement causes damage to the surfaces and make them comparatively rougher. Field tests, however, do not show corresponding behaviour. Tests by KRSMANOVIC and POPOVIC (1966) showed a drop in strength only for low normal stress (1.23 MPa) (12.5 kgf/cm^2) (178 lbf/in^2) and only for relatively smooth surfaces. Rough surfaces maintained their strength even at displacement of 30 mm (1.2 in) (0.107% of the specimen length). Perhaps, this displacement is too small to show any effect. Also, rough joints are never perhaps truely mating because of the displacement these have undergone and hence their true peak has already been exceeded during their geologic history and what is determined by the test is only the residual strength of the joint.

Thin filled joints are likely to show high shear strength at high normal loads due to their compression which is likely to cause interlocking of asperities. The increase in strength is likely to be smaller when the thickness of the filling is large due to the limit on closure of the joint with normal load. However, consolidation of the filling material contributes to increase in strength and may in certain cases be a dominating factor.

Except for smooth clean joints, the shear strength is not likely to increase linearly with increase in normal stress. For filled joints, the shearing behaviour of the filling is important and clays and sand + clay mixtures do not show a linear shear—normal stress envelope.

Total displacement at peak and rupture are dependent upon the shear stress sustained by the specimen (Fig. 8-123) and possibly the normal load acting upon it. The total displacement at peak (for schist) is about 15–30% of the displacement at rupture (TAKANO and FURUJO, 1966) and higher normal stress needs high horizontal displacement before rupture.

When a concrete block slides over rock, the following basic types of shear failure can be distinguished (Fig. 8-124):

(a) Shear and sliding of the concrete block with failure of the foundation rock. Sliding takes between blocks within foundation rocks (Fig. 8-124a).

(b) Shear at the interface followed by sliding of the block along the contact surface (Fig. 8-124b).
(c) Shear of the concrete with sliding of the block along concrete-concrete contact (Fig. 8-124c).
(d) Mixed shear (formation of cracks), with failure partly in concrete and partly along the contact surface (Fig. 8-124d).

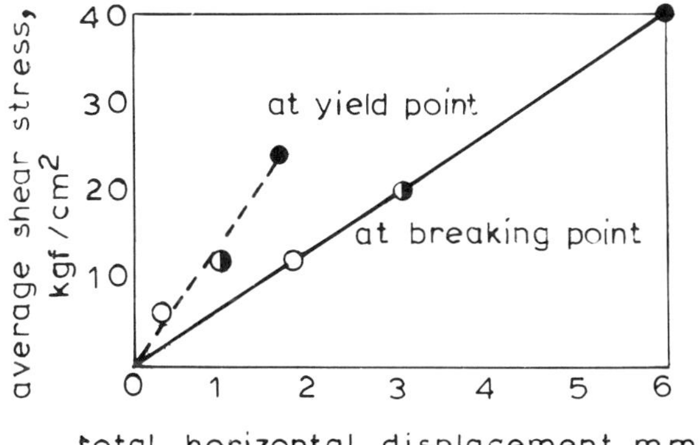

Fig. 8-123. Total horizontal displacement – average shear stress curves (after TAKANO and FURUJO, 1966).

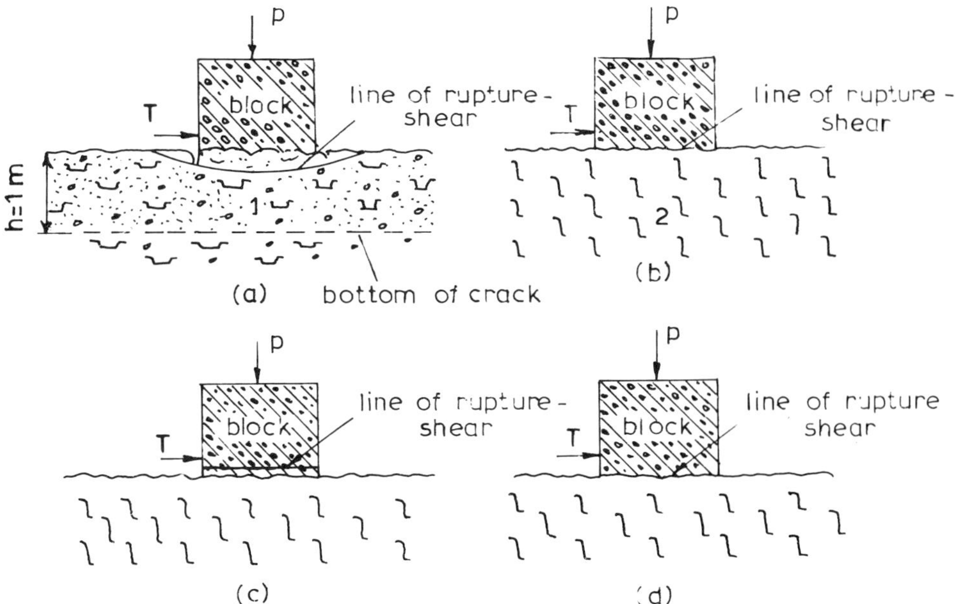

Fig. 8-124. Types of failure in shear and sliding of blocks (after EVDOKIMOV and SAPEGIN, 1967).

It is advisable to conduct the investigations to determine the type of failure by overturning the block and contouring the shear surface after the test.

In the case of fissured rock, the shear parameters obtained in sliding a concrete block on a rock surface shall invariably correspond to that of the rock with failure type obtained in Fig. 8-124a. In stronger unfissured rocks the type of failure may be that depicted by Fig. 8-124b and in very strong massive rocks, failure may occur through concrete, but this type of failure shall be extremely rare. With usual roughness of the concrete-rock surface, the possibility of sliding along the contact surface in a structure is much smaller and failure will unusually occur through the rock. The cohesion between rock and concrete usually does not exceed 0.5 MPa (5 kgf/cm^2) (70 lbf/in^2).

The peak shear resistance for a compact rock is larger than sliding (residual) resistance and the difference between the peak shear and residual values decreases as the normal stress increases. Fig. 8-125 shows these points very clearly.

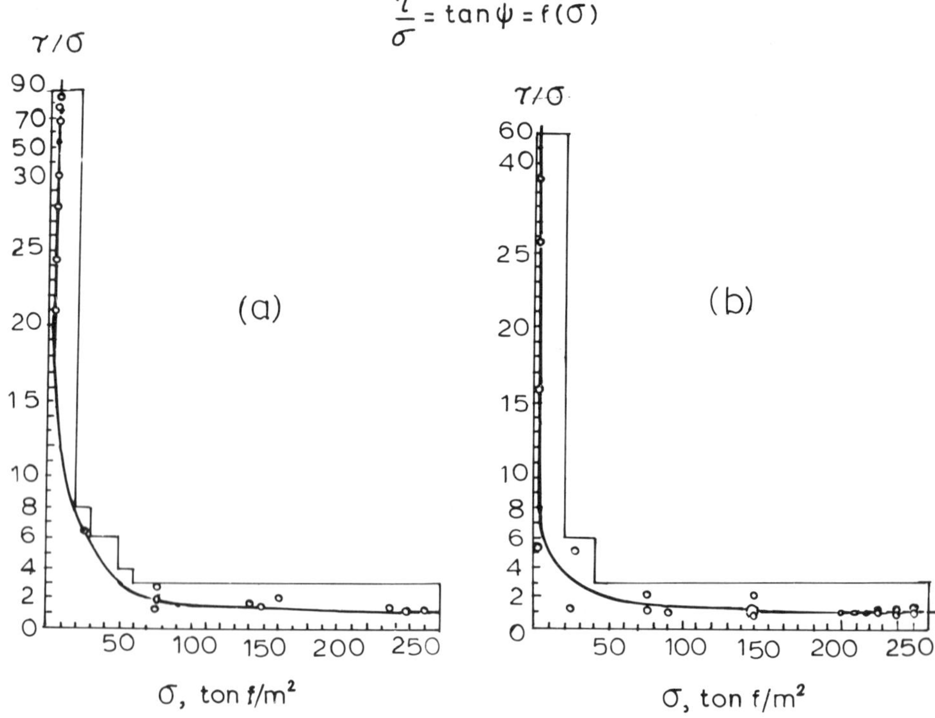

Fig. 8-125. Results of shear and sliding tests on concrete blocks in drift No. 1 and galleries of mine No. 2 (a) peak (b) residual (after Evdokimov and Sapegin, 1967).

The shear stress-shear displacement curves for a block sliding on a bedded rock are different depending upon whether it is loaded parallel to or at right angles to the bedding planes. Results of tests on shearing on a concrete block-rock interface are shown in Fig. 8-126. The angle of friction and cohesion are lower when loaded parallel to bedding plane (case C) than when loaded at right angles. The amount of displacement that blocks undergo before the peak shear stress is reached is higher when loaded at right angles than parallel to the bedding.

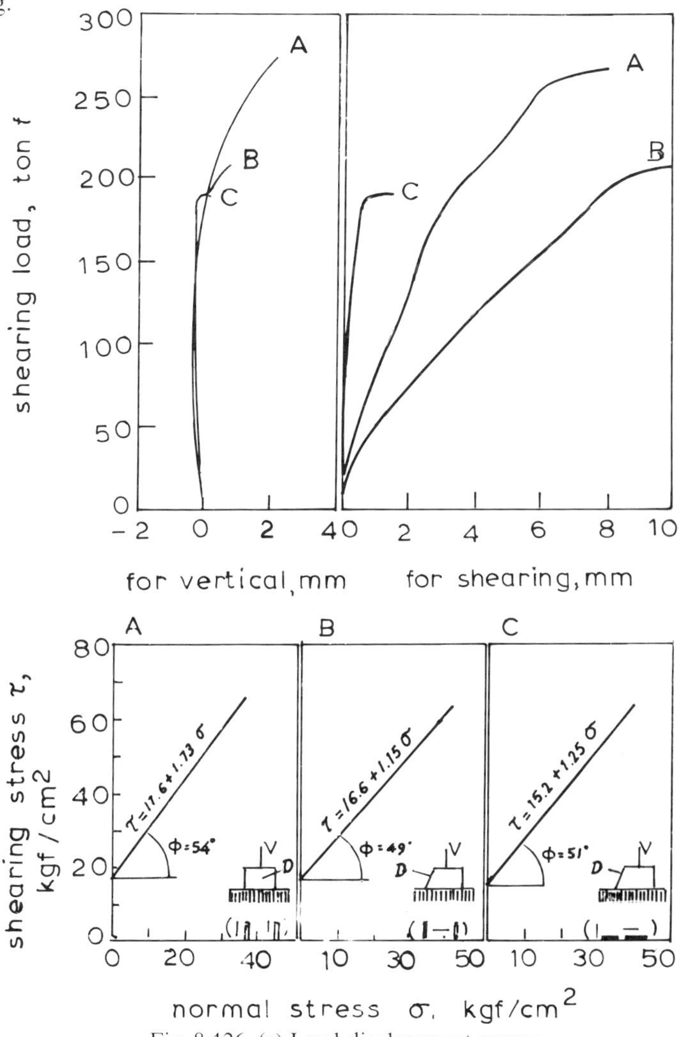

Fig. 8-126. (a) Load-displacement curves
(b) Correlation of τ and σ in bedded rock
(after Ishii et al, 1966).

When shearing stress is acting at an angle to the bedding, both cohesion and friction angle depend upon the angle of inclination. The angle of friction drops as the beds dip in the direction of shear stress (α-negative) or against the direction of shear stress (α-positive). The values more or less agree with the simple analysis using block sliding on an inclined surface. The cohesion values however are different. These are higher for negatively inclined blocks and lower for positively inclined blocks (Fig. 8-127).

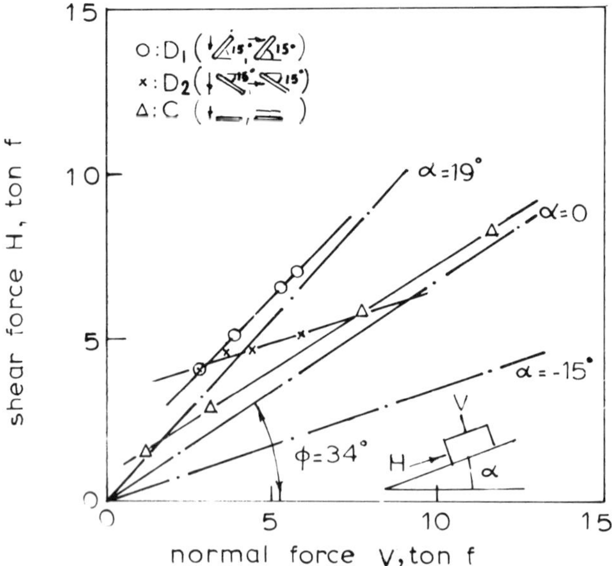

Fig. 8-127. Correlation of observed values H, V and dip of bedding plane (after Ishii et al, 1966).

An important point in in situ test data analysis is the acceptance of the criteria of rupture when no clear peak occurs.

Ruiz and De Camargo (1966) conducted in situ tests at the Jupia Dam, Parana River, for shear strength of basalt and sandstone. The results obtained in the 0.7 m × 0.7 m × 0.3 m (2.3 ft × 2.3 ft × 1.0 ft) rock blocks are given in Fig. 8-128 according to the following rupture criteria:

(a) Criterion of dilatancy represented by tangential and normal stresses corresponding to the inversion of vertical displacements of the rock block at the opposite side in relation to the jack.

(b) Criterion of maximum horizontal displacement represented by tangential and normal stresses at the bottom of rock block corresponding to a previously fixed horizontal displacement. In their tests, displacement was fixed at 1.0 mm (0.04 in).

(c) Criterion of ultimate strength represented by the maximum tangential and maximum normal stresses at the bottom of the rock block.

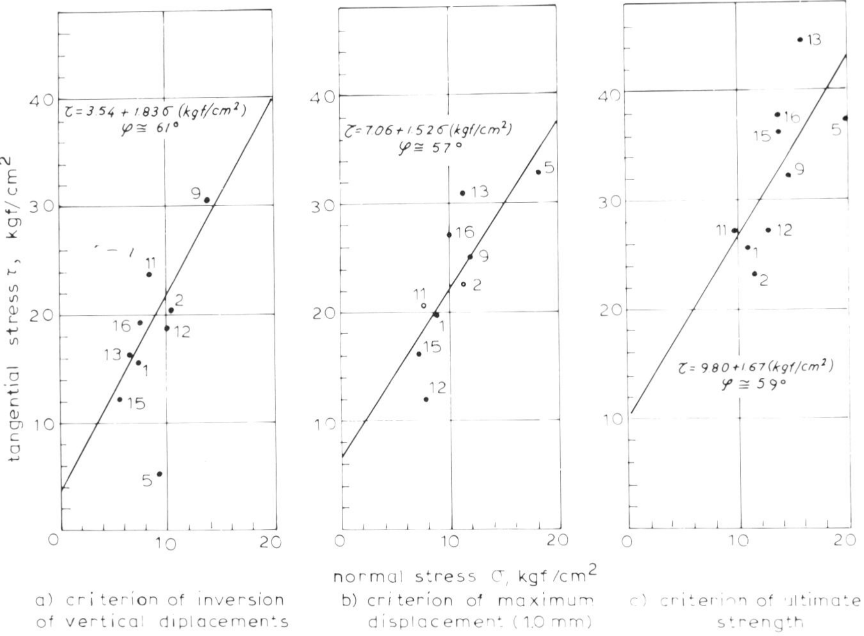

Fig. 8-128. Results of in situ small block shear tests according to rupture criteria (after RUIZ and DE CAMARGO, 1966).

Fig. 8-128 summarises the results obtained in the in situ shear tests. The numbers 1,2 and 11 correspond to vesicular basalts. The numbers 5, 9, 12, 15 and 16 correspond to vesicular and amygdaloidal sandstone- basalt breccia. Number 13 represents a silicified sandstone. Despite the difference in rock types a similar shear performance was obtained for the rocks tested.

Some more details on the shear behaviour of joints are discussed in Chapters 10 and 11 (Vol. IV).

8.12. Tensile Strength of Rock In Situ

The direct measurement of tensile strength of rocks in situ is very difficult and as yet no attempts have been made in this direction. Certain indirect methods have been devised, but these are more as index methods meant to determine certain properties such as workability of coal. The most common test is the pull test performed with the use of expanding bolt tester. In another application, tensile strength in bending of roof has been determined by uniformly loading it with compressed air. These methods are described below.

8.12.1. Pull Test

The principle of this test is to insert a bolt in a hole drilled in the face of the rock body and then to expand the head of the bolt and pull it out from the face, while the force required to pull it out is measured and related to the strength of the rock. FOOTE (1964) believes that the failure takes place due to shear but EVANS (1964) is of the opinion that the failure is a tensile one. For the purpose of the test, two devices have been developed:—(a) Russian device, and (b) M.R.E. device developed by the National Coal Board, U.K.

(a) *Russian device*: In the tests using the Russian device (Fig. 8-129) a hole is drilled on the face of the rock and the back end of the hole is reamed out to a large diameter (about 15.2–20.3 cm (6–8 in)) from the open end. A hydraulic mechanism is inserted which grips the rock in this enlarged region and pulls it away from the face and the force required for this purpose is measured. The pulling arrangement consists of a hydraulic ram and cylinder operated manually or mechanically. This test breaks a cone of rock from the face. The relationship between the breakage force F and the tensile strength of rock σ_t is given by:

$$\sigma_t = 1.2\, F/hr \tag{8.68}$$

where σ_t = tensile strength of rock
F = force at failure
r = radius of the fractured cone containing the bolt and
h = height of the cone extracted.

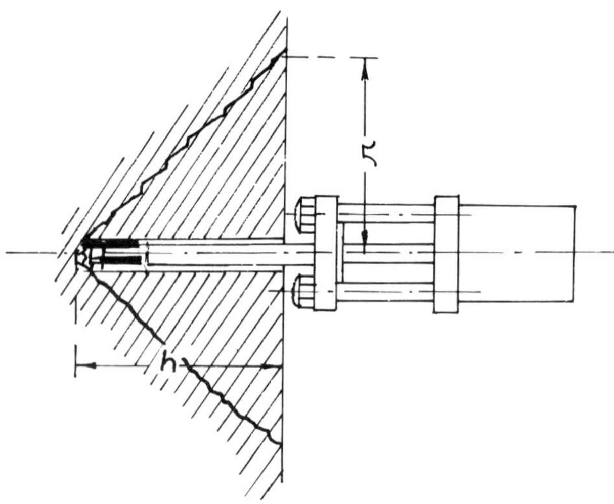

Fig. 8-129. Variant of determining the resistance of rock to breaking away from the pillar (after PROTODYAKONOV, 1961).

(*b*) *M.R.E. device*: The equipment used by Mining Research Establishment (N.C.B., U.K.) is shown in Fig. 8-130. The bolt is an expansion type and is made of En 25 steel. It is similar in construction to a 1.9 cm (0.75 in) rawl bolt and requires a 3.7 cm (1–$^{7}/_{16}$ in) diameter hole. A 15.2 cm (16 in) hole is drilled on a freshly exposed face and the bolt is inserted and fixed by expanding the segments to a standard distance. The bolt is attached to a hand operated hydraulic tripod with an extension rod consisting of 3 hydraulic rams forming the legs. The force required to pull out the bolt is measured with a dynamometer normally used in measuring the tension in the bolts. The tripod can apply a maximum load of 99.6 kN (22,400 lbf) to the bolt.

Two relationships are given to relate the force required to pull out the bolt and the constants of the test and the rock under test. There are differing opinions on the mechanism of failure.

FOOTE (1964) assumes that failure is due to shear and gives the following relationship:

$$F = \tau\pi\, h\,(r + 2\, r_h)/3 \qquad (8.69)$$

where F = force to pull out the bolt
τ = shear strength of the rock
h = height of the cone
r = radius of the cone and
r_h = radius of the hole.

The following relationship was derived by EVANS (1964) assuming tensile failure:

$$\sigma_t = \frac{F}{K r_h h} \qquad (8.70)$$

where σ_t = tensile strength of rock
F = force to pull out the bolt
K = constant
r_h = radius of the hole and
h = height of the cone.

The underground experiments indicated that the maximum force required to pull out is dependent upon the following factors:

(1) Pressure applied to the rock during the clamping of the bolt: The force increases with the increase in the clamping pressure.

(2) Depth of hole: The force increases with the increase in the depth;

$$F = K\,(\text{depth})^{1.5-2} \tag{8.71}$$

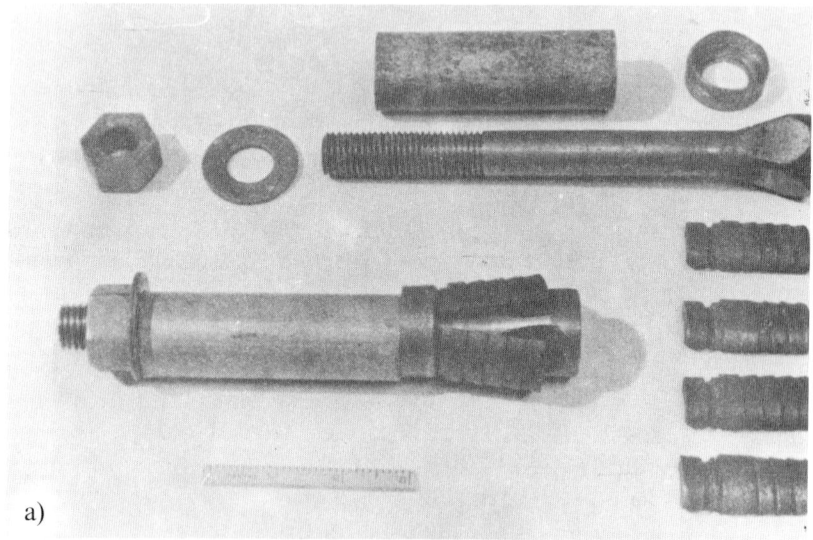

Fig. 8-130. M.R.E. expanding bolt seam tester
(a) Underground bolt
(b) Withdrawal apparatus
(after FOOTE, 1964).

(3) The force increases linearly with the increase in the radius of the hole.

(4) Overburden pressure.

(5) Direction of weakness planes.

8.12.2. Flexural Test

This method for the determination of flexural strength of laminated mine roof in situ was first used by MERRILL (1954) and MERRILL and MORGAN (1958). The method is based on bending of the roof beam under its own load (gravity-loaded) or loading it by artificial means (compressed air).

MERRILL (1954) carried out experiments in the oil shale mine at Rifle, Colorado, on a single layer of oil shale roof of thickness 50.8 cm (20 in). The room was widened successively till the span became so large that the roof layer could not stand its own weight and failed. The tensile stress at failure at the midpoint of the lower (convex) surface is calculated from the relationship:

$$\sigma_t = \frac{\rho w^2}{2t} \tag{8.72}$$

where σ_t = tensile stress at failure (tensile strength)
ρ = density of roof rock
w = width of span and
t = thickness of roof layer.

It was found that failure of the layer (ρ=24,430 N/m^3) (0.09 lbf/in^3) occurred when the room span was 24.4 m × 61.0 m (80 ft × 200 ft). The flexural strength determined was 14.1 MPa (2050 lbf/in^2).

The tensile strength value determined in laboratory on cores taken parallel to the bedding plane was 20.7 MPa (3000 lbf/in^2) while a centrifugal model study made by WRIGHT and BUCKY (1949) on samples taken from the mine gave a flexural strength of 12.8 MPa (1860 lbf/in^2).

In another study, an experimental room in a limestone mine was widened to 15.2 m (50 ft) and it was observed that bed separation had taken place and a 50.8 cm (20 in) thick layer had become detached from the overlying rocks freeing it from the overburden. Instead of continuing the extension of the room till the failure of this roof bed, compressed air was introduced on the top of the detached layer and was made to fail by applying a uniformly distributed load (MERRILL and MORGAN, 1958). This has the effect equivalent to increasing the unit weight of the beam. The maximum tensile stress developed at the mid-point of the beam at failure is calculated from the relationship:

$$\sigma_t = \frac{\rho w^2}{2t} + \frac{Pw^2}{2t^2} \tag{8.73}$$

where P = air pressure at failure.

Roof failed at an air pressure of 0.06 MPa (9 lbf/in^2) and the computed tensile strength was 34.1 MPa (4950 lbf/in^2) against a laboratory determined flexural strength of 32.1 MPa (4650 lbf/in^2). The uniaxial tensile strength was, however, much lower 8.2 MPa (1190 lbf/in^2).

In a modification of this method, the roof can be pulled down using rock bolts of length so that these remain imbedded only in the beds below the expected plane of bed separation. These bolts can then be pulled down by suitable anchoring system anchored to the floor of the excavation (WANG, 1975).

8.13. Triaxial Tests In Situ

The first triaxial tests in situ were conducted at Kurobe IV Dam in Japan (JOHN, 1961; MÜLLER, 1963) using prismatic specimens. The test arrangement consisted of a battery of hydraulic jacks (similar to the uniaxial compression tests) acting on the top and sides of a prismatic specimen cut from the rock with the surrounding rock forming the reaction surface.

GILG (1966) used cylindrical specimens with curved jacks placed around the specimen and a steel cylinder acting as a reaction. The arrangement is shown in Fig. 8-131.

The method consists of preparing a circular specimen (1) of required dimensions (0.7 m (2.3 ft) in this case). The specimen is surrounded by a steel cylinder (6) of appropriate stiffness and curved jacks (5) are placed between the specimen walls and steel cylinder (6). The space between the cylinder and curved jacks is cement grouted. The top surface of specimen is prepared plane with quick setting hard cement mortar and a distribution plate (2) is placed over it. Vertical pressure is applied using flat jacks (3) placed on the distribution plate (2) and the reaction frame (4). Vertical and horizontal deformations are measured using dial gauges (7 and 8).

LOGTERS and VOORT (1974) tested 1 m (3.3 ft) cube using flat jacks. Test specimens are prepared using skin to skin drill holes giving slots of 1.1 m (3.6 ft) length and 1.2 m (3.9 ft) depth (Fig. 8-132). Flat jacks of 1.0 m × 1.0 m (3.3 ft × 3.3 ft) are inserted in the slots 10 cm (3.94 in) below the rock surface.

The space between the flat jacks and rock is filled with cement grout. The top surface is loaded with hydraulic jacks of 100 tonnef (98.4 tonf) capacity. Deformation is measured using 3 point rod extensometers at the centre of the block. The lateral displacements can be measured either from volumetric displacement of flat jacks or preferably using L.V.D.T.s placed in the flat jacks (LNEC, Portugal flat jacks).

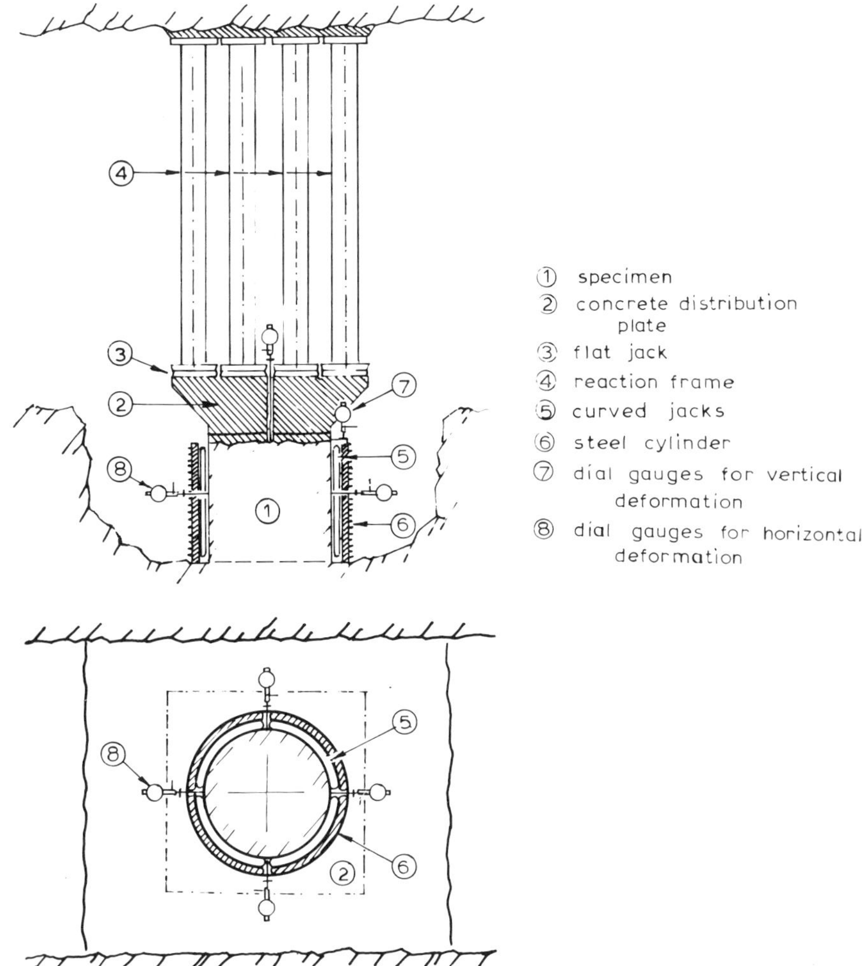

Fig. 8-131. Triaxial test in situ on cylindrical specimen (after GILG, 1966).

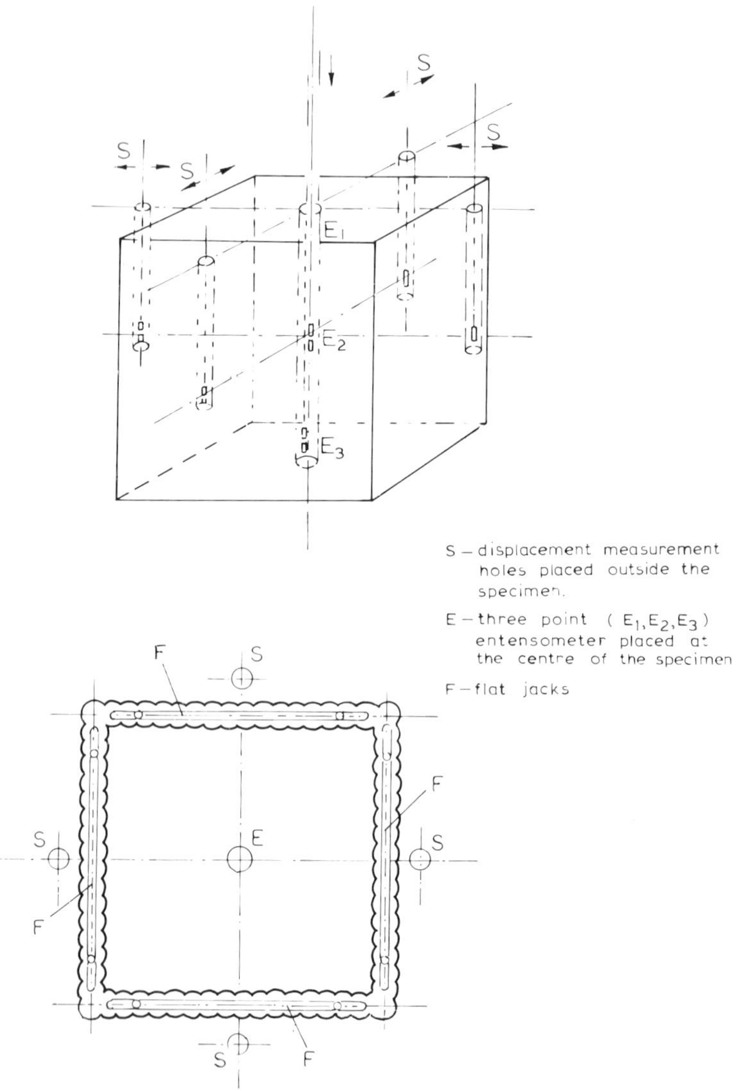

Fig. 8-132. Triaxial test in situ on cubical specimen (after LOGTERS and VOORT, 1974).

When the volumetric changes in flat jack are used to determine the lateral deformation of the specimen, it should be borne in mind that the total volumetric change is not only due to the expansion or contraction of the specimen but also due to the contraction and expansion of the surrounding rock. The exact value due to expansion (contraction) of the specimen can be measured if the lateral pressure is applied before the axial load is brought on (or reduced). The incremental expansion or contraction values can then be measured. The relationship between the expansion of the flat jack and the volumetric changes can be established by calibrating them in the laboratory. The method of analysing the strength of rock when tested to failure has been discussed in Chapter 5 (Vol. I).

In situ triaxial tests permit evaluation of in situ anisotropy of the rock mass. In situ triaxial tests in Tertiary limestones at Ghiona near Lidorikion (Greece) showed that the modulus of elasticity in situ may vary in the three directions by as much as 1 : 7 (Table 31). High values of POISSON's ratio observed in the vertical direction could be associated with the sliding of the inclined joints present in the specimen and anisotropy is more likely to be related with the bedding planes and joints.

TABLE 31

Moduli of elasticity (kgf/cm^2) for the three orthogonal directions x, y and z

(Rock type: Bedded limestone + shale + sandstone)

(after LOGTERS & VOORT, 1974)

Direction	Test 1	Test 2	Test 3
z (vertical)	14 000	12 500	100 000
x (horizontal)	9 500	4 900	7 500
y (horizontal)	1 500	9 500	7 000

GILG (1966) has compared the values of modulus of elasticity using uniaxial, triaxial and plate bearing tests (Table 32). While uniaxial and triaxial tests (70 cm (28 in) diameter) gave results close to each other, plate bearing test results were far higher.

TABLE 32

Comparison of results from different in situ tests at Punt Dal Gall dam

(after GILG, 1966)

	Type of Test					
	Uniaxial	P bars	Triaxial	P bars	Plate loading	P bars
Modulus of elasticity, MPa	2700	0–47	2400 to 6900	25 to 100	9900–10800 ($\nu = 0.15$)	0–143
	5500	12–72		side pressure	9200–10100 ($\nu = 0.3$)	
	1800	12–117				

Uniaxial and constrained tests on coal specimens have been conducted by LAMA (1966b). The constrained tests were conducted without separating the in situ specimen from the sides. The modulus of deformation and strength values obtained by him showed that constraining improved the deformation modulus of coal by 20% in well cleated coal. Improvement in non-cleated coal is much smaller.

8.14. Summary and Conclusions

The mechanical behaviour of rock mass for design of large structures should be determined by large scale in situ tests. Large scale in situ tests can be divided into 3 main categories:

(1) Deformability tests
(2) Shear tests
(3) Strength tests.

Table 1 gives the types of tests that are recommended for different rock structures. In situ tests must be carried out under simulated conditions keeping in view the variability of the rock mass. The in situ strength and deformation modulus values are lower than laboratory values. The size of the samples in in situ tests must contain at least 50–100 weakness planes. For coal, 1–1.5 m square specimen size is quite adequate. For shear tests, a specimen of size 0.7 × 0.7 m has been recommended by the ISRM, but the surface irregularities and their frequency must be taken into account in deciding upon the specimen size.

The test conditions influence the mode of failure and the strength values in in situ uniaxial compression testing. Tests on coal samples, in situ, showed that when load is applied through a flexible surface, failure takes place with development of vertical fractures (Fig. 8-10). When force is applied through rigid surfaces (using concrete cappings) failure occurs with development of cones (Fig. 8-11). The results representing relationship between size and uniaxial strength are given in Fig. 8-9. and Table 3.

The post failure stiffness drops with increase in width/height ratio of coal specimen in situ.

In the design of coal pillars, a safety factor of 1.5–1.7 should be used where the pillars are left for permanent support. If pillars are to be extracted at a later stage, a higher safety factor (~ 2.0) is suggested.

Three different types of tests are available to determine the deformability of rocks. They are:

(1) Plate bearing tests
(2) Pressure tunnel tests
(3) Borehole tests.

Plate bearing tests are most commonly conducted in foundation design, while pressure tunnel tests are preferred in tunnel design, particularly in high pressure hydraulic tunnels. Borehole tests are useful in testing areas not approachable from the test excavations or otherwise. Because of the small volume of rock involved, borehole tests are useful only for provisional delineation of zones.

In plate bearing tests, it is preferable, as far as possible, to conduct tests using circular plates with a flexible loading pad (e.g. flat jack) placed between the loading jack and the loaded surface. This ensures uniform loading. In other cases, an extremely rigid loading pad should be used. In this case the stiffness of the pad should be 5–10 times the stiffness of the loaded rock. While analysing results from plate bearing tests, the assumption of linear elasticity and anisotropy could lead to quite serious errors in the computed values. The errors in computed values may be of the order of 20–50 %. The results will be underestimated when $E_h/E_v >> G_h/G_v$ and over estimated when $E_h/E_v << G_h/G_v$. These results affect the surface deformation much more than any variation in load distribution under the plate. The errors are lower for stiff materials and larger for compressible materials. The deformation at depths greater than the radius of the loading plate are not influenced by the anisotropy of the material or the non-uniformity of load.

In layered systems, the results are dependent upon the relative thickness of the layers and the size of the plate.

The size of the plate in bearing tests should be as large as possible, and it is advisable to conduct tests using 3–4 different sized plates and extrapolate the results. It is advisable to conduct a large number of tests with a relatively smaller size plate than a few tests with a larger plate. Stiffness of the loading plate should be at least an order of magnitude higher than the rock. Displacements measured away from the plate are more representative of the rock mass. These are less sensitive to pressure distribution under the plate than those situated within the loaded area. In these tests deformations may be measured at the surface of the tunnel (for pressure tunnels) or at depth (for dam foundations). It is advisable to measure deformation with respect to a fixed point (e.g. centre-axis of tunnel) rather than radial displacement of two opposite points. Measurement of displacement at different depths permits calculation of modulus of different layers away from the tunnel wall as well change in state of the rock. Modulus values calculated assuming unfractured rock around the tunnel may be 50–100 % in error. Tensile stress develops in tunnel both due to loading as well as due to natural stress field and these may influence the results. The thickness of the lining used to load the tunnel surface should be as thin as possible.

A number of instruments have been developed for testing of boreholes. Table 15 gives a summary of the various devices. These are designed for use in small diameter holes (maximum 150 mm (6 in)). The results from these are greatly influenced by the presence of small fissures or cracks in boreholes. These devices are mainly used for preliminary study of the area and zonal classification for further investigations. Borehole testing devices can be used for determinations of modulus at greater depths normally not accessible. These can also be used to determine in situ tensile strength. These devices are simple, require much less site preparation and permit a large number of tests to be conducted in a shorter time at low costs. The disadvantage of borehole tests is small volume of rock mass involved so that test results may not be representative of the deformability of the rock mass. Because of tensile stresses developed in the borehole walls, the results may sometimes be lower than plate bearing tests. Higher values of borehole tests approach laboratory values.

Pressure displacement curves for rocks can be divided into 4 categories (Fig. 8-78) and a close look at these can give a fair amount of information about their behaviour. Deformation is strongly dependent upon joints, bedding planes and softer beds in the system. It increases with increase in load, indicating compaction of rocks. Only compact, massive rocks show no or very small hysteresis. Modulus of elasticity drops rapidly with weathering, though the E_h/E_v ratio tends to be maintained. Modulus increases with grouting except in cases where high grouting pressure has damaged rock due to crack propagation or decrease in friction along the sliding joints. Deformability is not the same in all directions and this ratio may, in some cases, be as high as 1 : 10 but usually 1 : 2; so also the ratio between E_e/E_d may not be the same in all directions. Anisotropy at low pressures may be due to induced anisotropy due to fracturing. Grouting may eliminate induced anisotropy but does not influence inherent anisotropy observed at high pressures.

The deformation modulus as a function of stress could be represented by some form of power law (Eqs. 8.54–8.56). Increase in hydrostatic pressure reduces anisotropy. There seems to be very poor correlation between *RQD* and field modulus values, though certain consistent trends have been noted between $(V_F/V_L)^2$ and (E_m/E_{seis}). (E_m/E_{core}) ratio is not influenced for drop in *RQD* below 0.60.

Load penetration curve of rocks obtained in bearing capacity determinations depends upon a number of factors such as presence or absence of fluids, temperature, indenter shape and size, rock porosity and brittle ductile nature of rock. In less deformable rocks failure occurs with initiation and development of Hertzian cracks. In comparatively more deformable rocks the initiation of Hertzian cracks is less important, and in highly porous rocks failure may be purely due to structural collapse. In calculating bearing capacities, the assumption of the mode of failure is important. Indications are that the expanding spherical bulb approach suggested by LADANYI (1967) is satisfactory for most rocks.

For the determination of the shear strength of joints and rock interfaces, the inclined load test has been most commonly used and is recommended by the International Society for Rock Mechanics. At least five tests should be performed at each place or on each type of plane. Sample dimensions with an area of 4000 cm^2 (620 in^2) has been found to be adequate for cohesion and frictional value determinations. It is important to recognise the plane of interest and align it with the shear plane of the test. Before deciding the location of the plane, it is advisable to identify these into different groups such as bedding planes, clean joints, filled joints > (1 mm (0.04 in) thick, 1–10 mm (0.04–0.4 in) thick and 10–30 mm (0.4–1.2 in) thick. The direction of shearing should correspond to the direction of anticipated shearing in the full structure. Besides the field test, laboratory test should be conducted on flat surfaces to obtain the residual angle of friction using flat-cut and ground-flat using No. 80 silicon carbide grit. After the test, the block should be inverted, and roughness, dip, direction and sheared area should be measured. Complete data about the behaviour of a joint can be obtained when normal and shear loads are applied in stages and displacements are measured as the block just begins to slide.

A few variations of the inclined shear test have been used to determine the shear strength and frictional behaviour of interfaces using a single jack and varying the angle of inclination of shear force.

Protection of moisture conditions during the test is important and test samples should be taken to ascertain moisture contents over the period. By protecting the top surface with a loading plate before the sides are excavated helps to protect the natural moisture content. If the rock or joint is saturated or has to be submerged in water, tests should be carried out under simulated conditions. Time required for saturation may take from a few days to several days depending upon the permeability. Water under pressure may be injected to speed up saturation.

The width (d) of the shear plane is important particularly when determining shear strength of rock material. The results are sensitive to the value of (d). When tests are conducted to determine cohesion and friction angle between rock and concrete, it is advisable to keep $d \cong 0$. In other cases Eq. 8.64 may be used.

Torsion test can be used to measure shear strength of discontinuities of inaccessible areas such as in boreholes. This method has also been recommended by the International Society for Rock Mechanics. The use of this method for determining shear strength of materials is doubtful (Volume 1, Section 4.2.).

Shear strength values are sensitive to moisture particularly in claystones and mudstones.

Punch test gives lower values than direct shear tests. If area of discontinuity present along the shear plane can be accounted for, the in situ shear strength values may be very close to laboratory values.

The shear strength of joints defined by the peak of stress-displacement curve increases with increase in normal stress. This peak strength consists of both cohesion and friction. The component of friction increases with increase in normal stress but the component of cohesion decreases with increase in normal stress. Rough surfaces in in situ tests maintain their strength even at large displacements—perhaps because they are never truely mating surfaces.

Total displacement at peak is about 15–30% of displacement at rupture and higher normal stress needs high horizontal displacement before rupture.

In sliding of concrete block over rock, shear may be limited to rock or concrete or transgress both materials. In fissured rock, shear parameters so obtained correspond to that of rock. The angle of friction and cohesion between a concrete block and rock interface are dependent upon bedding direction.

Measurement of in situ tensile strength using the M.R.E. or Russian device is rather an index test and determination of tensile strength using these techniques is dependent upon the type of failure. Flexural strength determinations may give values lower or higher than laboratory values depending upon the absence or presence of cracks and crack density.

In situ triaxial tests permit evaluation of in situ anisotropy better than plate bearing tests.

References to Chapter 8

1. Absi, E. and Seguin, M.: Le Noveau Geoextensomètre. Suppl. Annales de L'Institut Technique du Bâtiment et des Travaux Publics, No. 235–236, July–August 1967, pp. 1151–1158.
2. Ahlvin, R.G. and Ulery, H.H.: Tabulated values for determining the complete pattern of stresses, strains and deflections beneath a uniform circular load on a homogeneous half-space. Bull. Highway Research Board, No. 342, 1962, pp. 1–13.
3. Alas, M.C.: Dilatomètres de sondage. Proc. Inst. Symp. Rock Mech., Madrid, 1968, pp. 75–78.
4. Anonymous: Colloquium on in situ rock tests. Water Power, Vol. 15, 1963, pp. 155–156.
5. Barroso, M.: Contribution to Theme 8. Proc. 1st Cong. Int. Soc. Rock Mech., Lisbon, 1966, Vol. III, pp. 588–591.
6. Bicz, I.A.: Defining the susceptibility of coal seams to rock bursts. In Problems of Rock Bursts, WN1M1, Leningrad, 1962 (Translation in Polish by Central Mining Research Institute, Katowice).
7. Bieniawski, Z.T.: A design code for bord and pillar coal workings. Rep. S.African C.S.I.R., No. MEG 570, 1967.
8. Bieniawski, Z.T.: The effect of specimen size on compressive strength of coal. Int. J. Rock Mech. Min. Sci., Vol. 5, 1968, pp. 325–335.
9. Bieniawski, Z.T.: In situ large scale testing of coal. Proc. Conf. In situ Invest. Soils and Rocks, London, 1969, pp. 67–74.
10. Bieniawski, Z.T. and Van Heerden, W.L.: The significance of in situ tests on large rock specimens. Int. J. Rock Mech. Min. Sci. & Geomech. Abstr., Vol. 12, 1975, pp. 101–113.
11. Bieniawski, Z.T. and Vogler, U.W.: Load-deformation behaviour of coal after failure. Proc. 2nd Cong. Int. Soc. Rock Mech., Belgrade, 1970, Vol. I, pp. 345–351.
12. Bollo, M.F., Navalon, N. and Sanz-Saracho, J.M.: Considerations sur l'execution et l'interpretation d'essais de charge avec des pressiomètres plans de 2 a 4 m^2 de surface. Proc. 1st Cong. Int. Soc. Rock Mech., Lisbon, 1966, Vol. I, pp. 637–642.
13. Borowicka, H.: Influence of rigidity of a circular foundation slab on distribution of pressures over the contact surface. Proc. 1st Int. Conf. Soil Mech. Found. Eng., 1936, Vol. 2, pp. 144–149.
14. Brugman, B.J.: Deformation measurement in rock with eight-point extensometer. Proc. Int. Symp. Rock Mech., Madrid, 1968, pp. 65–69.
15. Bukovansky, M.: The determination of shearing strength of claystones and clays by means of in situ tests. (In Czech.) Univ. Bratislava, D. Sc. Thesis, 1962.
16. Bukovansky, M.: Three types of field shear tests on soft rocks. Proc. 1st Cong. Int. Soc. Rock Mech., Lisbon, 1966, Vol. I, pp. 343–346.
17. Burmister, D.M.: The general theory of stresses and displacements in layered soil systems. J. Appl. Phys., Vol. 16, 1945, pp. 89–95; pp. 126–127; pp. 296–302.
18. Burmister, D.M.: Application of layered system concepts and principles to interpretations and evaluations of asphalt pavement performances and to design and construction. Proc. Int. Conf. Structural Design of Pavements. Univ. Michigan, Ann Arbor, Mich., 1962, pp. 441–453.

19. CHAOUI, A., MARIOTTI, M. and ORLIAC, M.: In situ calcareous marls strain and shear strength: Comparison between different test characteristics. Proc. 2nd Cong. Int. Soc. Rock Mech., Belgrade, Vol. 2, 1970, pp. 357–365.

20. CHERCASOV, I.I.: The residual deformation caused by the deep penetration of a rigid plate into a brittle porous material. Proc. Geotechnical Conf. Shear Strength Properties of Natural Soils and Rocks, Oslo, 1967, Vol. I, pp. 179–180.

21. CLARK, G.B.: Deformation moduli of rocks. Proc. Symp. Testing Techniques for Rock Mech., Seattle, Wash., 1965, pp. 133–172.

22. COATES, D.F. and GYENGE, M.: Plate-load testing on rock for deformation and strength properties. Proc. Symp. Testing Techniques for Rock Mech., Seattle, Wash., 1965, pp. 19–35.

23. COMES, G.: Determination of mechanical characteristics of rock form foundation. In French. Travaux, No. 370, Nov., 1965, pp. 617–620.

24. COMES, G. and FOURNIER, G.: Measurement of in situ shear strength of marl. Proc. Int. Symp. Rock Mech., Madrid, 1968, pp. 195–199.

25. COOK, N.G.W.: Contribution to: A study of the strength of coal pillars by M.D.G. SALAMON and A.H. MUNRO. J.S. African Inst. Min. Metall., Vol. 68, 1967, pp. 192–195.

26. COOK, N.G.W., HODGSON, K. and HOJEM, J.P.M.: A 100-MN jacking system for testing coal pillars underground. J.S. African Inst. Min. Metall., Vol. 71, 1971, pp. 215–224.

27. CORDING, E.J., HENDRON, A.J. and DEERE, D.U.: Rock engineering for underground caverns. Proc. Symp. Underground Rock Chambers, Phoenix, Arizona, 1971, pp. 567–600.

28. COUETDIC, J.M. and BARRON, K.: Plate load testing as a method of assessing the in situ strength properties of Western Canadian Coal. Int. J. Rock Mech. Min. Sci. & Geomech. Abstr., Vol. 12, 1975, pp. 303–310.

29. DAVYDOVA, N.A.: Interpretation of pressiometric tests results with consideration of loaded section final length. Proc. Int. Symp. Rock Mech., Madrid, 1968, pp. 87–90.

30. DEERE, D.U.: 1963, unpublished.

31. DEERE, D.U., HENDRON, A.J., PATTON, F.D. and CORDING, E.J.: Design of surface and near surface construction in rock. Proc. 8th Symp. Rock Mech., Minneapolis, Minn., 1966, pp. 237–302.

32. DENKHAUS, H.G.: A critical review of the present state of scientific knowledge related to the strength of mine pillars. J.S. African Inst. Min. Metall., Vol. 63, 1962, pp. 59–75 and 143–150.

33. DIXON, S.J.: Pressure meter testing of soft bedrock. Proc. Symp. Determination of the In situ Modulus of Deformation of Rock. Denver, Colo., 1969. A.S.T.M. Spec. Tech. Pub. 477, 1970, pp. 126–136.

34. DODDS, D.J.: Interpretation of plate loading test results. Proc. Symp. Field Testing and Instrumentation. A.S.T.M. Spec. Tech. Pub. 554, 1974, pp. 20–33.

35. DODDS, R.K.: Suggested method of test for in situ shear strength of rock. A.S.T.M. Spec. Tech. Pub. 479, 1970, pp. 618–628.

36. DRYSELIUS, G.: Design of a measuring cell for the study of rock pressure. IV A Ingeniorsvetenskapskademiens Meddleland, Vol. 142, 1965, pp. 135–144.

37. DUFFAUT, P. and COMES, G.: Comparaison de la déformabilité statique d'un materiau de fondation mesurée en place sur le parament d'une excavation et la

paroi d'un sondage. Proc. 1st Cong. Int. Soc. Rock Mech., Lisbon, 1966, Vol. I, pp. 399–403.

38. EGOROV, K.E.: Concerning the question of the deformation of basis of finite thickness. I Mekhanika Gruntov, Gosstroiizdat, Moscow, 1958, No. 34.

39. EVANS, I.: The expanding bolt seam tester: A theory of tensile breakage. Int. J. Rock Mech. Min. Sci., Vol. 1, 1964, pp. 459–474.

40. EVANS, I. and POMEROY, C. D.: The strength of cubes of coal in uniaxial compression. Proc. Conf. Mechanical Properties of Nonmetallic Brittle Materials, London, Butterworths, 1958, pp. 5–28.

41. EVDOKIMOV, P.D.: Study of stability, shear strength and deformability of rock foundations of high concrete dams in the Soviet Union (in Russian). Gidrotechnicheskoe Stroitelstvo, No. 4, 1964, pp. 59–68.

42. EVDOKIMOV, P.D. and SAPEGIN, D.D.: Stability, shear sliding resistance and deformation of rock foundations. Israel Program for Scientific Translations, 1967, 145 p.

43. EVDOKIMOV, P.D. and SAPEGIN, D.D.: A large scale shear test on rock. Proc. 2nd Cong. Int. Soc. Rock Mech., Belgrade, 1970, Vol. 2, pp. 117–121.

44. EVERLING, G.: Ein Vorschlag zur Definition der Scherfestigkeit und damit zusammenhängender Begriffe. Rock Mech. Eng. Geol., Vol. I, 1963, pp. 181–185.

45. FERNANDEZ-BOLLO, M.: The influence of execution and the measurement methods on the deformability modulus obtained in rock masses tested with big size thin flat jacks. Proc. Int. Symp. Rock Mech., Madrid, 1968, pp. 95–97.

46. FISHMAN, YU. A.: Computing the stability and strength of rock foundations under concrete gravity dams. Hydrotechnical Construction, No. 5, May, 1976, pp. 450–461.

47. FOOTE, P.: An expanding bolt seam-tester. Int. J. Rock Mech. Min. Sci., Vol. 1, 1964, pp. 255–275.

48. FOSTER, C.R. and AHLVIN, R.G.: Stresses and deflections induced by a uniform circular bond. Proc. Highway Research Board, Vol. 33, 1954, pp. 467–470.

49. GALLICO, A.: A contribution to the design of foundations systems for arch dams. Water Power, Oct., 1974, pp. 323–329.

50. Geoprobe Instrument. Literature by Testlab. Corp., 216 North Clinton St., Chicago, Ill., 1967.

51. GEORGI, F., HÖFER, K.H., KNOLL, P., MENZEL, W. and THOMA, K.: Investigations about the fracture and deformation behaviour of rock masses. Proc. 2nd Cong. Int. Soc. Rock Mech., Belgrade, 1970, Vol. 2, pp. 299–305.

52. GERRARD, C.M., DAVIS, E.H. and WARDLE, L.J.: Estimation of the settlement of cross-anisotropic deposits using isotropic theory. Univ. Sydney, Sch. Civil Eng., Res. Rep. No. R-191, 1972.

53. GERRARD, C.M. and WARDLE, L.J.: Solutions for point loads and generalised circular loads applied to a circular anisotropic half-space. C.S.I.R.O., Australia, Div. Appl. Geomech., Tech. Paper 13, 1973a, 39p.

54. GERRARD, C.M. and WARDLE, L.J.: Solutions for line loads and generalised strip loads applied to an orthorhombic half-space. C.S.I.R.O., Australia, Div. Appl. Geomech., Tech. Paper 14, 1973b, 37p.

55. GIBSON, R.E. and HENKEL, D.J.: Influence of duration of tests at constant rate of strain on measured drained strength. Geotechnique, Vol. 4, 1954, pp. 6–15.

56. GICOT, H.: Measurements of deformation of the earth foundations of the Rossens dam. In French. Trans. 3rd Cong. Large Dams, Stockholm, 1948, Vol. 2, Paper R 56.

57. GILG, B.: Verformung und Bruch von Gesteinsproben unter dreiaxialer Belastung. Proc. 1st Cong. Int. Soc. Rock Mech., Lisbon, 1966, Vol. I, pp. 601–606.

58. GIMM, W.A.R., RICHTER, E. and ROSETZ, G.P.: A study of the deformation and strength properties of rocks by block tests in situ in iron ore mines (in German). Proc. 1st Cong. Int. Soc. Rock Mech., Lisbon, 1966, Vol. I, pp. 457–463.

59. GIROUD, J.P.: Settlement of a linearly loaded rectangular area. J. Soil Mech. Found. Div., Am. Soc. Civ. Eng., Vol. 94, 1968, pp. 813–831.

60. GOFFI, L.: La determinazione del modulo elastico di un ammasso roccioso mediante misure di deformazione per carichi concentrati in galleria sperimentale. Energia Elettrica, No. 9, 1963.

61. GOFFI, L.: Contribution to Theme 3. Proc. 1st Cong. Int. Soc. Rock Mech., Lisbon, 1966, Vol. III, pp. 271–272.

62. GONANO, L.P. and BROWN, E.T.: Stress gradient phenomenon and related size effects in brittle materials. Proc. 5th Australian Conf. Mech. Mater. and Struct., Melbourne, 1975, pp. 205–218.

63. GOODMAN, R.E.: Research in geological engineering at the University of California, Berkeley. Proc. 4th Annual Symp. Eng. Geol. and Soils Eng., Idaho Dept. Highways, Moscow, Idaho, 1966, pp. 155–156.

64. GOODMAN, R.E., VAN, T.K. and HEUZE, F.E.: Measurement of rock deformability in boreholes. Proc. 10th Symp. Rock Mech., Univ. Texas, Austin, Texas, 1968, pp. 523–555.

65. GREENWALD, H.P., HOWARTH, H.C. and HARTMANN, I.: Experiments on strength of small pillars of coal in the Pittsburgh bed. U.S.B.M. Tech. Paper 605, 1939, 22 p.

66. GREENWALD, H.P., HOWARTH, H.C. and HARTMANN, I.: Experiments on strength of small pillars of coal in the Pittsburgh bed. U.S.B.M. R.I. 3575, 1941.

67. GRISHIN, M.M., POSPELOV, V.B., CHUPRIKOV, I.K. and CHURAKOV, A.I.: Issledovaniya skalnogo osnovaniya charvakskoij plotiny. Proc. Moscow Civil Engineer Institute, No. 32, Gosenergoizdat, Moscow, 1961.

68. HAST, N.: Measuring stresses and deformations in solid materials. Stockholm, Centraltryckeriet, Esselte AB, 1943.

69. HENDRON, A.J., MESRI, G., GAMBLE, J.C. and WAY, G.: Compressibility characteristics of shales measured by laboratory and in situ tests. Proc. Symp. Determination of the In situ Modulus of Deformation of Rock. Denver, Colo., 1969, A.S.T.M. Spec. Tech. Pub. 477, 1970, pp. 137–153.

70. HEUZE, F.E.: The design of room and pillar structures in competent jointed rock. Example: The Crestmore mine, California. Univ. California, Berkeley, Ph. D. Thesis, 1970.

71. HOEK, E.: Rock mechanics: An introduction for the practical engineer. Part I. Min. Mag., Vol. 114, No. 4, April, 1966, pp. 236–255.

72. HULT, J.: On the measurement of stresses in solids. Trans. Chalmers Univ. Tech., Gothenburg, Sweden, No. 280, 1963.

73. Interfels, Salzburg, 1975.

74. International Society for Rock Mechanics. Suggested methods for determining shear strength. Feb., 1974.

75. International Society for Rock Mechanics. Recommendations on site investigation techniques, Final Report. July, 1975.

76. ISHII, F., IIDA, R., YUGETA, H., KISHIMOTO, S. and TENDOU, H.: On the strength characteristics of a bedded rock. Proc. 1st Cong. Int. Soc. Rock Mech., Lisbon, 1966, Vol. I, pp. 525–529.

77. JAEGER, C.: Rock mechanics and hydropower engineering. Water Power, Vol. 13, 1961, pp. 349–60; 391–396.

78. JAEGER, C.: Rock mechanics for dam foundations. Trans. 8th Cong. Large Dams, Edinburgh, 1964, Supplement, pp. 3–19.

79. JAEGER, J.C. and COOK, N.G.W.: Theory and application of curved jacks for measurement of stresses. Proc. Int. Conf. State of Stress in the Earth's Crust, Santa Monica, California, 1963, pp. 381–395.

80. JAEGER, J.C. and COOK, N.G.W.: Fundamentals of rock mechanics. London, Methuen & Co., 1969, 513 p.

81. JAHNS, H.: Messung der Gebirgsfestigkeit in situ bei wachsendem Massstabsverhältnis. Proc. 1st Cong. Int. Soc. Rock Mech., Lisbon, 1966, Vol. I, pp. 477–482.

82. JANOD, A. and MERMIN, P.: In situ measurement of rock properties using a dilatometer in a cylindrical jack. In French. Travaux, 1954, pp. 610–612.

83. JENKINS, J.D.: The bearing capacity of floors in Northumberland and Durham coalfield. Trans. Inst. Min. Eng., Vol. 117, Part 11, Aug., 1958.

84. JENKINS, J.D.: A laboratory and underground study of the bearing capacity of mine floors. Proc. 3rd Int. Conf. Strata Control, Paris, 1960, pp. 227–235.

85. JIMINEZ-SALAS, J.A. and URIEL, S.: Some recent rock mechanics tests in Spain. Trans. 8th Cong. Large Dams, Edinburgh, 1964, pp. 995–1022.

86. JOHN, S.W.: The technique of large scale tests on rock, illustrated by the example of the Kurobe No. 4 concrete dam in Japan. In German. Geol. u. Bauwesen, Vol. 27, 1961, pp. 9–14.

87. JUDD, W.R.: Some rock mechanics problems in correlating laboratory results with prototype reactions. Int. J. Rock Mech. Min. Sci., Vol. 2, 1965, pp. 197–218.

88. KAWABUCHI, K.: A study of strain characteristics of a rock foundation. Trans. 8th Cong. Large Dams, Edinburgh, 1964, Vol. I, Paper R-11.

89. KIMISHIMA, H.: A study of failure characteristics of foundation rock through a series of tests in situ. Rock Mech. in Japan, Vol. 1, 1970, pp. 91–93.

90. KONONG, H.: De spannungsverdeling in een homogeen, anisotroop elastisch half medium. LGM Mededelingen, Delft, Vol. 5, No. 2, 1960, pp. 1–19.

91. KRSMANOVIC, D. and MILIC, S.: Model experiments on pressure distribution in some cases of a discontinuum. Rock Mech. Eng. Geol., Suppl. 1, 1964, pp. 72–87.

92. KRSMANOVIC, D. and POPOVIC, M.: Large scale field tests of the shear strength of limestone. Proc. 1st Cong. Int. Soc. Rock Mech., Lisbon, 1966, Vol. I, pp. 773–779.

93. KUJUNDZIC, B.: Mesure de characteristique des roche en place. Ann. Inst. Tech. Bâtiment et Trav. Publ., Vol. 11, 1957, pp. 639–647.

94. KUJUNDZIC, B.: Survey of the methods of experimental investigation of mechanical characteristics of the rock masses in Yugoslavia. Jaroslav Cerni Institute for Development of Water Resources, Belgrade, Trans. No. 26, 1963, 25 p.

95. KUJUNDZIC, B., COLIC, B. and RADOSAVLJEVIC, Z.: Komplexe Untersuchung des mechanischen Verhaltens der Felsmasse für die Gründung der Bauwerke des Wasserkraft- und Schiffahrtsystems Derdap. Proc. 2nd Cong. Int. Soc. Rock Mech., Belgrade, 1970, Vol. 3, pp. 599–607.

96. KUJUNDZIC, B., RADOSAVLJEVIC, Z. and COLIC, B.: The methods and results of geotechnical investigations at the site of the proposed Derdap hydroelectric power plant. Jaroslav Cerni Institute for Development of Water Resources, Belgrade, Trans. Vol. XI, No. 31, 1964, pp. 1–16.

97. KUJUNDZIC, B. and STOJAKOVIC, M.: A contribution to the experimental investigation of changes of mechanical characteristics of rock masses as a function of depth. Trans. 8th Cong. Large Dams, Edinburgh, 1964, pp. 1051–1068.

98. KUTTER, H.K.: Stress distribution in direct shear test samples. Proc. Symp. Rock Fracture, Nancy, 1971, Paper II-6.

99. LADANYI, B.: Expansion of cavities in brittle media. Int. J. Rock Mech. Min. Sci., Vol. 4, 1967, pp. 301–328.

100. LADANYI, B.: Rock failure under concentrated loading. Proc. 10th Symp. Rock Mech., Univ. Texas, Austin, Texas, 1968, pp. 363–387.

101. LAMA, R.D.: A comparison of in situ mechanical properties of coal seams. Col. Eng., 1966a, pp. 20–25.

102. LAMA, R.D.: Elasticity and strength of coal seams in situ and an attempt to determine the energy in pressure bursting of road sides. D. Sc. Tech. Thesis, Faculty of Mining, Academy of Min. & Metall., Cracow, Poland, 1966b.

103. LAMA, R.D.: In situ and laboratory strength of coal. Proc. 12th Symp. Rock Mech., Rolla, Missouri, 1970, pp. 265–300.

104. LAMA, R.D.: Uniaxial compressive strength of jointed rock. Festschrift-Leopold Müller-Salzburg zum 65. Geburtstag, Institut für Felsmechanik und Bodenmechanik, Universität Karlsruhe, 1974, pp. 67–78.

105. LAMA, R.D. and GONANO, L.P.: Size effect considerations in the assessment of mechanical properties of rock masses. Proc. 2nd Symp. Rock Mech., Dhanbad, 1976.

106. LAMA, R.D., SIGGINS, A.F. and KHORSHID, M.: C.S.I.R.O., Australia, Division of Applied Geomechanics, (unpublished), 1977.

107. LANE, R.G.T.: Rock foundations: Diagnosis of mechanical properties and treatment. Trans. 8th Cong. Large Dams, Edinburgh, 1964, Vol. I, Paper R-8, pp. 141–166.

108. LAUFFER, H. and SEEBER, G.: Design and control of linings of pressure tunnels and shafts, based upon measurements of the deformability of the rock. Trans. 7th Cong. Large Dams, Rome, 1961.

108a. LAUFFER, H. and SEEBER, G.: Die Messung der Felsnachgiebigkeit mit der TIWAG – Radialpresse und ihre Kontrolle durch Dehnungsmessungen an der Druckschachtpanzerung des Kaunertal Kraftwerkes. Proc. 1st Cong. Int. Soc. Rock Mech., Lisbon, 1966, Vol. II, pp. 347–356.

109. LINK, H.: Evaluation of elasticity moduli of dam foundation rock determined seismically in comparison to those arrived at statically. Trans. 8th Cong. Large Dams, Edinburgh, 1964, Vol. 1, R-45, pp. 833–858.

110. LNEC (Laboratorio Nacional de Engenharia Civil), Portugal. Large flat jack. TE12, July, 1970.

111. LOCHER, H.G.: Some results of direct shear tests on rock discontinuities. Proc. Int. Symp. Rock Mech., Madrid, 1968.

112. LOGTERS, G. and VOORT, H.: In situ determination of the deformational behaviour of a cubical rock-mass sample under triaxial load. Rock Mech. Vol. 6, 1974, pp. 65–79.

113. LOTTI, C. and BEAMONTE, M.: Execution and control of consolidation works carried out in the foundation rock of an arch gravity dam. Trans. 8th Cong. Large Dams, Edinburgh, 1964, Vol. I, paper R-37, pp. 671–698.

114. MALINA, H.: Berechnung von Spannungsumlagerungen in Fels und Boden mit Hilfe der Elementenmethode. Veröffentlichung Nr. 40, Inst. f. Boden- u. Felsmechanik, Univ. Karlsruhe, 1969.

115. Manev, G. and Avramova-Tacheva, E.: On the valuation of strength and resistance condition of the rocks in natural rock massif. Proc. 2nd Cong. Inst. Soc. Rock Mech., Belgrade, 1970, Vol. I, pp. 59–65.

116. Martini, H.J., Durbaum, H. and Giesel, W.: Methods to determine the physical properties of rock. Trans. 8th Cong. Large Dams, Edinburgh, 1964, pp. 859–869.

117. McKinlay, D.G. and Anderson, W.F.: Determination of the modulus of deformation of a till using a pressuremeter. Ground Eng., Vol. 8, No. 6, 1975, pp. 51–54.

118. Ménard, L.: In situ measurements of physical properties in soil. In French. Annales des Ponts et Chaussées, Vol. 127, No. 3, May–June, 1957, pp. 358–377.

119. Ménard, L.: Rules for the calculation and design of foundation elements on the basis of pressuremeter investigations of the ground. Translated by B.E. Hartmann, Terrametrics, April, 1966.

120. Mencl, V.: Proportion of cohesion and of internal friction in the strength of rocks. Proc. Colloquium, Norwegian Geotechnical Inst., Oslo, 1962.

121. Mencl, V.: Factor of strength of rock material in the strength of rock mass. Proc. 1st Cong. Int. Soc. Rock Mech., Lisbon, 1966, Vol. I, pp. 289–290.

122. Merrill, R.H.: Design of underground mine openings, oilshale mine, Rifle, Colo., U.S.B.M. R. I. 5089, 1954, 56 p.

123. Merrill, R.H. and Morgan, T.A.: Method of determining the strength of a mine roof. U.S.B.M. R.I. 5406, 1958, 22 p.

124. Meyerhoff, G.G.: The ultimate bearing capacity of foundations. Geotechnique, Vol. 2, 1951, pp. 301–332.

125. Milovic, D.M.: Stresses and displacements in an anisotropic medium due to a circular load. Proc. 2nd Cong. Int. Soc. Rock Mech., Belgrade, 1970, Vol. 3, pp. 479–482.

126. Misterek, D.L.: Analysis of data from radial jacking tests. Proc. Symp. Determination of the In situ Modulus of Deformation of Rock, Denver, Colo., 1969. A.S.T.M. Spec. Tech. Pub. 477, 1970, pp. 27–38.

127. Mogilevskaya, S.: Effect of composition, features of structural fabric and water saturation on rock deformability. Proc. 2nd Cong. Int. Soc. Roch Mech., Belgrade, 1970, Vol. I, pp. 333–339.

128. Morocco, Direction de L'Hydraulique. Study of the characteristics of marl by means of laboratory and in situ tests. Proc. Int. Symp. Roch Mech., Madrid, 1968, pp. 201–220.

129. Müller, L.: Der Felsbau. Bd. I. Stuttgart, Ferdinand Enke, 1963, 623 p.

130. Muskhelishvili, N.I.: Some basic problems of the mathematical theory of elasticity. Groningen. Noordhoff, 1953.

131. Noel, G.: Mesure du module d'élasticité en profondeur dans les massifs rocheux. Cellule de mesure de l'Institut Technique du Bâtiment et des Travaux Publics, No. 85, 1963, pp. 533–540.

132. Nonveiller, E.: The determination of the deformation of loaded rock in tunnels. Proc. Yugoslav Soc. Soil Mech. Found. Eng., Ljubljana, No. 8, 1954.

133. Nose, M.: Rock tests in situ. Trans. 8th Cong. Large Dams, Edinburgh, 1964, pp. 219–252.

134. Oberti, G.: Experimentelle Untersuchungen über die Characteristika der Verformbarkeit der Felsen. ISMES Bull., No. 15, 1960.

135. Odemark, N.: Investigations as to the elastic properties according to the theory of elasticity. Statens Vaginstitut, Stockholm, Meddleland, 77, 1949.

136. Ohya, S., Ogura, K. and Tsuji, M.: The instrument of new type dilatometer "Elastmeter" for study. OYO Corporation, Tokyo, Japan, 1975.

137. Oliveira, E.R.A.: A utilizacaodos metodos numericos integrais na resolucao de problemas de elasticidade plana. Lab. Nac. de Eng. Civil, Lisbon, Internal Report, 1955.

138. Onodera, T.F.: Dynamic investigation of foundation rocks in situ. Proc. 5th Symp. Rock Mech., Minneapolis, Minn., 1962, pp. 517–533.

139. OYO Corporation. 2–1, Otsuka 3-chome, Bunkyo-ku, Tokyo, Japan, 1976.

140. Palmer, L.A. and Barber, E.S.: Soil displacement under a circular loaded area. Proc. High. Res. Board, Vol. 20, 1940, pp. 279–286; Discussion, pp. 319–332.

141. Panek, L.A.: Effect of rock fracturing on the modulus, as determined by borehole dilation tests. Proc. 2nd Cong. Int. Soc. Rock Mech., Belgrade, 1970, Vol. 1, pp. 383–387.

142. Panek, L.A. and Stock, J.A.: Development of a rock stress monitoring station based on flat slot of measuring existing rock stress. U.S.B.M. R.I. 6537, 1964.

143. Pomeroy, C.D.: A laboratory investigation of the penetration of coal by cylindrical indenters. MRE Rep. No. 2117, N.C.B., London, 1958.

144. Poulos, H.G. and Davis, E.H.: Elastic solutions for soil and rock mechanics. New York, Wiley, 1974.

145. Prandtl, L.: Uber die Härte plastischer Körper. Nachr. Kgl. Ges. Wiss. Göttingen, Math-Phys. Kl, 1920, pp. 74–85.

146. Pratt, H. R.: Personal Communication, 1976.

146a. Pratt, H. R., Black, A. D., Brown, W. S. and Brace, W. F.: The effect of specimen size on the mechanical properties of unjointed diorite. Int. J. Rock Mech. Min. Sci., Vol. 9, 1972, pp. 513–529.

147. Pratt, H.R., Brown, W.S. and Brace, W.F.: In situ determination of strength properties in a quartz diorite rock mass. Proc. 12th Symp. Rock Mech., Rolla, Missouri, 1970, pp. 27–43.

148. Prigozhin, E.S.: New design of pressiometer for rock massif strain modulus definition. Proc. Int. Symp. Rock Mech., Madrid, 1968, pp. 79–81.

149. Protodyakonov, M.M.: Methods of studying the strength of rocks, used in the U.S.S.R. Proc. Int. Symp. Min. Res., Rolla, Missouri, 1961, Vol. 2, pp. 649–668.

150. Rappoport, R.M.: Approximate methods for determining stresses and strains in layered rock foundations. Proc. 2nd Cong. Int. Soc. Rock Mech., Belgrade, 1970, Vol. 3, pp. 433–438.

151. Rice, L.O.: In situ testing of foundation and abutment rock for dams. Trans. 8th Cong. Large Dams, Edinburgh, 1964, Vol. I, Pap. R-5.

152. Richter, E.: Druckversuche in situ zur Bestimmung von Verformungs- und Festigkeitsparametern des klüftigen Gebirges. Bergakademie, Vol. 20, 1968, pp. 721–724.

153. Rocha, M.: Mechanical behaviour of rock foundations in concrete dams. Trans. 8th Cong. Large Dams, Edinburgh, 1964, Paper R-44, Q. 28, pp. 785–832.

154. Rocha, M.: New techniques in deformability testing of in situ rock mass. Proc. Symp. Determination of the In situ Modulus of Deformation of Rock, Denver, Colo., 1969. A.S.T.M. Spec. Tech. Pub. 477, 1970, pp. 39–54.

155. Rocha, M.: Present possibilities of studying foundations of concrete dams. Lab. Nac. de Eng. Civil (LNEC), Lisbon, Mem. No. 457, 1974.

156. ROCHA, M., DA SILVEIRA, A. F., GROSSMANN, N. and OLIVEIRA, E.: Determination of the deformability of rock masses along boreholes. Proc. 1st Cong. Int. Soc. Rock Mech., Lisbon, 1966, Vol. I, pp. 697–704.

157. ROCHA, M., SERAFIM, J. L. and DA SILVEIRA, A. F.: Deformability of foundation rocks. Trans. 5th Cong. Large Dams, Paris, 1955, Vol. III, pp. 531–561.

158. ROESLER, F. C.: Brittle fractures near equilibrium. Proc. Phys. Soc., Vol. B 69, 1956, pp. 981–992.

159. ROMERO, S. U.: In situ direct shear test with saturation of the rock, and interpretation of the great shear test in the nature. Proc. 1st Cong. Int. Soc. Rock Mech., Lisbon, 1966, Vol. I, pp. 353–357.

160. ROMERO, S. U.: In situ direct shear tests on irregular surface joints filled with clayey material. Proc. Int. Symp. Rock Mech., Madrid, 1968, pp. 189–194.

161. ROUSE, G. C. and WALLACE, G. B.: Rock stability measurements for underground openings. Proc. 1st Cong. Int. Soc. Rock Mech., Lisbon, 1966, Vol. II, pp. 335–340.

162. RUIZ, M. D.: Anisotropy of rock masses in various underground projects in Brazil. Proc. 1st Cong. Int. Soc. Rock Mech., Lisbon, 1966, Vol. I, pp. 263–267.

163. RUIZ, M. D. and DE CAMARGO, F. P.: A large-scale field shear test on rock. Proc. 1st Cong. Int. Soc. Rock Mech., Lisbon, 1966, Vol. I, pp. 257–261.

164. RUIZ, M. D., DE CAMARGO, F. P., MIDEA, N. F. and NIEBLE, C. M.: Some considerations regarding the shear strength of rocks. Proc. Int. Symp. Rock Mech., Madrid, 1968, pp. 159–169.

165. SALAMON, M. D. G.: A method of designing bord and pillar workings. J. S. African Inst. Min. Metall., Vol. 68, No. 2, Sept., 1967, pp. 68–78.

166. SALAMON, M. D. G.: Elastic moduli of a stratified rock mass. Int. J. Rock Mech. Min. Sci., Vol. 5, 1968, pp. 519–527.

167. SALAMON, M. D. G.: Personal communication, Oct., 1976.

168. SEEBER, G.: Evaluation of static strain measurements in rock. In German. Geol.- u. Bauwesen, Vol. 26, 1960, pp. 152–175.

169. SEEBER, G.: Some rock mechanics measurement results in the pressure chamber of the Kaunertal power plant. In German. Rock Mech. Eng. Geol., Supplement 1, 1964, pp. 130–148.

170. SELLERS, J. B.: Measurement of in situ shear strength using the torsional shear method. Proc. Symp. Field Testing and Instrumentation, A.S.T.M. Spec. Tech. Pub. 554, 1974, pp. 147–155.

171. SERAFIM, J. L. and GUERREIRO, M.: In situ tests for the study of rock foundations of concrete dams. Proc. 1st Cong. Int. Soc. Rock Mech., Lisbon, 1966, Vol. II, pp. 549–556.

172. SERAFIM, J. L. and GUERREIRO, M.: Shear strength of rock masses at 3 Spanish dam sites. Proc. Int. Symp. Rock Mech., Madrid, 1968, pp. 147–157.

173. SERAFIM, J. L. and LOPES, J. J. B.: In situ shear tests and triaxial tests of foundation rocks of concrete dams. Lab. Nac. de Eng. Civil, Lisbon, Tech. Pap. No.190, 1962, 7 p.

174. Shannon and Wilson, Inc.: Report on in situ rock tests, Dworshak dam site, for U.S. Corps of Engineers, District Walla Walla, Seattle, Wash., Dec., 1964.

175. SHIELD, R. T.: On Coulomb's law of failure in soils. J. Mech. Phys. Solids, Vol. 4, 1955, pp. 10–16.

176. SILVEIRA, A.: Contribution to Theme 2. Proc. 1st Cong. Int. Soc. Rock Mech., Lisbon, 1966a, Vol. III, pp. 237–239.

177. SILVEIRA, A.: Contribution to Theme 8. Proc. 1st Cong. Int. Soc. Rock Mech., Lisbon, 1966b, Vol. III, pp. 578–582.

178. STAGG, K.G.: In situ tests on the rock mass. In Rock Mechanics in Engineering Practice (Editors – K.G. STAGG and O.C. ZIENKIEWICZ), London, Wiley, 1968, pp. 125–156.

179. STEARS, J.H.: Evaluation of penetrometer for estimating roof bolt anchorage. U.S.B.M. R.I. 6646, 1965, 23 p.

180. STEINBRENNER, W.: Tafeln zur Setzungsberechnung. Die Strasse, Vol. I, 1934, p. 121.

181. STUCKY, A.: Research centre for studies on dams. In French. Centenaire de l'Ecole Polytechnique de Lausanne, Lausanne, Switzerland, 1953.

182. TAKANO, M. and FURUJO, I.: Deformation and resistance in in situ block shear test on a block schist and a characteristic loading pattern. Proc. 1st Cong. Int. Soc. Rock Mech., Lisbon, 1966, Vol. I, pp. 765–768.

183. TAKANO, M. and SHIDOMOTO, Y.: Deformation test on mudstone enclosed in a foundation by means of tube-deformeter. Proc. 1st Cong. Int. Soc. Rock Mech., Lisbon, 1966, Vol. I, pp. 761–764.

184. TALOBRE, J.: Rock mechanics. In French. Paris, Dunod, 1957.

185. TALOBRE, J.: Experimental determination of resistance of rocks in dams and in the walls of underground structures. In French. Trans. 7th Cong. Large Dams, Rome, 1961, Vol. 2, pp. 269–293.

186. TALOBRE, J.: La mesure in situ des propriétés mecaniques des roches et la sécurité des barrages de grande Hauteur. Trans. 8th Cong. Large Dams, 1964, pp. 397–399.

187. TALOBRE, J., FOX, P. and MEYER, A.: Foundations of the Pahlavi dam on the Dez river. Trans. 8th Cong. Large Dams, Edinburgh, 1964, pp. 1–22.

188. TERZAGHI, K.: Theoretical soil mechanics. New York, Wiley, 1943.

189. THENN DE BARROS, S.: Deflection factor charts for two- and three-layer elastic systems. High. Res. Rec. No. 145, 1966, pp. 83–108.

190. TIMOSHENKO, S. and GOODIER, J.N.: Theory of elasticity. 2nd ed., New York, McGraw Hill, 1951, 506 p.

191. TRANTER, C.J.: On the elastic distorsion of a cylinder hole by a localised hydrostatic pressure. Q. Appl. Maths., Vol. 4, No. 3, 1946, pp. 298–302.

192. TSYTOVICH, N.A., UKHOV, S.B., KUBETSKY, V.L., PANENKOV, A.S. and TERNOVSKY, I.N.: To the problem of determining the modulus of deformation of in situ fissured rock masses. Proc. 2nd Cong. Int. Soc. Rock Mech., Belgrade, 1970, Vol. I, pp. 301–306.

193. UESHITA, K. and MEYERHOF, G.G.: Deflection of multilayer soil systems. J. Soil Mech. Found. Div., Am Soc. Civ. Eng., Vol. 93, 1967, pp. 257–282.

194. UKHOV, S.B. and TSYTOVICH, N.A.: Some principles of mechanical properties of chloritic schists. Proc. 1st Cong. Int. Soc. Rock Mech., Lisbon, 1966, Vol. I, pp. 781–786.

195. U.S. Bureau of Reclamation. Morrow Point Dam and Power Plant foundation investigation. Water Resources Tech. Pub., Denver, Colo., 1965.

196. VAN HEERDEN, W.L.: In situ determination of complete stress-strain characteristics for 1.4 m square coal specimens with width-to-height ratio of up to 3.4. Rep. S. African C.S.I.R., No. ME 1265, 1974.

197. VAN HEERDEN, W.L.: In situ complete stress-strain characteristics of large coal specimens. J.S. African Inst. Min. Metall., Vol. 75, No. 8, March, 1975, pp. 207–217.

198. Vesic, A.S.: Discussion-Session III. Proc. 1st Int. Conf. Structural Design of Asphalt Pavements, Univ. Michigan, Ann Arbor, Mich., 1963, pp. 283–290.

199. Voight, B.: On the functional classification of rocks for engineering purposes. Proc. Int. Symp. Rock Mech., Madrid, 1968, pp. 131–135.

200. Wagner, H.: Determination of the complete load deformation characteristics of coal pillars. Proc. 3rd Cong. Int. Soc. Rock Mech., Denver, 1974, Vol. 2, pp. 1076–1081.

201. Wallace, G.B., Slebir, E.J. and Anderson, F.A.: Radial jacking test for arch dams. Proc. 10th Symp. Rock Mech., Univ. Texas, Austin, Texas, 1968, pp. 633–660.

202. Wallace, G.B., Slebir, E.J. and Anderson, F.A.: In situ methods for determining deformation modulus used by the Bureau of Reclamation. Proc. Symp. Determination of the In Situ Modulus of Deformation of Rock. Denver, Colo., 1969. A.S.T.M. Spec. Tech. Pub. 477, 1970, pp. 3–26.

203. Wang, C.: Mechanical roof pulling technique for evaluating the effectiveness of roof bolting systems. Trans. Soc. Min. Eng. A.I.M.E., Vol. 258, 1975, pp. 65–68.

204. Westergaard, H.M.: On the elastic distortion of cylindrical hole by a localised hydrostatic pressure. Theodore von Karman Anniv. Vol., Calif. Inst. Tech., 1941.

205. Wilkins, J.: Contribution to Theme 8. Proc. 1st Cong. Int. Soc. Rock Mech., Lisbon, 1966, Vol. 3, pp. 586–588.

206. Worotnicki, G., Enever, J.R. and Spathis, A.: A pressiometer for determination of modulus of rock in situ. Proc. Symp. In situ Testing of Rock Parameters, Victorian Geomech. Soc., Melbourne, 1976.

207. Wright, F.D. and Bucky, P.B. Determination of room and pillar dimensions for the oil shale mine at Rifle, Colorado. Trans. A.I.M.E., Vol. 181, 1949, pp. 352–359.

208. Zienkiewicz, O. C. and Stagg, K. G. Cable method of in situ rock testing. Int. J. Rock Mech. Min. Sci., Vol. 4, 1967, pp. 273–300.

Uncited References to Chapter 8

1. Bernaix, J.: Properties of rock and rock masses. Proc. 3rd Cong. Int. Soc. Rock Mech., Denver, Colorado, Vol. I-A, 1974, pp. 9–38.
2. Brady, B.T., Hooker, V.E. and Agapito, J.F.: Laboratory and in situ mechanical behaviour studies of fractured oil shale pillars. Rock Mech., Vol. 7, No. 2, 1975, pp. 101–120.
3. Brandon, J.R.: Rock mechanics properties of typical foundation rock types. U.S.B.R. Engng. Res. Center, Rep. No. REC-ERC-74-10, July 1974, 99 pp.
4. Chan, S.S.M.: A proposed method to obtain actual strength parameters of mine rocks and rock masses. Proc. 2nd Cong. Int. Soc. Rock Mech., Belgrade, Vol. 2, 1970, pp. 83–88.
5. Habetha, E.: Large-scale shear tests for the Waldek II (Germany) pump-fed storage station construction project. Proc. 2nd Int. Conf. Int. Assoc. Engng. Geology, Sao Paulo, Brazil, August, 1974, Vol. I, Theme IV-35.
6. Herrmann, L.R. and Taylor, M.A.: Characterization of the structural behaviour of rock masses. Vol. I & II. OFR 67 (1) – 75,67 (2) – 75. USBM. 1975.
7. Hoek, E. and Londe, P.: Surface workings in rock. Proc. 3rd. Cong. Int. Soc. Rock Mech., Denver, Colorado, Vol. I-A, 1974, pp. 613–654.
8. Oliveara, R., Esteves, J. M. and Rodrigues, L. F.: Geotechnical studies of the foundation rock mass of Valhelhas dam (Portugal). Memoria No. 463, LNEC, Lisbon, Portugal, 1975.
9. RISO, R. De: Field survey of rock mass deformability of dam foundations. Experience in South Italy (in Italian) Memoire Nole, Istituto. Geol., Univ. Napoli, Vol. XII, 1972, pp. 1–56.
10. Roman, J., Coll, M., Sanz-Saracho, J.M. and Peironcely, J.M: Essais de compression radialle en galérie. Proc. 2nd Cong. Int. Soc. Rock Mech., Belgrade, 1970, Vol. II, pp. 469–74.
12. Tsytovich, N. A., Ukhov, S. B. and Burlakov, V. N.: Failure mechanism of a fissured rock base upon displacement of a loading plate. Proc. 2nd Cong. Int. Soc. Rock Mech., Belgrade, Vol. 2, 1970, pp. 89–93.
13. Ward, W.H. and Burland, J.B.: Assessment of the deformation properties of jointed rock in the mass. Proc. Int. Symp. Rock Mechanics, Madrid, 1968, pp. 37–44.
14. Weir-Jones, I. and Lang, H.: The use of in situ moduli determinations as an aid to U.G. opening design and their relationship to laboratory determined values. 9th Canadian Symp. Rock Mech., Montreal, Dec. 1973, pp. 1–20.

CHAPTER 9

Time-Dependent Properties of Rock

9.1. Introduction

The understanding of time-dependent effects or creep behaviour is important in further development of knowledge in the field of rock mechanics, rock bursts, ground control, seismology and other geological and geophysical phenomena occurring in the earth's crust.

In this chapter, the concepts concerning the deformational process occuring in rocks are given. The various mechanical models used to describe the time-dependent properties of solids are discussed together with applications of these models to rocks. A comprehensive review is given of the phenomenon of creep including creep laws, factors influencing creep and the mechanism of creep. In addition the equipment used to study creep in rocks is described and the methods for the determination of the time-dependent strength of rocks are included.

The subject of creep in rocks has been studied now over fifty years and a considerable amount of data is available under different loading conditions. Time-dependent behaviour in several rocks has been studied both in the laboratory under compression, bending and torsion (MICHELSON, 1917, 1920; PHILLIPS, 1931, 1932, 1948; GRIGGS, 1936, 1939, 1940; EVANS and WOOD, 1937; LOMNITZ, 1956; POMEROY, 1956; NISHIHARA, 1957a and b; HARDY, 1959, HARDY et al, 1969; MATSUSHIMA, 1960; ROBERTSON, 1960, 1963; PRICE, 1964; LECOMTE, 1965; RUMMEL, 1965, 1969; HOBBS, 1970; DREYER, 1972) and in mines (HOFER, 1958; REYNOLDS and GLOYNA, 1961; BARRON and TOEWS, 1963; BRADSHAW et al, 1964; POTTS, 1964; HEDLEY, 1967; LANGER, 1969). In spite of the extensive work done in this field under a variety of test conditions, the results are more qualitative than quantitative; it is only in a few instances that the elastic and viscous constants have been determined. Besides, no standard test procedures have been evolved as yet to conduct the tests.

9.2. Elasticity and Plasticity in Rocks

It is known that rocks do not follow Hooke's law. The deviation of the behaviour of rocks from the Hooke's law has been termed "viscoelastic". This has been confirmed mostly from the stress-strain curves. To explain this complex behaviour and the deformational behaviour of rocks under the effect of time, MAXWELL, KELVIN-VOIGT models and many other combinations of springs and dashpots have been introduced. High deformation observed under constant load in uniaxial or triaxial compression has been termed "creep" or "plasticity". VON KARMAN's (1911) experiments on sandstone specimens showed that whilst under uniaxial stress conditions deformation was 0.4%, under triaxial conditions deformation was 3% (at lateral stress of 26.8 MPa (3891 lbf/in^2)). There is, however, no proof to indicate whether the high deformation was due to plastic flow in the true sense or to opening and propagation of cracks. True plastic deformation is possible if shifting of the particles takes place in such a way that broken bonds are re-established thus eliminating the possibility of crack development. The density and strength of the rock mass must remain constant (as long as no structural changes take place). These conditions have not been confirmed, but indications are that strength and density are lowered. ROBINSON (1959) showed that the density of rocks decreases as a result of creep. This indicates the opening of minute internal cracks. His work on Indiana limestone showed that there was a decrease in the pore volume at initial deformation whilst at the point of yielding and thereafter liquid had to be supplied to the rock interstices, indicating an increase in the pore volume. Obviously, any plastic yield reported is due to destruction of the internal structure and opening of cracks rather than true plastic deformation. OBERT and DUVALL (1942, 1945a and b), ANTONIDES (1955), CHUGH et al (1967), HARDY et al (1969) and many others have shown that microcracks develop well before their ultimate failure.

The barrel shape of the specimens as reported many times under triaxial compression loading conditions is due to high friction between the restrained cracked particles. This restraint does not allow them to disintegrate. Friction at specimen ends introduces high lateral stresses which are comparatively small at the centre of the specimen. Therefore, widening of cracks is more pronounced at the centre giving relatively high lateral deformation. JONES (1958) reported that pyramidal portions near to the press platens of the specimens fractured in compression remained free from cracks.

Geologists, while concentrating on bending and twisting of rocks, have paid some attention to the flow properties of rocks. GÜMBEL (ADAMS and NICOLSON, 1901) as early as 1880 subjected small cylinders of orthoclase, quartz, Iceland spar and alabaster in steel collars to pressures of 2231 to 2536 MPa (323620 to 367750 lbf/in^2). The cylinders of orthoclase and quartz crushed to an incoherent powder but calcite retained its coherence but became perfectly opaque. The calcite was observed to have been forced into small depressions in the collar where it was found in a powdery condition. This material had lost its cleavage although the rest of the cylinder had retained the cleavage and conchoidal fracture. Alabaster behaved in a similar manner. GÜMBEL considered that these experiments proved the absence of plasticity. He also showed that folded rocks have distinct microscopic fractures.

PFAFF (ADAMS and NICOLSON, 1901) experimented on lithographic limestone samples enclosed in cylinders filled with wax and subjected to a pressure of 1011 MPa (146659 lbf/in^2) maintained for seven weeks. From the results he concluded that pressure alone was incapable of making rocks flow.

ADAMS and NICOLSON (1901) subjected marble specimens of 3.81 cm (1.5 in) height and 2.54 cm (1.0 in) diameter to triaxial pressure conditions and found that the strength of the deformed specimens was much smaller than that of the undeformed specimens. The results showed that when deformation was carried out very slowly, the resulting specimens were comparatively stronger than when deformed rapidly (Table 33). Using scanning microscope, MONTOTO (1974) has shown that cracks propagate in specimens and these join together before final disintegration of specimens.

TABLE 33

Strength of deformed specimens

(after ADAMS and NICOLSON, 1901)

Specimen	Original		Greatest diameter after deformation cm (in)	Deformation time	Crushing strength of deformed specimens MPa(lbf/in^2)
	Height cm (in)	Diameter cm (in)			
A	4.04 (1.594)	2.54 (1.00)	3.57 (1.407)	64 days	36.9 (5350)
O	4.04 (1.594)	2.54 (1.00)	3.06 (1.203)	1.5 hours	27.6 (4000)
P	3.82 (1.505)	2.54 (1.00)	3.53 (1.388)	10 mins.	19.1 (2776)

(The specimens had strength before deformation as high as 78.8 to 82.9 MPa (11430 to 12026 lbf/in^2)).

Although rocks do not follow the laws of elasticity, plasticity or viscoelasticity, it is still important to determine the stress-strain curve for the rocks and the time-dependent strain so that the mechanical behaviour of rocks can be predicted. Due to the varied nature of rocks, their exact behaviour cannot be easily defined in mathematical terms. Several mechanical models have been suggested that may have direct or indirect application to the description of the behaviour of rock. These are described here. Creep and the mechanism of creep are discussed in more detail in later sections.

9.3. Rheological Models—Simple Behaviour

The theory of rheological models uses a combination of several mechanical elements (springs, dashpots, frictional resistance to movement) connected in series or parallel or both. These systems represent only the mechanical behaviour of the material under conditions of testing, but have nothing in common with the real material. The following are three basic elements forming mechanical models.

1. A perfectly elastic spring representing truly elastic deformation.
2. A dashpot representing viscous deformation (truly Newtonian material).
3. A mass resting on a plane with a frictional force equal to the yield limit which prevents the movement of the block under the action of the forces below the yield limit. This represents truly plastic deformation.

Some simple models obtained from these elements representing ideal materials are described here.

9.3.1. Perfectly Elastic or Hookean Material

For materials which obey Hooke's law perfectly (ideal materials), the relationship between stress σ and strain ε can be represented as

$$\sigma = E\varepsilon \tag{9.1}$$

where $E =$ Young's modulus of the material. The mechanical behaviour of such materials can be represented by a simple spring (Fig. 9-1). In this model, σ represents the force acting on the spring and ε represents the deformation produced in the spring while E is termed the spring constant.

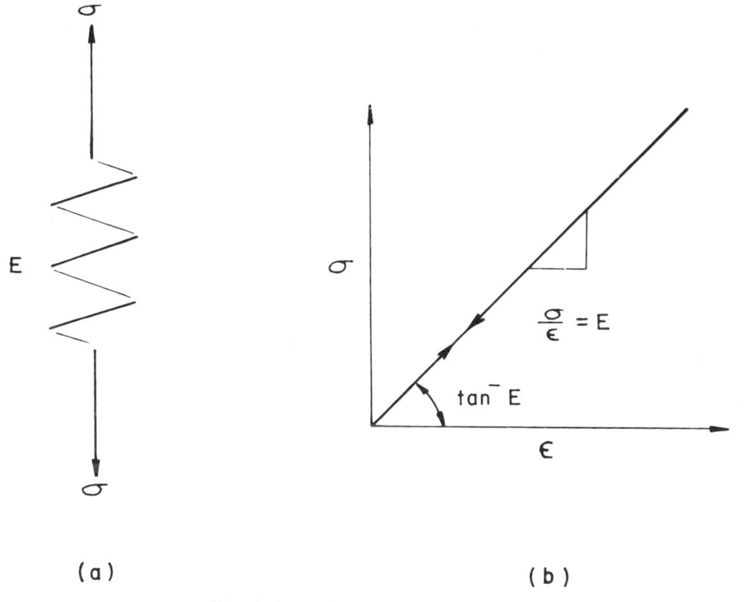

Fig. 9-1. Perfectly elastic material
(a) Model
(b) Stress-strain curve.

9.3.2. Perfectly Plastic Material

A perfectly plastic material is that which will not deform at all as long as the stress applied is less than a particular well defined limit σ_0; if the applied stress is equal to this limit the material will continue to deform without supporting any further increase in stress. The model of such a material is a load W lying on a surface having a fixed coefficient of friction μ (Fig. 9-2). The model thus represents a relationship

$$\sigma_0 = \mu W \tag{9.2}$$

and graphically the characteristic behaviour can be represented by the stress-strain curve given in Fig. 9-2b.

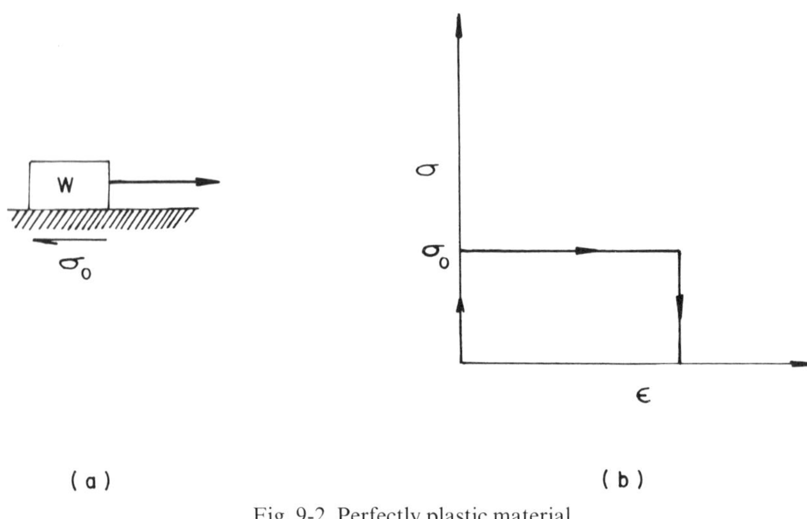

Fig. 9-2. Perfectly plastic material
(a) Model
(b) Stress-strain curve.

9.3.3. Perfectly Elastoplastic or St. Venant Material

St. Venant material is the one which behaves perfectly elastically below a certain defined stress level σ_0 and perfectly plastically as soon as this stress level is reached. Hence this material is a combination of a perfectly elastic element E and a perfectly plastic element W coupled in series (Fig. 9-3a) and its mechanical behaviour is represented by Fig. 9-3b.

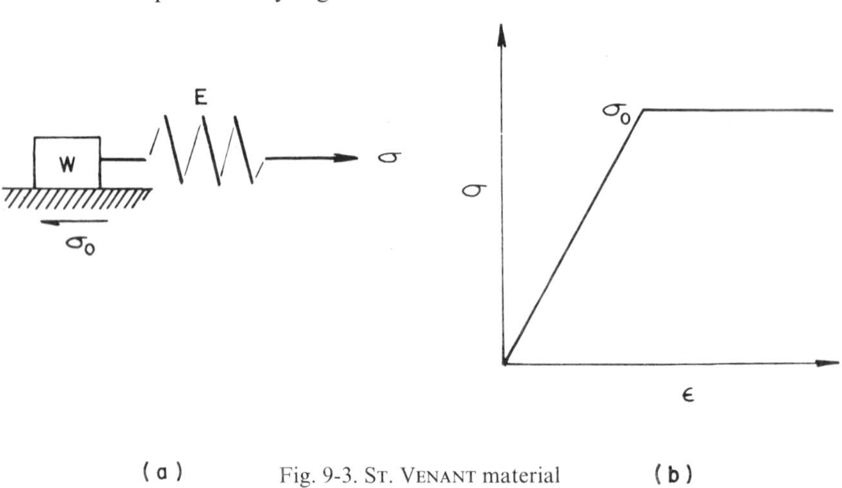

Fig. 9-3. St. Venant material
(a) Model
(b) Stress-strain curve.

9.3.4. Perfectly Viscous or Newtonian Material

A perfectly viscous material develops shear stresses proportional to the rate of change of shear strains. Mathematically it can be represented by

$$\tau = \eta \dot{\gamma} \tag{9.3}$$

where τ = shear stress

$\dot{\gamma}$ = rate of shear strain $\left(\frac{d\gamma}{dt}\right)$ and

η = coefficient of viscosity $\left(\frac{1}{\text{fluidity}}\right)$.

A dashpot represents the mechanical model of such a material (Fig. 9-4a).

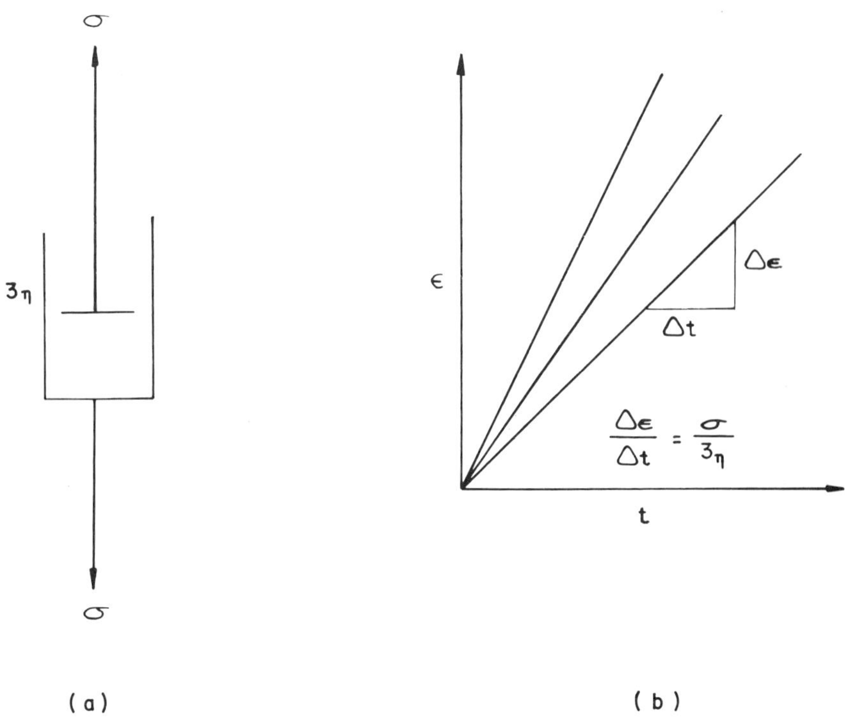

Fig. 9-4. NEWTONIAN material
(a) Model
(b) Stress-strain curve under constant stress.

When a tensile rod composed of a Newtonian material is being acted upon by a constant normal tensile stress σ_1 and the rate of change of normal strain is $\dot{\varepsilon}$ the maximum shear stress induced in the rod can be given by

$$\tau = \frac{\sigma_1}{2} \tag{9.4}$$

Since the Poisson's ratio of the material is 0.5, lateral strain will be related to the normal strain by the relationship

$$-\varepsilon_3 = -\varepsilon_2 = \frac{\varepsilon_1}{2} \tag{9.5}$$

and the maximum shear strain

$$\gamma = \varepsilon_1 - \varepsilon_2 = \frac{3\varepsilon_1}{2} \tag{9.6}$$

From Eqs. 9.3, 9.4 and 9.6, the normal stress is related to the normal strain by

$$\sigma = 3\,\eta\dot{\varepsilon}$$

or

$$\sigma = 3\,\eta\frac{d\varepsilon}{dt} \tag{9.7}$$

Upon integration of Eq. 9.7

$$\varepsilon = \frac{\sigma t}{3\eta} \tag{9.8}$$

Graphically, Eq. 9.8 depicting the behaviour of a Newtonian material can be represented by Fig. 9-4b.

9.3.5. Viscoelastic or MAXWELL Material

When two elements, a spring and a dashpot are coupled in series, they produce a model termed the MAXWELL model (Fig. 9-5a) and the material which has such a characteristic is called MAXWELL material.

When such a material is subjected to normal tensile stress σ, the total strain produced is the sum of the instantaneous (elastic) strain ε' of the spring, and the viscous strain ε'' of the dashpot, a function of time.

Thus

$$\varepsilon = \varepsilon' + \varepsilon'' \tag{9.9}$$

At zero time, i.e. immediately when stress is applied,

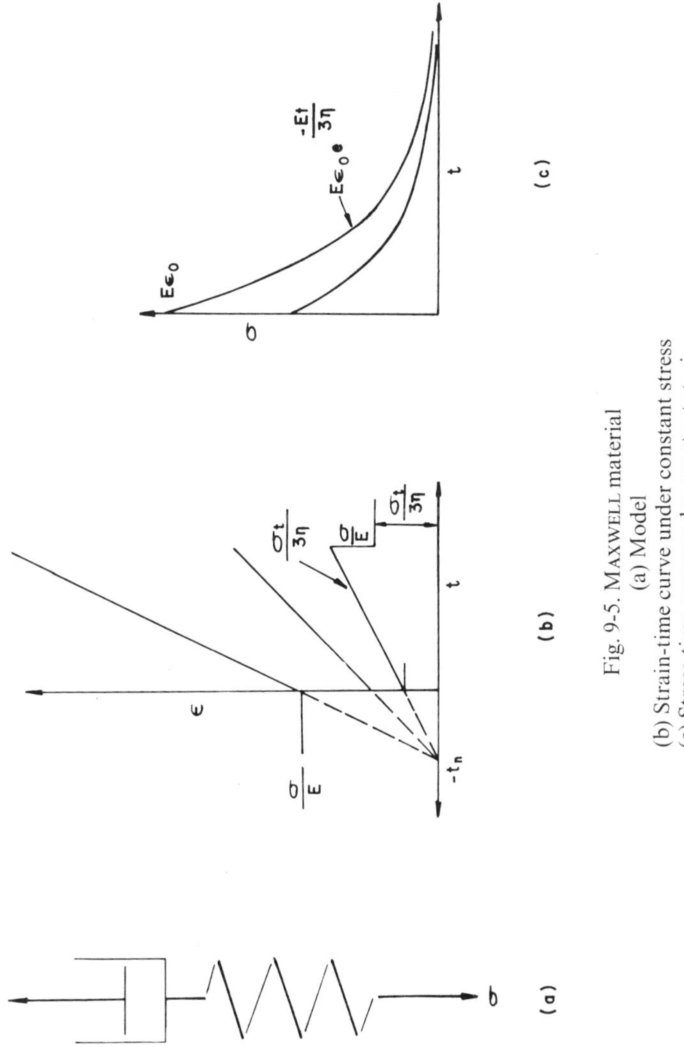

Fig. 9-5. MAXWELL material
(a) Model
(b) Strain-time curve under constant stress
(c) Stress-time curve under constant strain.

$$\varepsilon' = \frac{\sigma}{E}, \; \varepsilon'' = 0$$

After time t,

$$\varepsilon'' = \frac{\sigma t}{3\eta}$$

Therefore, total strain after time t,

$$\varepsilon_t = \frac{\sigma}{E} + \frac{\sigma t}{3\eta} \tag{9.10}$$

The mechanical behaviour can be represented by Fig. 9-5b.

If strain $\varepsilon = \varepsilon_0$ is suddenly applied, then at time $t = 0$ (Fig. 9-5c),

$$\varepsilon' = \frac{\sigma}{E}$$

$$\varepsilon'' = 0$$

If ε is kept constant, with time a certain reduction in stress must take place to compensate for the strain produced in the dashpot. This is equivalent to the reduction of strain in the spring. Therefore, after differentiation of Eq. 9.10

$$\frac{\dot{\sigma}}{E} = \frac{-\sigma}{3\eta}$$

$$\text{or} \quad \frac{d\sigma}{\sigma} = \frac{-E dt}{3\eta}$$

Upon integration

$$\sigma = E\varepsilon_0 e^{-\frac{Et}{3\eta}} \tag{9.11}$$

The term $E\varepsilon_0$ represents the stress σ (initial stress), at time $t = 0$. The stress relaxes from the initial value of σ to $\left(\frac{\sigma}{e}\right)$ in time t_n (called the MAXWELL relaxation time) and can be given by $t_n = \frac{3\eta}{E}$ (Fig. 9-5b).

Relaxation time t_n is also the time required to produce, under a sustained stress, an inelastic strain equal to the elastic strain. Thus the family of creep curves of the MAXWELL material forms a family of straight lines which meet at a common point on the time axis where $t = -t_n$.

A relationship for shear stress can be derived simultaneously and the resulting equations can be represented by:

For constant shear stress

$$\gamma = \frac{\tau}{G} + \frac{\tau t}{\eta} \tag{9.12}$$

For constant shear strain

$$\tau = \tau_0 e^{-\left(\frac{Gt}{\eta}\right)} \tag{9.13}$$

The relaxation time for shear stress can be given by $t_t = \frac{\eta}{G}$. For an incompressible material, $E = 3G$, the value of t_t and t_n will be equal.

This model was introduced by MAXWELL to describe the behaviour of materials such as pitch. The model has also been applied to the study of earth's mantle. It is essentially the rheid model introduced by CAREY (1953).

9.3.6. Firmo-viscous or KELVIN or VOIGT Material

When a spring and a dashpot are connected in parallel (Fig. 9-6a) it is called the KELVIN model. The material having the properties of this model is called a firmo-viscous, KELVIN or VOIGT material. When such a material is subjected to a normal tensile stress, the strain in the elastic element (spring) must be equal to the strain in the viscous element (dashpot). The total stress acting will be the sum of the stress distribution in the two elements such that

$$\begin{aligned} \sigma &= \sigma' + \sigma'' \\ &= E\varepsilon + 3\eta\dot{\varepsilon} \end{aligned} \tag{9.14}$$

At the moment of application of stress ($t = 0$), total strain $\varepsilon = 0$, and if the applied stress σ_0 is constant; integrating Eq. 9.14,

$$\varepsilon = \frac{\sigma_0}{E}\left(1 - e^{-\frac{Et}{3\eta}}\right) \tag{9.15}$$

The term $e^{-\frac{Et}{3\eta}}$ approaches 0 as $t \to \infty$. Hence a firmo-viscous material attains a strain exponentially equal to the Hookean material after a very large lapse of time (Fig. 9-6b).

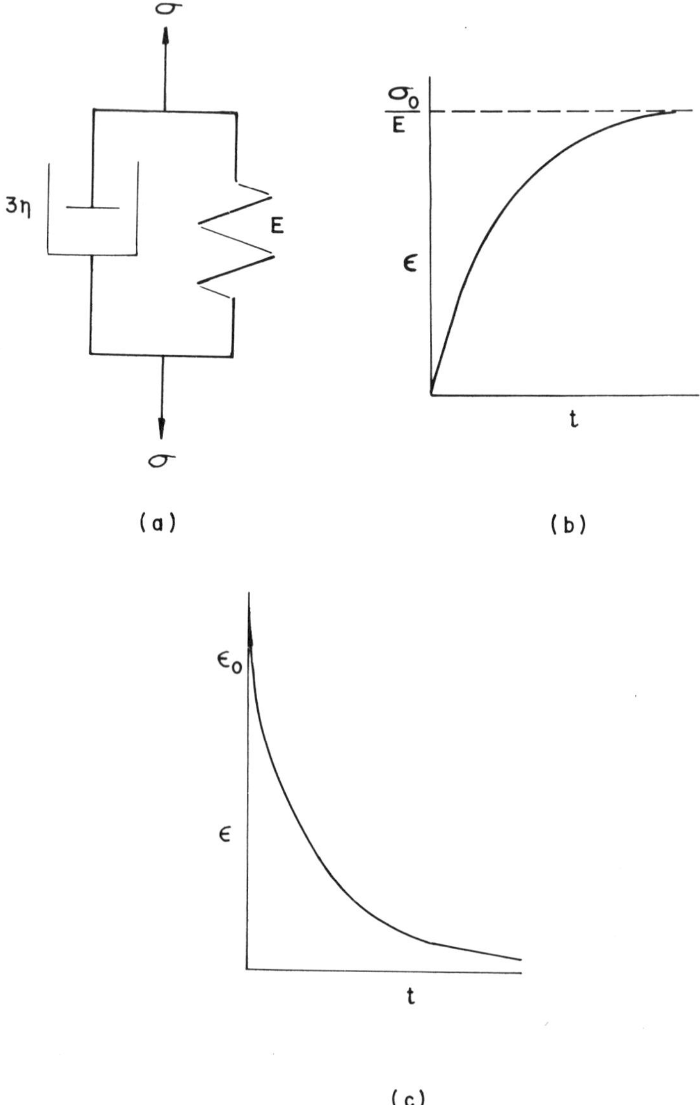

Fig. 9-6. KELVIN material
(a) Model
(b) Strain-time curve under constant stress
(c) Strain-time curve upon removal of stress.

If a strain ε_0 is produced in the KELVIN material and then destressed completely, $\sigma_0 = 0$, Eq. 9.14 becomes

$$\frac{d\varepsilon}{\varepsilon} = -\frac{E\,dt}{3\eta}$$

and on integration

$$\varepsilon = \varepsilon_0 e^{-\frac{Et}{3\eta}} \tag{9.16}$$

Thus the strain decays exponentially from its initial value ε_0 and the relaxation time for strain to fall to $\frac{\varepsilon_0}{e}$ is given by $\frac{3\eta}{E}$.

9.4. Rheological Models—Complex Behaviour

The mechanical behaviour of real materials can rarely be represented by any of the simple models described above. Their behaviour often requires applying two or more simple models in series or parallel. Some complex rheological models are described here.

9.4.1. Generalised KELVIN Model

The KELVIN model described above suffers from the drawback that it does not give an instantaneous elastic strain so often observed in the creep curves of materials. The deficiency is overcome by introducing a spring E_2 in series with the simple KELVIN model (Fig. 9-7a). This is called a generalised KELVIN model or NAKAMURA model (NAKAMURA, 1949). From Eqs. 9.1 and 9.14 it follows:

$$\sigma = E_2 \varepsilon_2$$

also

$$\sigma = E_1 \varepsilon_1 + 3\eta_1 \dot{\varepsilon}_1 \tag{9.17}$$

and

$$\varepsilon = \varepsilon_1 + \varepsilon_2$$

Eliminating ε_1 and ε_2 from Eq. 9.17,

$$\sigma = E_1\left(\varepsilon - \frac{\sigma}{E_2}\right) + 3\eta_1\left(\dot{\varepsilon} - \frac{\dot{\sigma}}{E_2}\right)$$

or

$$3\eta_1 \dot{\sigma} + (E_1 + E_2)\sigma = E_2(3\eta_1 \dot{\varepsilon} + E_1 \varepsilon) \tag{9.18}$$

On the application of a stress σ_0 at time $t=0$, the instantaneous strain produced will be

$$\varepsilon_2 = \frac{\sigma_0}{E_2} \tag{9.19}$$

and after a time t, strain at sustained stress σ_0 will be

$$\varepsilon = \frac{\sigma_0}{E_2} + \frac{\sigma_0}{E_1}\left(1 - e^{-\frac{E_1 t}{3\eta_1}}\right) \tag{9.20}$$

Fig. 9-7b represents the strain-time character of such a material.

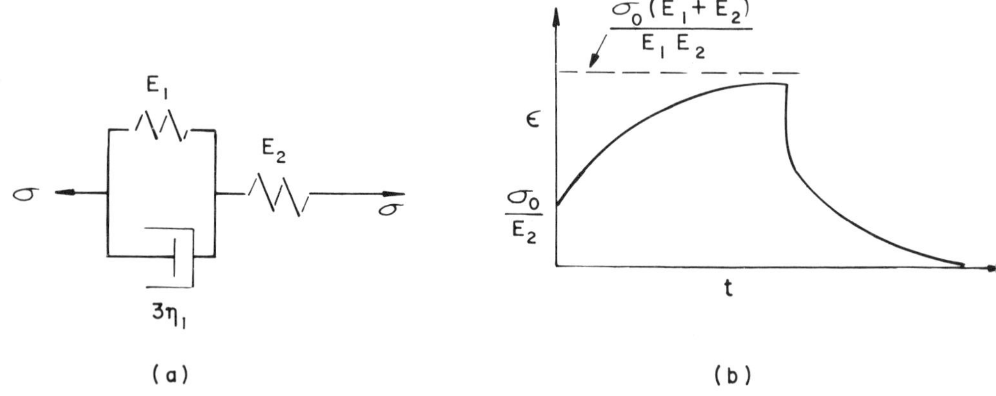

Fig. 9-7. Generalised KELVIN material
(a) Model
(b) Strain-time curve.

9.4.2. BURGER'S Model

By combining a MAXWELL model and a KELVIN model in series, a standard BURGER's model is obtained (Fig. 9.8a). When such a model is subjected to a constant stress σ_0 maintained for a time t, the strain produced will be the sum of the strains in the MAXWELL and KELVIN units and can be given by sum of Eqs. 9.10 and 9.15,

$$\varepsilon_t = \varepsilon_m + \varepsilon_k$$

where ε_t = strain in BURGER's model

ε_m = strain in MAXWELL unit and

ε_k = strain in KELVIN unit.

or

$$\varepsilon_t = \frac{\sigma_0}{E_m} + \frac{\sigma_0 t}{3\eta_m} + \frac{\sigma_0}{E_k}\left(1 - e^{-\frac{E_k t}{3\eta_k}}\right) \tag{9.21}$$

The Eq. 9.21 shows that the total strain under a stress is made up of 3 parts, i.e. instantaneous elastic strain $\left(\frac{\sigma_0}{E_m}\right)$, secondary creep $\left(\frac{\sigma_0 t}{3\eta_m}\right)$ and primary crepp $\frac{\sigma_0}{E_k}\left(1 - e^{-\frac{E_k t}{3\eta_k}}\right)$.

If at this time t, stress is released, the elastic part will immediately recover, the primary creep* will be recovered gradually and the secondary creep will remain permanently in the MAXWELL unit (Fig. 9-8b).

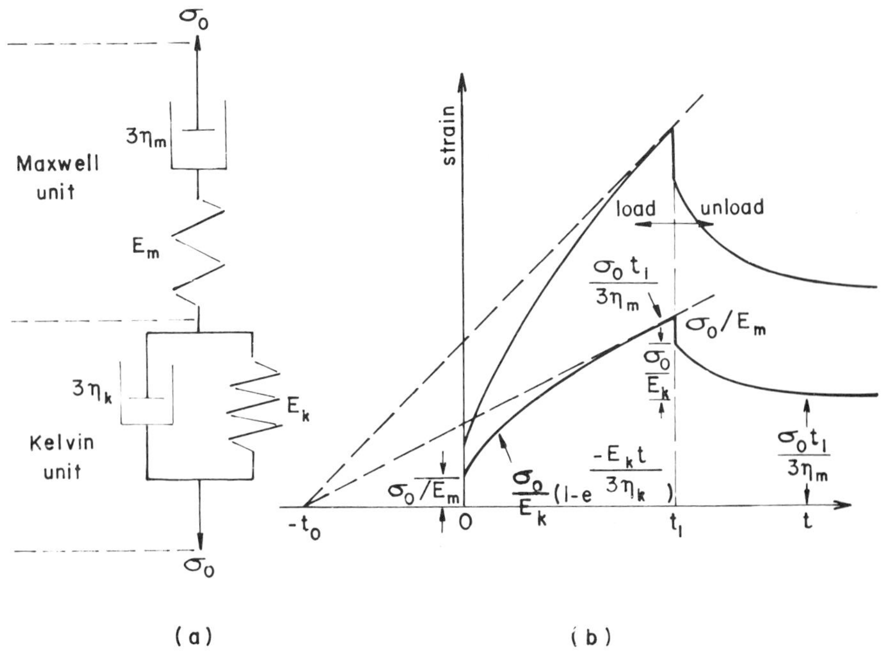

Fig. 9-8. BURGER's material
(a) Model
(b) Strain-time curve under constant stress.

* The terms primary creep and secondary creep are explained in Section 9.6.

It can be seen that secondary creep lines originate from a common point on the time axis ($-t_0$), as in the MAXWELL model, the value of this point is given by

$$t_0 = -3\eta_m\left(\frac{1}{E_m}+\frac{1}{E_k}\right) \tag{9.22}$$

This model gives the simplest representation of rock materials (NISHIHARA, 1957a and b; MATSUSHIMA, 1960; PRICE, 1964; HARDY, 1967a and b).

The complex behaviour of real materials very often requires the introduction of a number of, rather than two, constants (elastic modulus and coefficient of viscosity). This is achieved by parallel coupling of a number of MAXWELL units or by series coupling of a number of KELVIN units (Fig. 9.9). Each unit has its constants E_{mi}, η_{mi} and E_{ki}, η_{ki}, the subscript i taking all values from 1 to n consecutively, where n represents the number of units.

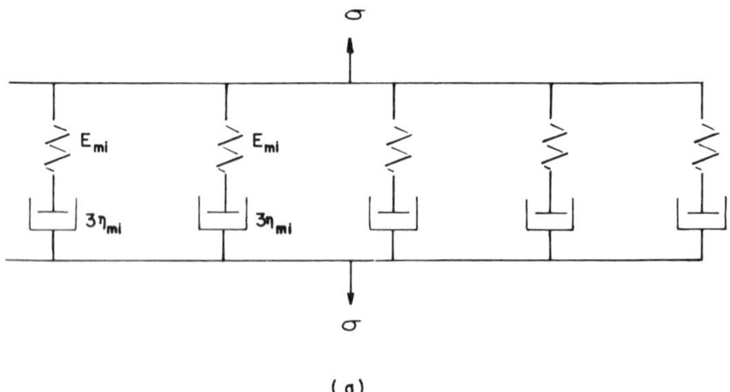

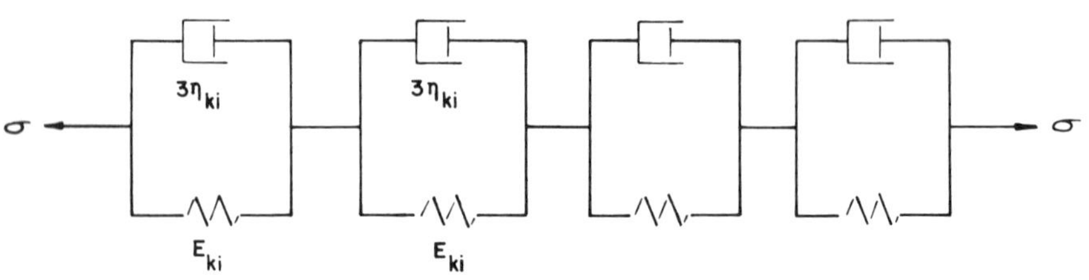

Fig. 9-9. (a) MAXWELL units coupled in parallel
(b) KELVIN units coupled in series.

The coupling of MAXWELL units in parallel is important for materials with different particle sizes or for materials consisting of particles with different activation energies in their respective equilibrium positions. These give rise to phases of different relaxation times in materials with unordered structure.

MAXWELL models with multiple relaxation times can reproduce various primary creeps, the larger the number of units coupled, the more extended the period of primary creep and better the approximation to given creep curves.

TERRY and MORGANS (1958) and HARDY et al (1969) have shown that a generalised form of the BURGER's model (Fig. 9.10) is more suitable for coal and a large number of rocks. In the generalised model, a series of KELVIN units is incorporated together with a MAXWELL unit. When a constant stress σ_1 is applied for any time t, the strain ε_t is given by,

$$\varepsilon_t = \sigma_1 \left[\frac{1}{E_m} + \frac{t}{3\eta_m} + \Sigma_1^n \frac{1}{E_k} (1 - e)^{\left(\frac{-E_k t}{3\eta_k}\right)} \right] \tag{9.23}$$

where E_m, η_m are the parameters for the MAXWELL unit and E_k, η_k are the parameters for the KELVIN unit.

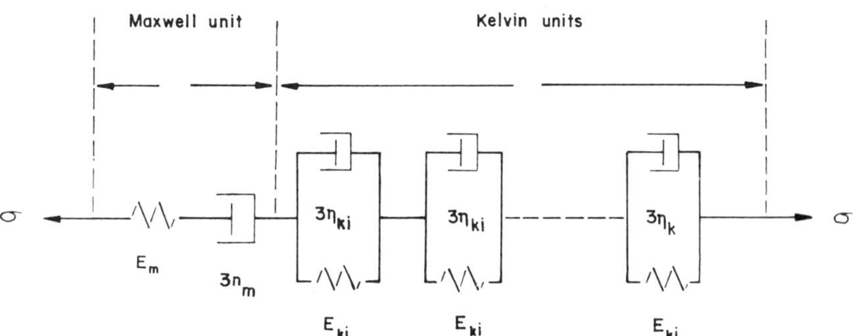

Fig. 9-10. Generalised BURGER's model

The type of time-strain curve (Fig. 9-20) can be represented by a BURGER's model (Fig. 9-8a). Alternatively, provided the applied load is very high, this can be represented by a complex BINGHAM model (Fig. 9-11). To determine which of these models is better able to represent the rheological behaviour of rock, it is necessary to produce a series of time-strain curves in which the successive tests are conducted at different constant levels of stress. By determining the slope of the secondary creep curve for different loads and plotting this information, one may distinguish between the two complex models. The results of such a test in uniaxial compression on coal-measure siltstone are given in Fig. 9-12, which shows that this rock body behaves as a BINGHAM body. About a dozen of rock types tested by PRICE (1969) gave similar data.

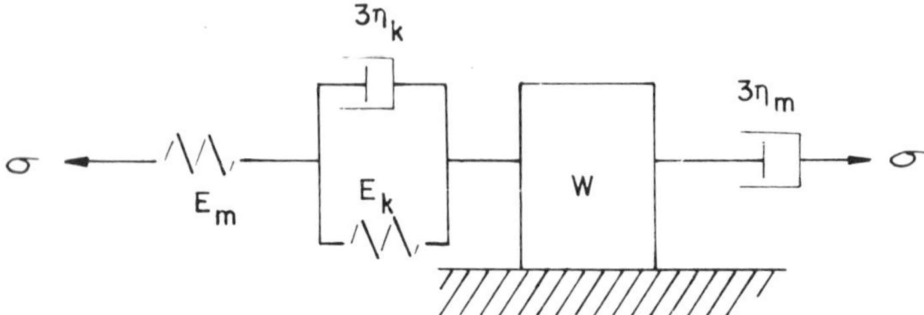

Fig. 9-11. BINGHAM model

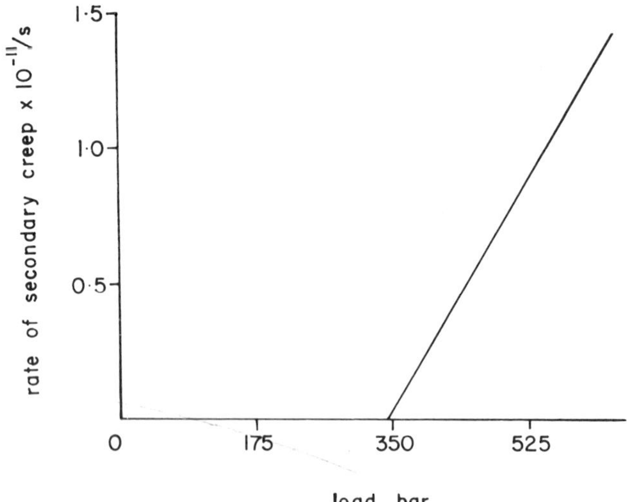

Fig. 9-12. Rate of secondary creep against applied load for siltstone (after PRICE, 1969).

The analysis of the behaviour of a multiple element model even for a constant load is quite cumbersome due to complex redistribution of stress in the units. For example, in a multiple MAXWELL unit combination, on immediate removal of stress, units with short relaxation times release their loads more quickly but with long relaxation times release their loads slowly and hence carry some of the load released by other units. As such, their load will first increase and then decrease while in units with short relaxation it continuously decreases to zero. In springs it increases continuously to the final value associated with full stress.

Sliding weight type model (BINGHAM model) has been used to describe hysteresis loop in cyclic load tests (NISHIMATSU and HEROESEWOJO, 1974).

Some different types of models that have been used to explain the behaviour of rock materials are given in Table 34.

TABLE 34

Rheological models for different rock types

Rock type	Rheological model	Behaviour	Source
Solid hard rock	HOOKEAN	Elastic	OBERT and DUVALL, 1967
Rock in general	KELVIN	Visco-elastic	SALUSTOWICZ, 1958
Rock at greater depths	MAXWELL	Visco-elastic	SALUSTOWICZ, 1958
Rock loaded for short interval	Generalised KELVIN or NAKAMURA	Visco-elastic	NAKAMURA, 1949
Sandstone, limestone and other rocks	HOOKE's model parallel with MAXWELL	Visco-elastic	RUPPENEIT and LIBERMANN, 1960
Coal	Modified BURGER	Visco-elastic	HARDY, 1959; BOBROV, 1970
Dolomite, claystone, anhydrite	HOOKE's model + number of KELVIN models in series	Visco-elastic	LANGER, 1966, 1969
Carboniferous rocks	KELVIN	Visco-elastic	KIDYBINSKI, 1966
Carboniferous rocks	ST. VENANT parallel with NEWTONIAN	Elasto-viscoplastic	LOONEN and HOFER, 1964

9.5. Creep Test Equipment

Creep studies have been conducted at constant load conditions and under different stress systems (bending, torsion, compression, tension and shear). There has been a number of measurements in compression while work in bending, torsion and tension has been very limited. No work has so far been reported on creep of rocks in direct shear.

Creep test equipment must satisfy two conditions in addition to the type of loading; firstly, the equipment must be of sufficient capacity and be capable of maintaining constant load on the specimen over long periods that may be days, months and even years; and secondly, the strain measuring instrument should be stable and must not show any drift during long periods of testing.

It is essential that any drift in the measuring instrument is less than the minimum increment in measurement.

For loading purposes a dead weight or weight through levers, springs, or hydraulic systems can be applied. In spring loaded systems, it is essential that stiffness of the spring is less than that of the specimen. In hydraulic systems pneumatic regulator-boosters are helpful in producing and controlling high pressures for long term tests.

Dial gauges [0.001 mm (0.00004 in) sensitivity], optical-mechanical gauges (TUCKERMAN, MARTENS) and linear variable differential transformers (LVDT) are quite suitable for long term tests. Electromechanical transducers, especially strain gauges bonded to the specimens, are less stable over long periods and are not recommended unless special care is taken in their selection and in bonding and curing of the glue.

For accuracy in creep testing, the loading and the strain-measuring system must be checked for creep by placing a dummy specimen of hardened steel and loading it to a value at least 1.5 times the maximum load to which the specimens are to be subjected.

Many materials and particularly the testing equipment using pneumatic regulator-boosters are affected by changes in temperature, humidity, dust etc. It is essential that these be controlled within limits.

A timing device which actuates in case of failure of specimens must be incorporated in the equipment in order to indicate the period elapsed between loading and failure.

The test environment must be rigidly controlled for the duration of the test period. It is important that temperature of the surroundings is kept within ± 1 °C and humidity $\pm 5\%$. The test specimens should preferably be stored in the test room for at least two weeks in advance in order to acquire stability.

When tests are required to be conducted under conditions different from the ambient conditions, special chambers are required where the necessary conditions can be produced. Special furnaces are available for high temperature work. The natural humidity of the specimens can be maintained by enclosing the specimen either in a rubber membrane or by waxing or coating the surface with some other impervious layer; however, care must be taken that the coating material does not alter the properties.

Some different types of apparatus used in practice are described in detail.

a) Testing in Bending

PHILLIPS (1931) conducted creep tests in bending. The apparatus (Fig. 9-13a) consists of a cast-iron table (AB), 61 cm (24 in) long, slotted from end to end and supported by two cast-iron legs (CD). The rock beam (EF) rests on two adjustable knife-edge supports (GH), which are clamped to the table from below. The beam specimen is loaded by placing loads on the hanger (J), which is supported from a collar on the beam. (K) is an optical lever. Fig. 9-13b shows a section of the apparatus at the point of application of the load. The optical lever (K) has two point supports on a fixed platform (N), the third point support passes through a hole in the collar (M) and rests on the upper surface of the beam.

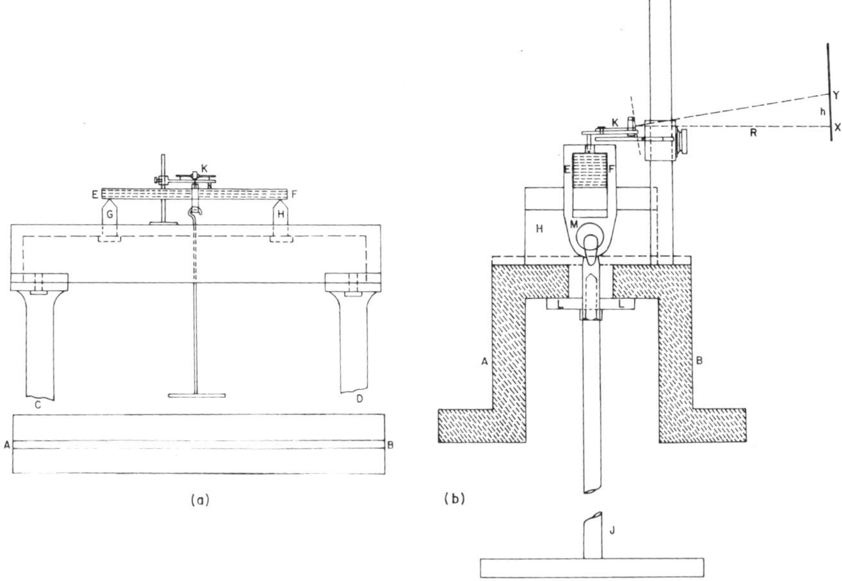

Fig. 9-13. (a) Elevation of apparatus and plan of table
(b) Cross-section of beam-testing apparatus, showing the optical lever in position.
(after PHILLIPS, 1931).

On loading, any deflection of the beam at the point of application of the load will cause the mirror to tilt slightly. As the deflection of the beam is very small, it is magnified to a considerable extent by the arrangement shown. A telescope, to which a vertical scale is attached, is fixed at a distance R away. In consequence of the deflection of the beam, the mirror support moves through a distance d, and the mirror turns through an angle ϕ. If r is the distance between the front and back point supports of the optical lever, $d = r \sin \phi$. At the same time, the line of sight reflected from the mirror moves a distance h from X to Y on the scale through an angle 2ϕ, and Y is the scale reading on the telescope. Then $h = R \tan 2\phi$; and $\dfrac{d}{h} = \dfrac{r \sin \phi}{R \tan 2\phi}$.

Since ϕ is very small

$$d = \frac{rh}{2R}$$

The equipment is calibrated by placing the deflecting support of the optical lever on a micrometer and noting the scale readings h for different micrometer readings d. The system is simple and stable.

b) Testing in Torsion

LOMNITZ (1956) conducted creep tests in torsion and the apparatus used by him is shown in Fig. 9-14. The equipment consists of a steel frame which can accommodate 45.72 cm (18 in) long specimens. The torsional couple is applied at the lower end of specimen through a 38.1 cm (15 in) diameter pulley. The strain measuring system is optical and consists of two facing mirrors one of which is mounted on the specimen while the other is stationary. A light source throws light onto the movable mirror. The light is reflected back and forth between the mirrors and is magnified (10–12 reflections). The reflected light is optically concentrated to a luminous point and is recorded photographically on a standard seismograph drum having a trace speed of 1.5 cm/min (0.59 in/min). The equipment is capable of giving a reading where every mm on drum corresponds to shear strain at the specimen surface of 1.4×10^{-6} radian.

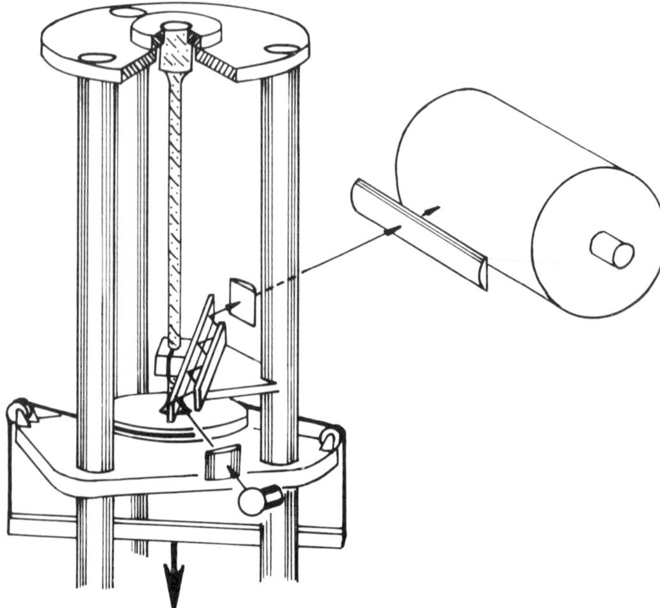

Fig. 9-14. Schematic diagram of apparatus for creep in torsion (after LOMNITZ, 1956).

c) Testing in Compression/Tension

MISRA and MURRELL (1965) carried out creep tests on rocks at high temperatures. The schematic diagram of the apparatus used by them is shown in Fig. 9-15. The apparatus has two levers which give an overall ratio of approximately 100 : 1. Hardened steel knife edge pivots (K1, K2, K3 and K1′, K2′, K3′) are used and the pull exerted by the weight pan is converted to compressive load by the load reverser (Fig. 9-15b). The spring and jack assembly below the reverser helps to adjust the position of reverser while on load so that lever number 2 is horizontal. At this position, the rod connecting the spring and jack assembly to the load reverser is locked to the platform through a nut. The equipment is calibrated using a proving ring or a metal specimen fitted with wire resistance strain gauges. In the latter case a subsidiary calibration is made using a testing machine of known accuracy. L.V.D.T. or Martens mirror extensometer is used for the measurement of strain. The heating furnace encloses the specimen and temperature is measured by thermo-couples.

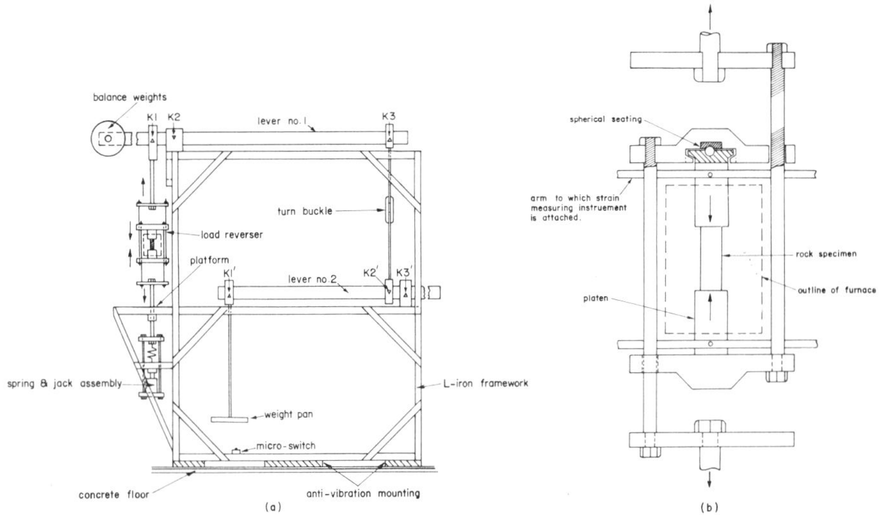

Fig. 9-15. Schematic diagrams of
(a) Compression creep testing machine
(b) Load reversing device
(after MISRA and MURRELL, 1965).

The use of compound levers is probably the best and most inexpensive method of obtaining constant load over a long period. But, for specimens of large diameter and high strength, the loading frame tends to become so bulky and the time taken to apply the weights becomes so large that the initial strain measurements can be lost. An alternative arrangement is to load the specimen through plate springs which have a horizontal force-displacement characteristics but this arrangement has the disadvantage that the equipment requires periodical resetting of the load.

More suitable and simple equipment uses the liquid/air interface coupled with hydraulic cylinders serving as a pressure intensifier and control unit. Such equipment has been extensively used (HOOPER, 1967; WAWERSIK and BROWN, 1971, 1973).

WAWERSIK and BROWN (1971, 1973) describe a testing system for conducting creep tests in uniaxial compression/tension and under confining pressure. Hydraulic loading machines (Fig. 9-16) were used for all creep tests in uniaxial compression. Each loading machine consists of a precision, double acting hydraulic cylinder, four tie rods, and cross heads. The hydraulic rams have capacities of 222 kN (50,000 lbf) at 21 MPa (3000 lbf/in^2) pressure. Accurate alignment of the loading frames is assured by carefully machined spacer tubes which fix the relative positions of the upper and lower machine cross heads.

Fig. 9-16. Loading apparatus
(after WAWERSIK and BROWN, 1971).

The displacement of and the force generated by the loading pistons of the hydraulic rams is controlled by regulating the hydraulic pressure in each cylinder. The hydraulic pressure in turn is generated optionally by means of an air pump or by means of a high-pressure gas accumulator. To limit the piston travel during tests which may remain unattended over long periods, solenoid valves are placed between the pressure source and each hydraulic cylinder.

These solenoid valves are actuated and release the cylinder pressure when the piston travel has reached a predetermined amount.

To maintain a constant force, a high pressure gas accumulator is pressurised in series with each hydraulic cylinder for subsequent load control. As soon as the desired load (stress) value is reached, the pressure source is isolated from the loading system and the load is then held constant to within plus 178 N (40 lbf) by means of the gas accumulator.

For experiments under confining pressure, axial load is generated by means of an actuator in a standard loading frame. Confining pressure is applied in the pressure vessel depicted in Fig. 9-17. It accommodates specimens 2.54 cm (1 in) in diameter by 7.00 cm (2.75 in) long. The specimens are loaded between two pistons of indentical diameter and jacketed in loosely fitting polyurethane membranes. To test water-saturated specimens at reasonably constant (atmospheric) pore pressure, all specimens are vented through small holes in the loading pistons.

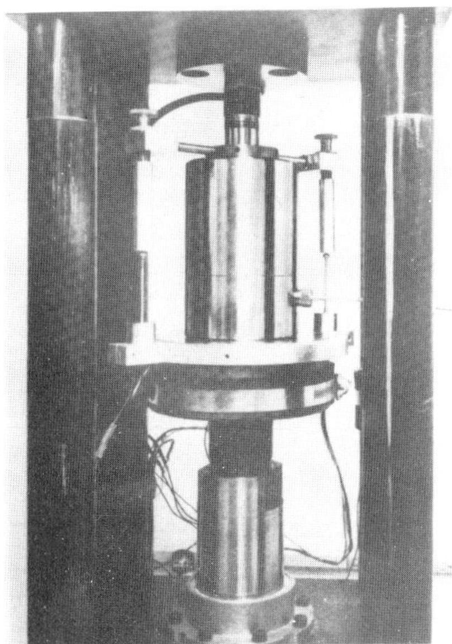

Fig. 9-17. Pressure vessel
(after WAWERSIK and BROWN, 1973).

Pressure is applied by means of a hand pump and/or by means of a screw-driven pressure regulator. To maintain constant pressure in long-term creep experiments, the action of the pressure regulator is servo-controlled using the output of a specially fabricated differential pressure gauge as the feedback signal.

The tension apparatus consists of a double acting hydraulic cylinder, Spherco rod-end bearings and a ring-shaped load cell (Fig. 9-18). Each specimen is cemented with epoxy resin to the two cylindrical end pieces. The end pieces are then connected to the fixed and movable cross heads of the testing machine by means of a short, threaded stud. Two accumulators which are placed in series with the actuator serve to apply and maintain a constant load in creep experiments. Specimen alignment is of course a problem. Hydraulically operated grips or cast metal grips are useful for accurate alignment (MTS, 1975).

Fig. 9-18. Tension apparatus
(after WAWERSIK and BROWN, 1973).

Force is measured either by means of strain gauge mounted load cells or with a temperature compensated pressure gauge. Axial strain is measured either indirectly by monitoring end-to-end specimen displacements (compression tests) or directly by means of strain gauges (tension experiments). In uniaxial compression and in tension, the lateral or tangential strain is measured with the aid of strain gauges. In confining pressure tests, the lateral (radial) strain is determined from measurements of the integrated radial specimen deformation. This technique was first employed by CROUCH (1970). It is based on the fact that the volume adjustments to the confining pressure medium (needed to maintain a fixed confining pressure during radial deformation of specimen) directly proportional to the (average) radial specimen strain. In this system, the volume adjustments to the confining pressure medium are determined simply by tracking the motion of the servo-controlled pressure regulator. This indirect technique for monitoring radial strains in confining pressure experiments is very economical. The method is particularly useful where strain gauges cannot be employed, for example on wet specimens or when large strains are encountered (see Section 6.5.5).

Servo-hydraulic testing machines are commercially available which permit maintenance of constant load or constant strain (for relaxation work) very conveniently with very high accuracy.

In geological studies, the interest on creep is mostly centred at high temperatures (up to 1000 °C) and ultra high pressures (up to 20 kbars and more). Testing at such high pressures requires extreme care and special designs. The usual triaxial cells are not suitable for the purpose. At such high pressures and temperatures, sealing of the fluid becomes a serious problem. High pressures (up to 20 kb) can be achieved using a graphite furnace embedded in a solid medium as used by GRIGGS (1967). Such a system is convenient for qualitative work and study of structural changes. Accurate deformation measurements require a fluid pressure medium or the use of inert gases (CO_2—up to 800° C at 5 kb, GRIGGS, TURNER and HEARD, 1960; argon—1000° C and 10 kb, HEARD and CARTER, 1968). BAIDYUK (1967), BUTCHER, CORNISH and ROUFF (1964) give details of different techniques for measurement of creep strain at high hydrostatic pressures.

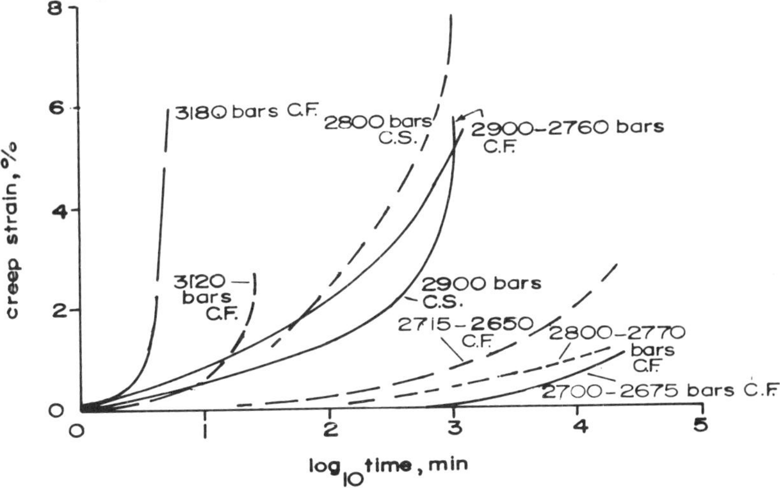

Fig. 9-19. Creep strain versus log time curves for wet Solenhofen limestone tested at 20° C, 600 bars confining pressure in the constant force (C. F.) and constant stress (C. S.) testing modes. Differential stress or range of differential stress is given with each curve. (after RUTTER, 1972).

Since most of the tests on rocks are conducted at low total strain, the influence of change in the cross-sectional area of the specimen with deformation has not been accounted for. RUTTER (1972) has shown that if the load on the specimen is monitored keeping the stress constant taking the centre of the specimen as a reference (which is an approximation since the lateral deformation of the specimen throughout its height is not uniform), the creep curves obtained for Solenhofen limestone are considerably different than those obtained in the conventional test. This (Fig. 9-19) is particularly true at higher differential pressures. Also depending upon the rock type there may be suddenly very large variations for an increase in differential pressure by a few percent.

Two approaches are usually adopted for determining the creep parameters of a rock. In the first case specimens are subjected to different load values ranging from 30 to 90% of the short term strength; deformation is measured with time. In the second case, the same specimen is loaded to different load values starting from a small load and increasing it in steps. The deformation is measured with time at each step.

TABLE 35

Observations of acceleration of transient creep of rocks at 15° C and in compression

(after MURRELL, 1969)

Rock	Ratio[a] σ/σ_c	Duration of experiment (minutes)	Final strain (10^{-6})	Creep acceleration observed after: time (minutes)	strain[b] (10^{-6})
Anhydrite	0.465	1600	84	—	—
	0.64	1500	125	600	94
	0.89	235	156	12	65
		296	c		
Dolomite	0.56	2000	116	—	—
	0.68	2200	204	1000	185
Sandstone	0.63	700	18	—	—
(Darley	0.73	650	28	—	—
Dale)	0.86	645	41	—	—
Marble	0.61	2100	156	—	—
(Lime-hillock)	0.775	2200	230	1000	200
Micro-	0.55	1500	188	—	—
grano-diorite	0.65	1500	228	?1000	?200[d]
Peridotite	0.54	3000	124	—	—

a σ_c = uniaxial compressive strength
b approximate figures
c specimen fractured
d evidence of acceleration uncertain

In both these cases, it is important that deformation is measured until the steady state creep is reached. In practice it is very difficult to say when this steady state creep has been reached and if this will not be followed by tertiary creep with the passage of time. It is dependent upon the long term strength of the material and that is very difficult to determine. It is possible that acceleration will take place even though steady state creep has been in existence for a long time. Table 35 gives some idea of the times measured.

9.6. Time-Strain Curve

An idealised creep curve for rock at constant stress is given in Fig. 9-20. The curve consists of four sections:

1. The instantaneous elastic strain (AB),
2. primary or transient creep (BC),
3. secondary or steady-state creep (CD) and
4. tertiary or accelerated creep (DE).

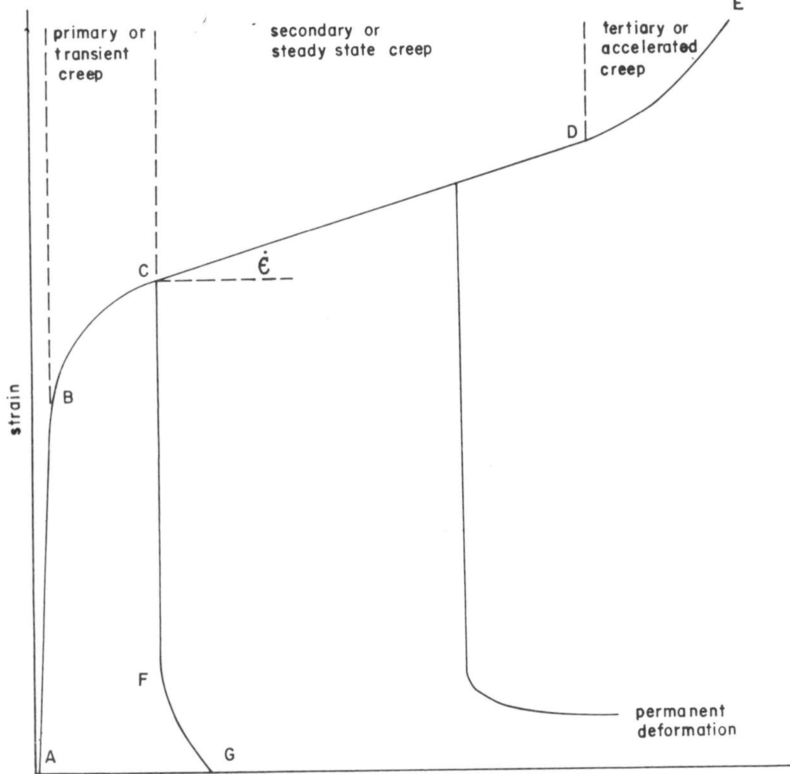

Fig. 9-20. Theoretical time-strain curve at constant stress.

The instantaneous elastic strain (AB) occurs immediately on application of load and is followed by primary creep in which the rate of strain decreases with time. Primary or transient creep is sometimes called delayed elastic deformation for if the specimen is destressed at the point before (C), there will be instantaneous recovery (CF) followed by delayed elastic recovery (FG).

If the stress is allowed to persist beyond (C), secondary creep sets in where the strain-rate is constant and the specimen undergoes permanent deformation. If specimen is destressed at any point between the range (CD), the permanent deformation takes place. The amount of permanent deformation is governed by steady strain rate $\dot{\varepsilon}$ and the time elapsed.

If the specimen is not destressed at the point (D), the rate of strain increases at an increasing rate and the specimen eventually fails. This phase (DE) is called tertiary creep. This creep phase, if once initiated, is usually so short that failure cannot be arrested. The tertiary creep does not represent a pure deformation process but progressive damage and is fundamentally different from the two preceding stages of deformation. All the three stages of creep are observed only at proportionally high stress levels.

Most of the work on time-dependent strain has been conducted on primary and secondary creep phases only and the tertiary creep phase has not been investigated in appreciable detail, because it is probably not important from the design point of view.

The constant stress creep curve for a number of materials can be expressed by a general relationship of the form,

$$\varepsilon = \varepsilon_e + \varepsilon(t) + At + \varepsilon_T(t) \tag{9.24}$$

where ε = total strain

ε_e = elastic strain

$\varepsilon(t)$ = function expressing transient creep

At = linear function of time t representing steady state creep, the constant A depends upon test conditions, and

$\varepsilon_T(t)$ = function expressing tertiary creep.

A number of empirical equations have been developed to express transient and steady state creep but no simple equation has been found so far for tertiary creep.

For primary creep (transient creep) the simplest curve fitting expressions are power laws of the form,

$$\varepsilon(t) = At^n \tag{9.25}$$

ANDRADE (1910, 1914) used the one-third power law to describe the creep in soft metals in the form,

$$\varepsilon(t) = At^{1/3} \tag{9.26}$$

COTTRELL (1952) has shown that the equation,

$$\varepsilon(t) = At^{n};\ o < n < 1 \tag{9.27}$$

is applicable to a variety of cases. The constants are dependent upon stress level, temperature and structure of the material.

Certain other investigators have suggested logarithmic or exponential laws to represent primary creep. GRIGGS (1939) suggested

$$\varepsilon = B \log t \tag{9.28}$$

HARDY (1967 a) used BURGER's model which gives primary creep in the form

$$\varepsilon = B(1 - \exp(-Ct)) \tag{9.29}$$

In general, the laws of primary creep can be divided into two main groups, the exponential law

$$\dot{\varepsilon}_t = A_1 \exp(-A_2 t) \tag{9.30}$$

and the power law

$$\dot{\varepsilon}_t = B_1 t^{B_2} \tag{9.31}$$

where $\dot{\varepsilon}_t$ = strain rate at a particular time t and A_1, A_2, B_1, B_2 = constants. A_2 and B_2 determine the rate of decrease of strain rate $\dot{\varepsilon}_0$ at time $t = 0$. The values of strain rate $\dot{\varepsilon}_t$ decrease with elapse of time depending upon the values of A_2 and B_2 in both the cases. A_2 and B_2 therefore can be regarded as strain hardening parameters.

The Eqs. 9.30 and 9.31 can be compared further by taking their logarithms and writing them in the form

$$\log \dot{\varepsilon}_t = \log A_1 - A_2 t \tag{9.32}$$

$$\log \dot{\varepsilon}_t = \log B_1 + B_2 \log t \tag{9.33}$$

Differentiating the Eqs. 9.32 and 9.33 with respect to time t, we get

$$d(\log \dot{\varepsilon}_t)/dt = -A_2 \tag{9.34}$$

$$d(\log \dot{\varepsilon}_t)/dt = B_2/t \tag{9.35}$$

From Eq. 9.34, it is clear that the rate of change of logarithm of strain rate for the exponential law of primary creep remains constant with time while that for a power law (Eq. 9.35), the rate of change of the logarithm of strain rate is inversely proportional to time i.e. at longer times, the rate of change is less rapid than at short times and will be smaller than the exponential law. At $t=0$, the rate of strain hardening in a power law of primary creep is infinite. At intermediate values of t, both laws behave similarly.

Most of the investigators have used the graphical method as a technique to determine the form of the law. Recently, numerical methods involving computers have been used (CRUDEN, 1969). These methods have definite advantage for exactly deciding the type of law which fits the data best. CRUDEN (1971 a) has shown that power law of transient creep fits most of the data published on rocks and minerals satisfactorily without the addition of any steady state creep component. The exponential law even with a component of the steady state creep is inadequate for the description of that data.

9.7. Creep in Rocks and Minerals

MURRELL and MISRA (1962), ROBERTSON (1963) and LAMA (1972) have given comprehensive reviews on creep in rocks and minerals.

Some of the earliest work has been done by MICHELSON (1917, 1920) who measured creep in some rocks subjected to torsion and expressed his results in the form

$$\phi_t = F_1 + F_2\,(1-\exp(-\alpha t^{1/2})) + F_3\,(t)^{\beta} + F_4 \tag{9.36}$$

where ϕ_t = angle of twist at any time t
F_1, F_2, F_3 and F_4 are of the form $F = cT\exp(hT)$ and
T = applied torque.

The ten constants in the MICHELSON equation are α, β, c_1, c_2, c_3, c_4, h_1, h_2, h_3 and h_4. The value of β is approximately 1/3.

MICHELSON found that part of the strain

$$F_2\,(1-\exp(-\alpha t^{1/2})) \tag{9.37a}$$

is recoverable. He did not find steady state or secondary creep.

PHILLIPS (1931, 32) conducted tests on creep in bending of coal measure rocks by line loading of 25.4 cm (10 in) long beams at the centre. He found that "time yield" for 24 hours increased with the increased load up to a load of 71.2N (16 lbf) but the rate of yield from 62.3 to 71.2N (14 to 16 lbf) was less than for previous loads while there was no "time yield" at the end of 24 hours although 9.6 divisions which were recovered, were observed during the first few hours. This is, however, abnormal and PHILLIPS attributes it to the inability of the material to adjust itself further and that fracture condition may be imminent.

EVANS (1936) conducted tests on marble, granite, sandstone, slate and concrete and found that part of the creep is recoverable. According to him creep and recovery could be described by the equation

$$\varepsilon = A(1-\exp(B-Ct^{n})) \tag{9.37b}$$

where A, B, and C = constants and $n \simeq 0.4$.

GRIGGS (1936, 1939, 1940) carried out creep measurements on limestone, talc, shale and on many mineral crystals in compression and observed both primary and secondary creep. He expressed the results in the form

$$\varepsilon = \varepsilon_e + B\log t + Dt \tag{9.38}$$

where ε_e = elastic strain
$B\log t$ = primary creep and
Dt = secondary creep
($D = \frac{\sigma}{3\eta}$ in MAXWELL's model)
B and D = constants and depend upon stress σ.

LOMNITZ (1956) tested slender cylindrical specimens of granodiorite and gabbro of 45.72 cm (18 in) length, 2.22 cm (7/8 in) central diameter in torsion at room temperature and atmospheric pressure. He found that creep for constant torque up to one week can be represented by the equation

$$\gamma(t) = \frac{\tau}{G}\,[1 + q\,ln(1+\alpha t)] \tag{9.39}$$

where $\gamma(t)$ = total shear strain in radians
τ = constant shear stress
G = rigidity modulus
t = time in seconds
q = constant and
α = coefficient such that $\alpha t > 1$.

The above relationship is applicable only in the range of small strains. The results obtained by him are given in Fig. 9-21. The value of q for recovery is only half that for creep under load. The values (P) on the curves are the surface stresses.

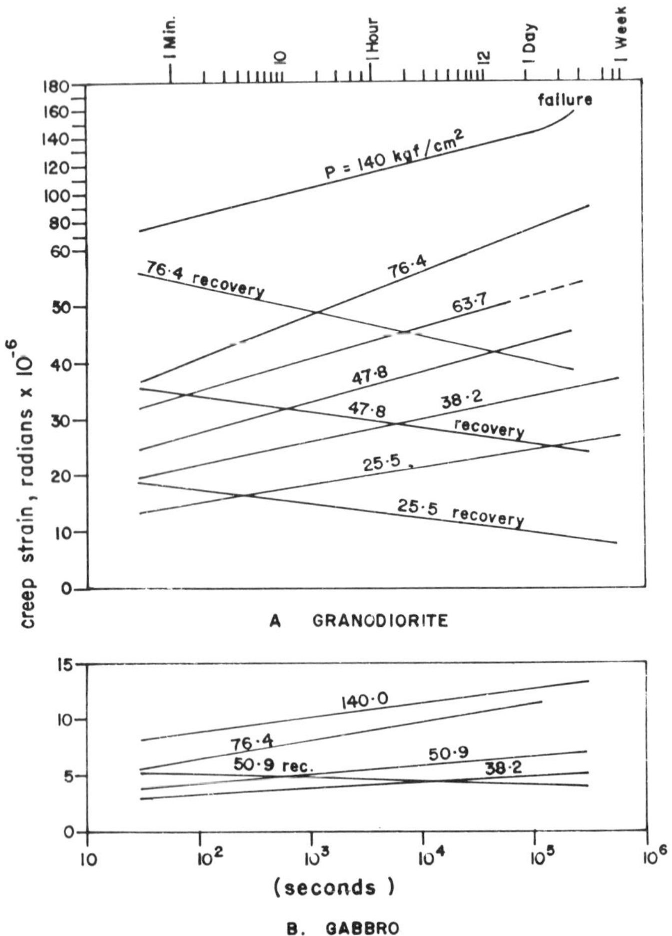

Fig. 9-21. Creep curves for granodiorite and gabbro at room temperature and atmospheric pressure (after LOMNITZ, 1956).

MATSUSHIMA (1960) conducted creep tests on igneous rocks at atmospheric pressure and room temperature and found that longitudinal creep for granite could be represented by the relationship

$$\varepsilon = A_0 + A_1 e^{-\alpha_1 t} + A_2 e^{-\alpha_2 t} + A_3 e^{-\alpha_3 t} + B \log t + Ct \tag{9.40}$$

where t = time

A, B & C = constants

$\alpha_1, \alpha_2, \alpha_3$ = reciprocals of which represent the retardation times of KELVIN's model and are of the order of 10, 10^2, 10^4 seconds respectively.

HARDY et al (1969) studied creep in Indiana limestone, Crab Orchard sandstone and Barre granite and fitted the curves to the generalised BURGER's model, using a computer programme developed for the purpose, and found that curves fit excellently to the model which incorporates two KELVIN units.

SINGH (1970, 1975) conducted creep tests in compression on a number of rocks including Sicilian marble and Wombeyan marble. He observed all the 3 stages of idealised creep. Specimens were loaded beyond their yield point limit. The results for Sicilian marble are given in Fig. 9-22. It was found that primary stage axial creep could be represented by the equation

$$\varepsilon = 0.4395\, t^{0.4929} \times 10^{-4} \tag{9.41}$$

and secondary creep

$$\varepsilon = (0.1817\, t - 0.8022) \times 10^{-4} \tag{9.42}$$

for this marble.

The following equation was found to represent both primary and secondary creep:

$$\varepsilon = 0.4205 t^{0.5044} \times 10^{-4} \tag{9.43}$$

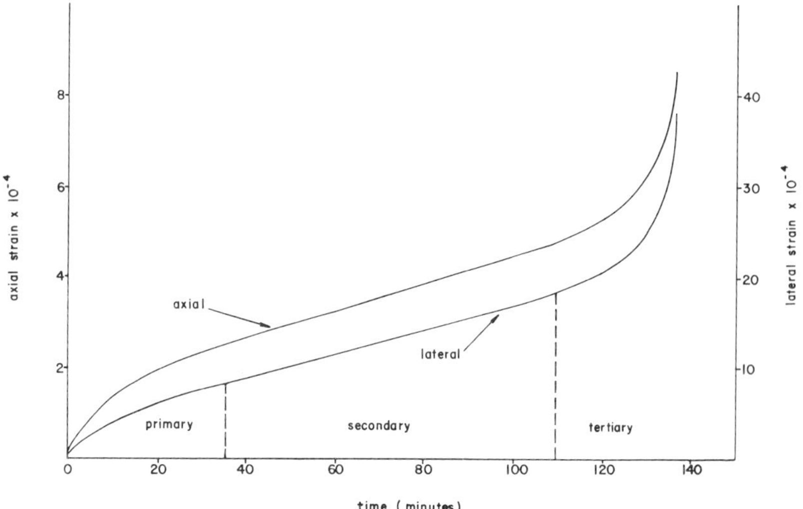

Fig. 9-22. Axial and lateral creep curves of a Sicilian marble specimen at a constant pressure of 12,500 lbf/in^2 (after SINGH, 1970).

The equation of the lateral creep curve up to the secondary stage region is

$$\varepsilon = 1.1610t^{0.569} \times 10^{-4} \tag{9.44}$$

WAWERSIK and BROWN (1971) conducted creep tests in compression on three rock types, namely, Westerly granite, Nugget sandstone and Tennessee marble. They also observed all the three stages of idealised creep. For Westerly granite and Nugget sandstone, strains are proportional to t^n during primary creep and to t during secondary creep where t denotes time. Specifically, it was found that

$$\varepsilon_I = 10^c t^n \tag{9.45}$$

$$\varepsilon_{II} = 10^{c'} \tag{9.46}$$

where ε_I and ε_{II} denote the strains during primary and secondary creep, respectively. c and c' are functions of stress. In general, the creep of marble appeared to lack the secondary creep stage even where time-dependent deformations terminated in creep fracture. Primary creep was approximated by an expression of the form

$$\varepsilon_I = 10^c \log t \tag{9.47}$$

HOBBS (1970) conducted incremental creep tests on different coal measure rocks such as siltstones, shales, mudstones, limestones and sandstones of strength varying from 61 to 206 MPa and subjected to compressive stresses to 26.4 to 41.4 MPa with tests extending to 40,000 minutes. He found that longitudinal strain-time behaviour can be approximated by the following

$$\varepsilon = \frac{\sigma}{E_c} + g\sigma^f t + K_2 \sigma \log(t+1) \tag{9.48}$$

where E_c = mean incremental modulus
σ = stress level
t = times in minutes
g, K_2 and f = constants.

The values of these parameters are given in Table 36. The strain rates predicted by this relationship are less than observed when t is small. HOBBS also found that after unloading the instantaneous recovery was followed by time-dependent recovery which almost ceased after about 15,000 minutes. The primary creep recovery for Ormonde siltstone and Portland limestone was similar to primary creep under load and recovery rate was higher than the comparable creep rates under load.

TABLE 36 Values of parameters of Eq. 9.48
(after HOBBS, 1970)

Rock type	E_c GPa	σ_c MPa	g	f	K_2	Ratio of primary/ secondary creep
Ormonde siltstone	17.7	68.9	1.04×10^{-61}	32.2	3.07×10^{-6}	4.0
Lea Hall sandstone	9.6	47.6	1.12×10^{-35}	18.4	6.8×10^{-6}	2.0
Hucknall shale	9.5	58.7	—	—	7.56×10^{-6}	1.0
Portland limestone	40.4	85.0	0	—	3.34×10^{-6}	10.0

TABLE 37 Rheological characteristics of various types of rock salt
(after ALBRECHT and LANGER, 1974)
$\left(\text{a measured, b estimated, }^{+}\text{) equivalent Viscosity } \eta = \frac{\Delta\sigma}{3\,\delta\varepsilon/\delta_t}\right)$

Rocktype	Plasticity η_{pl} bar · sec	KELVIN-Viscosity η_i bar · sec	Retardation time $T_{ret} = \frac{\eta_i}{\mu_1}$	loading time $t = t_B$ (test)	$h = \frac{t_B}{T_{ret}}$
Carnallit	$_a2.4\times10^7$	1.4×10^7	1 min	10 min	10
Hartsalz	$_a$— —	1.4×10^{12}	140 days	911 days	6.5
Gips/ Anhydrit	1×10^{12}	4×10^7	8 min	60 min	7.5
	$_a1\times10^{12}$	1×10^{10}	3 days	20 days	6.7
	— —	10	4×10^{-5} sec	3×10^{-4} sec	7.5
rock salt	$_b7\times10^{12}$	7×10^{10}	116 days	—	—
rock salt	$_b3\times10^{12}$	9×10^{11}	1 year	—	—
rock salt	$_a3.5\times10^{13\,+})$	—	—	3 days	—
rock salt (artificial)	$_a4\times10^{12\,+})$	—	—	3.5 days	—

An extensive amount of work has been done on the creep of rock salt and different models have been proposed. SCHUPPE in 1963 (HOFER, 1964) suggested a combination of KELVIN and NEWTONian models placed in series. Extensive work done by DREYER (1972) showed that creep of halite can be represented by a power law of the form Eq. 9.26.

Many authors represent the creep behaviour of rock salt by KELVIN models and their results are given in Table 37.

MENZEL and SCHREINER (1975) have examined the stability safety of excavations in carnallitite and found that the following relationship can be used successfully; for constant stress

$$\varepsilon = \frac{\sigma}{E} + [(1+S)\ B]^{\frac{1}{(S+1)}} \cdot \sigma^{\frac{\gamma}{(S+1)}} \cdot t^{\frac{1}{(S+1)}} \qquad (9.49)$$

for constant stress rate

$$\varepsilon = \frac{\sigma}{E} + [(1+S)\ \mathrm{B}]^{\frac{1}{(S+1)}} \cdot \sigma^{\frac{\gamma}{(S+1)}} \cdot \left(\frac{t}{1+\gamma}\right)^{\frac{1}{(S+1)}} \qquad (9.50)$$

The values for various carnallitite salts are given in Table 38. The factor $\frac{\gamma}{(S+1)}$ as the exponent of stress in certain carnallitites varies and lies between 2 and 5.

TABLE 38

Values of constants for carnallitite (t in hours, σ in kgf/cm^2)

(after MENZEL and SCHREINER, 1975)

	E kgf/cm^2	S	γ	$B \times 10^{48}$
Trummer carnallitite Südharz				
A	60000	2	15	5.2
B	65000	2	15	18
C	40000	2	15	5.9
Trummer carnallitite Werra	50000	2	15	12
Weißer carnallitite Werra	50000	2	15	6.6
Roter carnallitite Werra	62000	2	15	42
Carnallitite Calvorden Scholle	50000	2	15	4.7

9.8. Factors Influencing Creep

The time-dependent deformation of rocks is dependent upon a number of factors. Some important ones are discussed here.

9.8.1. Nature of Stress

Most of the work on creep has been done under uniaxial compression, but some workers have also used bending (PHILLIPS, 1931, 1932, 1948; POMEROY, 1956; WACHTMAN and MAXWELL, 1956; PRICE, 1964), torsion (MICHELSON, 1917, 1920; LOMNITZ, 1956) and uniaxial tension (WAWERSIK and BROWN, 1973).

Specimens subjected to bending and torsional stresses show a given creep rate at comparatively lower stresses as compared to specimens subjected to compressive stresses.

TABLE 39
Physical properties and coefficients of logarithmic creep equation for the three rocks under uniaxial tension and uniaxial compression
$[\varepsilon_t = A\sigma + B \log t + Ct]$
(after CHUGH, 1974)

Physical property	Indiana limestone	Tennessee sandstone	Barre granite
Unconfined tensile strength, lbf/in^2	750 – 800	850 – 900	1200 – 1300
YOUNG's modulus in tension, lbf/in^2	4.8×10^6	1.5×10^6	6.9×10^6
POISSON's ratio in tension	0.27	0.09	0.11
Unconfined compressive strength, lbf/in^2	7000 – 11,000	18,000 – 20,000	29,000 – 31,000
YOUNG's modulus in compression, lbf/in^2	5.2×10^6	5.1×10^6	10.5×10^6
POISSON's ratio in compression	0.30	0.20	0.22
Apparent porosity, %	17.2	6.7	0.7
Permeability, darcy	5.2×10^{-2}	3.7×10^{-4}	5.0×10^{-6}
Moisture content, %	0.02	0.50	0.32

Rock	Uniaxial tension				Uniaxial compression			
	Stress lbf/in^2	A $\times 10^{-6}$	B $\times 10^{-6}$	C $\times 10^{-6}$	Stress lbf/in^2	A $\times 10^{-6}$	B $\times 10^{-6}$	C $\times 10^{-8}$
Tennessee sandstone	205.61	0.670	8.91	—	9,383	0.212	20.37	—
	312.15	1.080	14.02	—	12,725	0.182	17.45	—
	418.00	1.330	22.44	—	17,709	0.161	19.37	—
					19,011	0.189	32.04	3.6
					20,662	0.164	30.09	5.3
Barre granite	881.04	0.163	1.33	—	11,565	0.103	1.81	—
	991.58	0.161	1.40	—	13,394	0.096	3.39	—
	1103.23	0.157	1.68	0.9	21,650	0.089	3.41	—
	1213.78	0.142	2.39	3.3	26,674	0.082	4.42	0.6
					28,200	0.076	4.42	1.3
Indiana limestone	203.0	0.228	—	—	6,392	0.185	4.31	—
	308.05	0.225	0.73	—	7,174	0.192	6.51	—
	412.47	0.227	—	—	7,955	0.188	8.16	—
	516.89	0.215	1.05	—	8,737	0.182	14.15	—
	621.32	0.215	0.66	—				
	725.74	0.230	1.54	—				
	830.16	0.204	—	—				

WAWERSIK and BROWN (1973) carried out creep tests on Westerly granite in uniaxial compression and uniaxial tension. In compression, creep accelerated gradually during tertiary creep providing some warning that creep fracture was impending. In uniaxial tension the tertiary creep was of very short duration. Consequently, creep rate of granite subjected to uniaxial tension was extremely high resulting in sudden failure. CHUGH (1974) has investigated the creep of Indiana limestone, Tennessee sandstone and Barre granite both under tension and compression. The results are given in Table 39. The creep strain in tension in Tennessee sandstone is about 6 times higher than under compression for the same percentage of failure stress. In Indiana limestone and Barre granite the creep rate in tension is only slightly greater than in compression.

9.8.2. Level of Stress

The rate of strain and the value of strain at any particular time depends upon the relative level of stress in relation to the yield limit of the rock. GRIGGS (1939, 1940) carried out detailed measurements on alabaster immersed in water (Fig. 9-23) and Solenhofen limestone and found that the value of the coefficient of viscosity represented in the MAXWELL model depends upon stress level. The value of steady state creep constant D (Eq. 9.38) was found to be

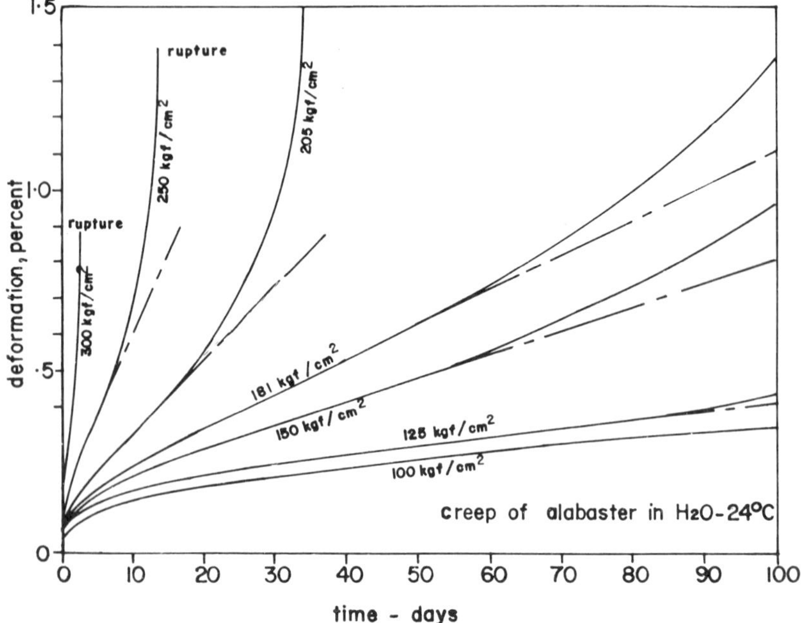

Fig. 9-23. Assembled creep curves of alabaster. Seven specimens, all in distilled water, at the same temperature, but loaded differently (after GRIGGS, 1940).

$$D = \dot{\varepsilon}_s = a \sinh b(\sigma - s) \tag{9.51}$$

where a, b and s = constants and

σ = compressive stress.

Since, according to MAXWELL model $D = \frac{\sigma}{3\eta}$,

$$\frac{dD}{d\sigma} = \frac{1}{3\eta} = \text{constant.}$$

GRIGGS found that for wet alabaster

$$\frac{dD}{d\sigma} = a b \cosh b(\sigma - s) \tag{9.52}$$

which increases as stress increases.

HARDY et al (1969) and SINGH (1970) have also found that strain rate increases with the stress level.

To relate the influence of stress level on strain rate, the following empirical equation has been used (OBERT and DUVALL, 1967):

$$\dot{\varepsilon} = A\sigma^{n} \tag{9.53}$$

The values of n for different rocks are given in Table 40.

TABLE 40

Values of the exponent n in Eq. 9.53 for different rocks

(after OBERT and DUVALL, 1967)

Rock	Maximum strain $\times 10^{-3}$	Maximum stress MPa	($lbf/in^2 \times 10^3$)	Exponent n
Slate	1.0	351.6	(51.0)	1.8
Granite	1.0	344.8	(50.0)	3.3
Alabaster (in water)	5.0	19.3	(2.8)	2.0
Limestone	7.0	137.9	(20.0)	1.7
Halite	10.00	29.0	(4.2)	1.9
Gabbro	0.01	9.7	(1.4)	1.0
Granodiorite	0.20	9.7	(1.4)	1.0
Granite	3.00	96.5	(14.0)	3.0
Shale	3.00	9.7	(1.4)	2.7

LOMNITZ (1956) and ROBERTSON (1963) have given the value of $n = 1$ and PRICE (1964) found a straight line between the steady state creep rate and applied stress for sandstone specimens subjected to bending stresses. WAWERSIK and BROWN (1973) reported that creep in granite was, however, related to stress in a nonlinear manner.

LECOMTE (1965) found that for rock salt a change of axial stress from 6.9 to 13.8 MPa (1000 to 2000 lbf/in^2) increased the creep rate by a factor of 7 at confining pressure of 100 MPa (14,500 lbf/in^2) and 29°C (84.2°F). At higher temperature, 104.5°C (220.1°F) by doubling the axial stress from 3.45 to 6.9 MPa

TABLE 41
Average development of the instantaneous strain and the secondary creep rate in rocks at various stress levels
(after AFROUZ and HARVEY, 1974)

Rock type	σ_c (kN/cm^2)	Stress level (kgf/cm^2)	Instantaneous strain $\times 10^{-6}$	Secondary strain rate ($\times 10^{-6}$/hr)
Air-dried coal	0.56	25.7	1120	6.6
		51.4	1850	11.9
Air-dried underclay	1.90	25.7	274	5.4
		51.4	450	10.0
Water-saturated coal	0.33	6.6	900	5.5
		12.9	1650	8.5
		19.5	2290	11.6
		25.7	3300	13.5
		32.4	3600	19.8
		39.0	4330	22.9
		51.4	5600	30.1
Water-saturated underclay	1.14	19.5	2100	15.4
		25.7	2690	20.0
		39.0	3220	41.1
		51.4	3880	65.5
		58.5	4110	78.2
		77.1	4600	115.0
		102.8	4880	166.6
Water-saturated underclay with clay band	0.95	6.6	1050	28.0
		12.9	1700	50.0
		13.4	1800	54.5
		19.5	2400	102.2
		25.7	2880	195.0
Calcitic limestone	4.55	25.7	2800	2.0
		51.4	4850	2.5
Dolomitic limestone	6.88	25.7	660	1.9
		51.4	940	2.3
Sandstone	9.42	25.7	1800	0.7
		51.4	3300	1.0

(500 to 1000 lbf/in^2), creep rate increased 12 fold. The influence is more marked for coarse-grained specimens of rock salt than fine-grained specimens.

AFROUZ and HARVEY (1974) tested different rocks under air-dried and saturated conditions at different stress levels and found that on doubling the stress level, the secondary creep rate was slightly less than double (increase by about 90%) while instantaneous creep increased by about 50 to 80%. The values obtained by them are given in Table 41.

STAVROGIN and LODUS (1974) have shown that strain rate and stress level are related by the equation

$$\dot{\varepsilon} = \dot{\varepsilon}_0 \cdot e^{\alpha\sigma} \tag{9.54}$$

where $\dot{\varepsilon}_0$ and α are constants. The value of α for potash salts was about 0.045 to 0.047.

9.8.3. Confining Pressure

GRIGGS (1936) when testing Solenhofen limestone under high confining pressures observed creep when a particular differential pressure was reached (Fig. 9-24). This was because confining pressure raised the differential stress at fracture and permitted considerable deformation before failure.

ROBERTSON (1960) studied transient creep of Solenhofen limestone specimens in compression at confining pressures up to 400 MPa (58,000 lbf/in^2) at room temperature. Experiments were conducted only for a short time in a bomb without a provision for maintaining constant pressure. Therefore, the results are only qualitative, but they permit the important conclusion that hydrostatic pressure greatly decreases transient creep rate to be drawn. In ROBERTSON's tests creep rate per unit stress difference decreased a hundred fold for an increase in hydrostatic pressure from 100 to 200 MPa (14,500 to 29,000 lbf/in^2).

LECOMTE (1965) conducted creep tests on artificial polycrystalline rock salt specimens and found that increase in differential stress increased the creep rate considerable. An increase in confining pressure decreased creep rate somewhat.

THOMPSON and RIPPERGER (1964) conducted triaxial creep tests on sylvinite and rock salt. They showed that for these rocks, the constant creep rate in the secondary stage of creep under a uniaxial compressive stress was the same as the creep rate induced by the same magnitude of differential stress applied in triaxial loading. The investigation conducted by SERATA (1968), into the plastic yielding and viscoelastic flow of rock salt in polyaxial state of stress, also appeared to give support to the phenomenon described.

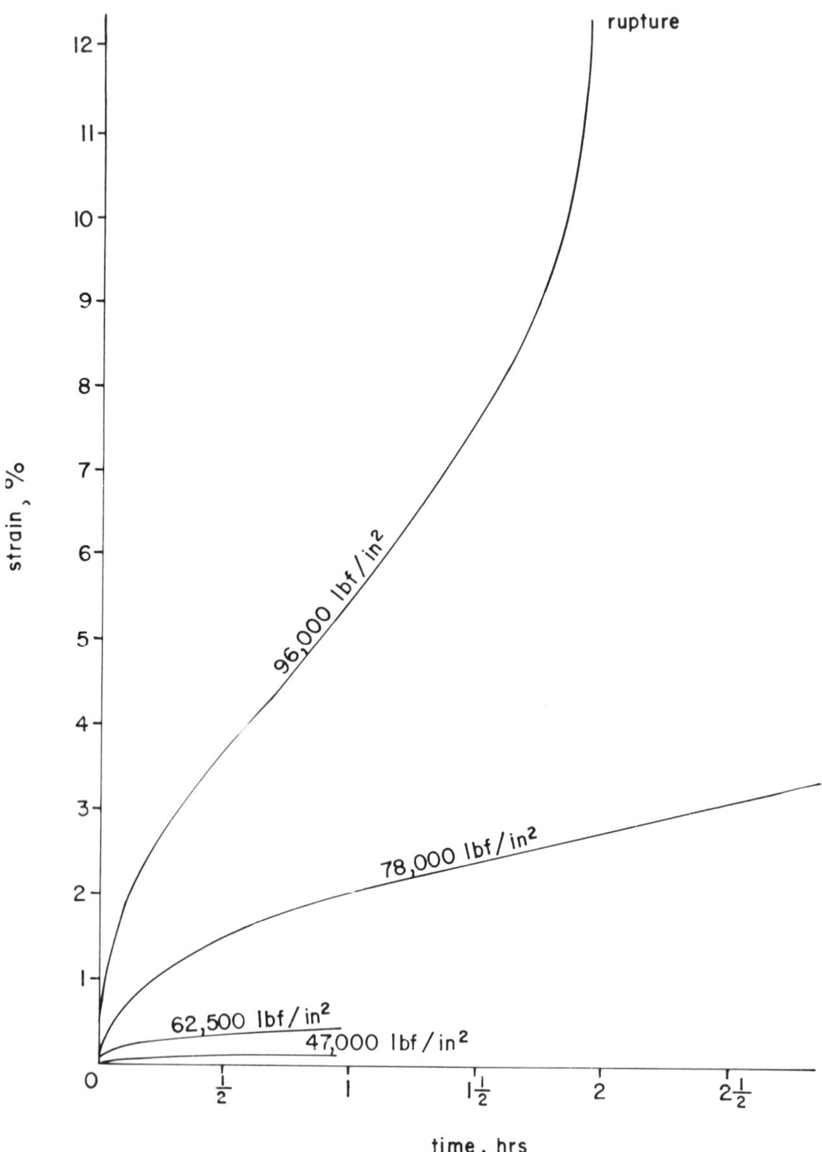

Fig. 9-24. Creep of Solenhofen limestone subjected to different loads and at confining pressure of 1000 lbf/in^2 (after GRIGGS, 1936).

WAWERSIK and BROWN (1973) measured creep on Westerly granite and Nugget sandstone in uniaxial compression and at 6.9 MPa (1000 lbf/in^2), 34.5 MPa (5000 lbf/in^2) and 69.0 MPa (10,000 lbf/in^2) confining pressure. In case of the granite, shear strains and volumetric strains (during creep) increased rapidly with increases in the shear stress; these decreased as the confining pressure was raised. Creep and creep fracture observations on granite indicate that the times-to-failure at fixed shear stresses will increase by orders of magnitude with increasing confining pressure. Hence creep fracture is much less likely to occur under confining pressure than it is in uniaxial compression. This conclusion is based on measured changes of the secondary creep rates as a function of confining pressure.

9.8.4. Temperature

MISRA and MURRELL (1965) tested a number of rocks of different types up to 750°C (1382°F) and found that at temperatures below $0.2T_m$ (where T_m is the melting point of rock in degrees Kelvin), creep strain depends logarithmically on time and is proportional to stress and temperature. At higher temperatures an additional term appears which is a function of t^m, where t is the time and $0 < m < 1$. The value of m increases as temperature increases and is 1/3 at $0.5\ T_m$. This type of creep increases very rapidly with stress (perhaps following a power law, with an exponent ~2 to 3) and with temperature. In the latter case it follows the exponential law

$$\varepsilon \propto \exp\left(-\frac{U}{KT}\right) \tag{9.55}$$

where U = activation energy
K = BOLTZMANN's constant and
T = absolute temperature.

LECOMTE's (1965) work on artificial rock salt showed that at 100 MPa (14,500 lbf/in^2) confining pressure and 6.9 MPa (1000 lbf/in^2) axial pressure, an increase in temperature from 29°C to 104.5°C (84.2°F to 220.1°F) increased creep rate by a factor of 4–5, whereas an increase from 29°C to 198.2°C (84.2°F to 388.7°F) increased the creep rate by a factor of about 22. At lower confining pressures but at the same temperature difference, creep rate increased considerably.

RUMMEL (1969) studied the role of temperature on creep of granite and eclogite at stresses much below strength and at temperatures up to 400°C (752°F) and found that creep is governed by the equation

$$\varepsilon = C_1 \log C_2 t + C_3 t \tag{9.56}$$

where C_1, C_2 and C_3 are constants and C_1 increases linearly with temperature.

LOMENICK and BRADSHAW (1965, 1969) conducted laboratory tests on model salt pillars and reported that deformation of mine pillars at ambient temperatures can be given by the relationship

$$\dot{\varepsilon} = 0.39 \times 10^{-37} \cdot \sigma^3 T^{9.5} t^{0.7} \tag{9.57}$$

where $\dot{\varepsilon}$ = vertical (convergence) strain rate of pillar (in/in/hr)
T = absolute temperature (°K)
σ = average total pillar stress (lbf/in^2) and
t = time (hr).

MCCLAIN (1966) and MCCLAIN and BRADSHAW (1970) investigated the effect of thermal stresses on the convergence of pillars of salt as a result of heating. They found that heating increased the strain rate by a factor of nearly 100. According to them, the high convergence rate is due to added thermal stresses caused in the pillars and that the change in creep rate due to rise in temperature alone is rather small.

OKSENKRUG and SAFARENKO (1975) have found very good agreement between experimental data and the theoretical curve for salt using the following equation

$$\sigma_i = G\varepsilon_i - [1 - \exp(-\bar{a}t^{\bar{\delta}})] g \cdot \varepsilon_i^c \tag{9.58}$$

where $c = \dfrac{G\varepsilon_i'}{G\varepsilon_i' - \sigma_i'}$

$g = \dfrac{G}{c\left[\dfrac{c\sigma_i^{\infty}}{G(c-1)}\right]^{c-1}}$

σ = stress level
ε_i = strain at any time
G = shear modulus
σ_i^{∞} = ultimate time-dependent strength
ε_i' = strain at rupture
σ_i' = stress corresponding to strain ε_i'
$\bar{a}$ and $\bar{\delta}$ = parameters depending upon the stress where
$\bar{a} = a\sigma_i + \gamma$
$\bar{\delta} = \delta\sigma_i + \rho$
$\bar{\delta}$ = constant and
$\bar{a}$ is determined from the above equation.

All the stages of rock deformation up to fracture are described by a single equation (Eq. 9.58) and the time-dependent stress σ_i' has a linear relationship with the strain represented by (Fig. 9.25)

$$\sigma_i' = \frac{c-1}{c} G\varepsilon_i' \tag{9.59}$$

The curves using the Eq. 9.58 are shown in Fig. 9-26. The various parameters were established in laboratory testing and agree very well with the equation.

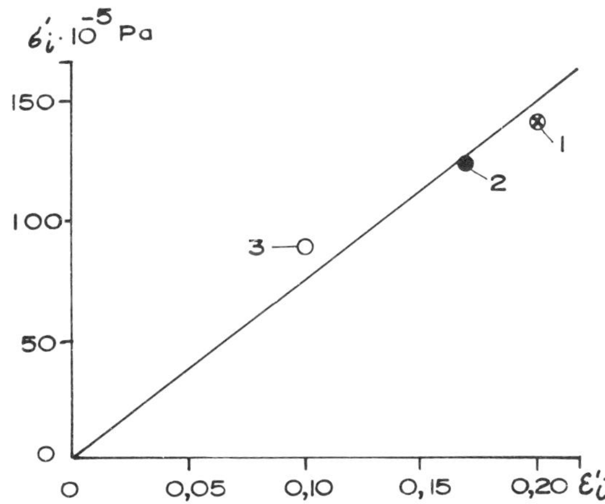

Fig. 9-25. Relation of limiting deformation of rock salt and the acting stresses. (1) Fracture at $t = 235$ h; (2) fracture at $t = 739$ h; (3) fracture at $t = 5770$ h. (after OKSENKRUG and SAFARENKO, 1975).

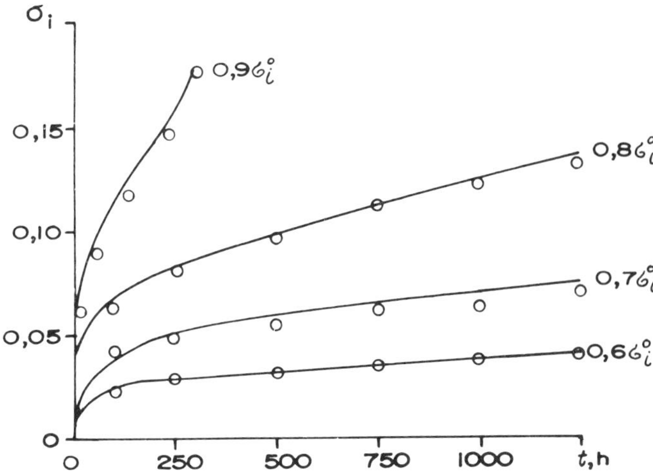

Fig. 9-26. Creep curves of rock salt specimens. Circles represent experimental data; curves are approximations of Eq. 9.58 with $G = 1.27 \times 10^9$ Pa, $c = 1.057$, $g = 13.7 \times 10^9$ Pa, $\bar{a} = 0{,}89 \times 10^{-7}\ \sigma_i + 1.51$, $\bar{\delta} = -0.38 \times 10^{-8}\ \sigma_i + 0.094$ (after OKSENKRUG and SAFARENKO, 1975).

KUZNETSOV and VASHCHILIN (1970) tested sandstone specimens in bending at different temperatures 16°, 43° and 58°C at 70 to 80% of the short term strength and found that both the instantaneous and secondary creep deformations increased with temperature (Fig. 9-27). The creep rate increased markedly during the first three days. When the specimens were unloaded and their strengths determined, specimens III and IV which were creep tested at 16°C showed strengthening by 1.2%, whereas specimens I and II tested at 58°C and 43°C showed strength increase from 25 to 29%—an increase in strength in spite of the creep deformation.

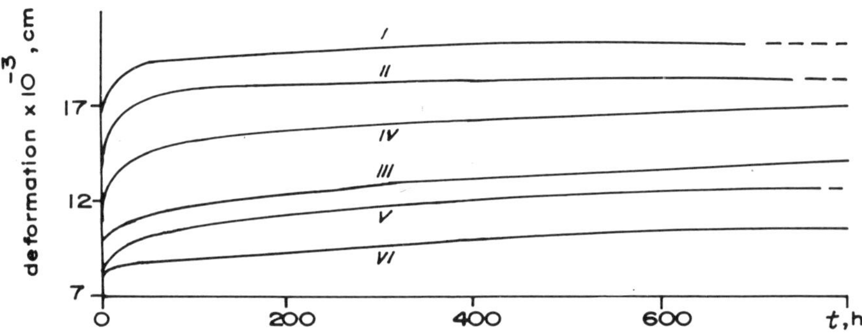

Fig. 9-27. I–III) At stresses $\sigma_t = 0.7\ \sigma_b$ and at 58, 43 and 16°C, respectively; IV=VI) at σ of 0.8, 0.6 and 0.5 of σ_b, respectively, at T=16°C (after KUZNETSOV and VASHCHILIN, 1970).

MURRELL and CHAKRAVARTY (1971) found that rocks such as microgranodiorite, dolerite and dunite remained brittle even up to 1000°C (1832°F) and that creep strain was proportional to a fractional power of the time.

9.8.5. Cyclic Loading

PHILLIPS (1932, 1948) while investigating creep in bending found that when a beam of rock previously loaded by bending in one direction was subjected to the same load in the reverse direction, the rate of creep increased.

ROBERTSON (1955) found that the influence of cyclic loading on Solenhofen limestone was to increase the yield point if the confining pressure was raised after each loading cycle (Fig. 9-28).

LOMNITZ (1956) conducted cyclic loading tests (torsion) on a granodiorite specimen. The specimen was subjected to 2 weeks loading at 4.7 MPa (681.6 lbf/in^2) followed by 2 weeks recovery and 5 cycles of loading and unloading of 4 minutes each at the same stress. He found that the creep constant q (Eq. 9.39) decreased appreciably after the first unloading but after a few repetitions the value of q remained constant and was equal in direct creep and recovery. According to him, a smaller creep rate must be assumed in rocks subjected to alternating stresses than under unidirectional forces of long duration, a conclusion in contradiction to that of PHILLIPS'. More details are given in Section 9.11.

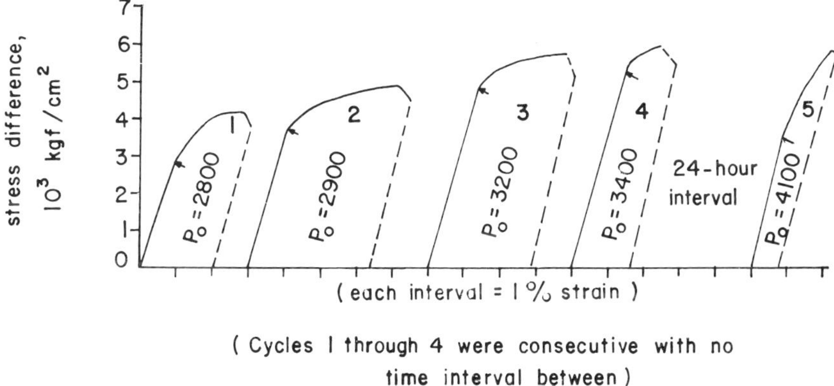

Fig. 9-28. Shift in the yield point of cyclically loaded Solenhofen limestone (after ROBERTSON, 1955).

9.8.6. Moisture and Humidity

PHILLIPS (1931) observed that creep rate increased with wet rocks. The work of GRIGGS (1940) on alabaster immersed in solutions is important. He found that immersion in solutions markedly increases the creep rate (Fig. 9-29). Steady state creep appeared in alabaster immersed in water or dilute HCl solution at stresses which were insufficient for dry alabaster. It has been suggested that dissolution would occur at the most highly stressed regions and deposition at the free regions. The creep would thus take place by recrystallisation. GRIGGS, however, found that for alabaster creep rate in HCl solution was more than in water although the solubility was less in the former than in the latter. GRIGGS also found no difference in creep between specimens soaked in solution for 68 days and those immersed immediately. Creep rate of alabaster specimens increased when the soaking solution was subjected to pressure (Fig. 9-30). It was suggested that the rate of creep of alabaster is not due to simple recrystallisation but may be a function of ionic mobility of Ca^{++} and SO_4^{--} in solvent (GORANSON, 1940).

The work of TURNER (1949) on Yule marble has shown that marble has the capacity to absorb water along grain boundaries (the rate of penetration for Yule marble is 38 mm (1.5 in) in 70 minutes with a porosity 0.15%). Its deformation at different degrees of saturation (GRIGGS and MILLER, 1951;

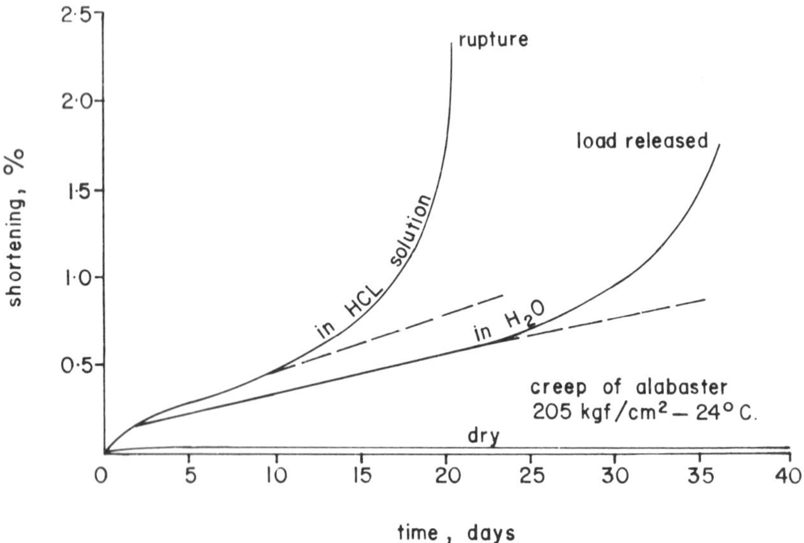

Fig. 9-29. Creep of alabaster in different chemical environments. Three specimens, all loaded to 203 kgf/cm^2, all at the same temperature, but one dry, one in distilled water, and one in dilute HCl. (after GRIGGS, 1940).

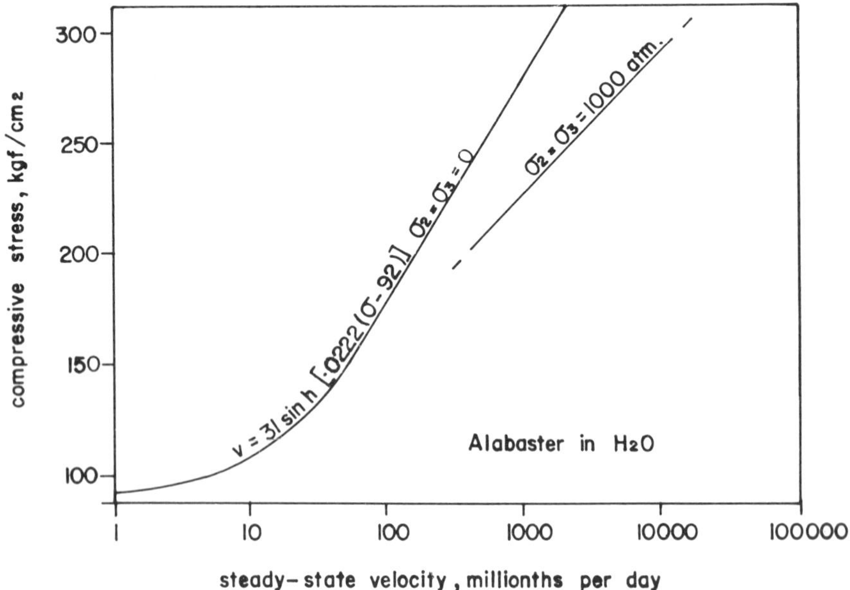

Fig. 9-30. Creep rate of alabaster in water as a function of confining pressure (after GRIGGS, 1940).

GRIGGS et al, 1951, 1953; HANDIN and GRIGGS, 1951) and at a confining pressure of 5000 kgf/cm^2 (490 MPa) (71,000 lbf/in^2) and temperature up to 300°C (572°F) shows that interstitial fluid does not affect lattice deformation directly (as temperature does) but operates through the conditions at the grain boundaries which facilitate external rotation. Microstructural study of initial and deformed material has shown that physico-mechanical effect is absent.

Creep strain measurements on slate and porphyrite under dry and wet conditions by KANAGAWA and NAKAARAI (1970) showed that initially strain rates for wet samples were 2 to 5 times higher, but after about 20 to 100 days the secondary creep rate tended to be more or less the same. The stabilisation period depended upon the rock type, stress level and perhaps the moisture content also. The results are given in Fig. 9-31.

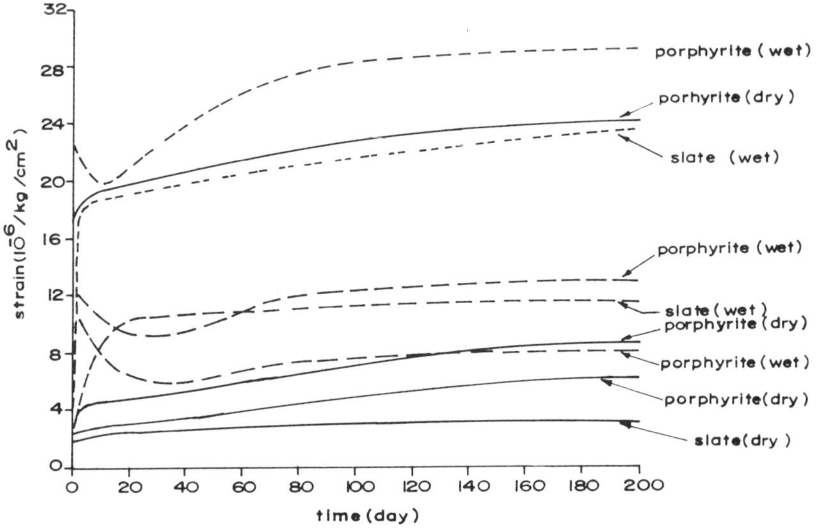

Fig. 9-31. Creep strain under dry and wet condition (after KANAGAWA and NAKAARAI, 1970).

RENZHIGLOV and PAVLISHCHEVA (1970) calculated the viscosity coefficient* of rocks and found that water saturation increases the viscosity a few to several fold depending upon the rock type. The various values obtained by them at different effective load values and test intervals are given in Table 42.

* The viscosity coefficient (η) is calculated using the following relationship:

$$\eta = \frac{\tau - \tau_{\lim}}{\dot{\varepsilon}}$$

where τ = shear stress; $\tau_{\lim}$ = threshold stress when $\dot{\varepsilon} = 0$.

TABLE 42
Viscosity coefficients of rocks
(after RENZHIGLOV and PAVLISHCHEVA, 1970)

State of the rock (tensile strength, kgf/cm²)*		Effective stress, kgf/cm²	Viscosity ($\eta \cdot 10^{17}$), P in a period of (h)			
			0–12	48–96	96–144	240–720
		Sandy-clay shale				
Air-dried	(600)	450	0,24	3,56	8,25	12,9
		250	0,23	4,74	9,55	32,6
Water-saturated	(340)	250	0,026	1,14	2,47	6,75
		150	0,028	2,22	2,22	25,8
		Clay shale				
Air-dried	(860)	625	0,90	18,5	57,0	112,0
		250	0,76	13,0	30,0	103,0
Water-saturated	(600)	375	0,098	3,4	9,6	18,0
		150	0,170	5,3	1,13	22,0
		Sandstone				
Air-dried	(720)	300	0,152	3,12	6,65	10,3
		150	0,180	8,75	17,6	30,2
Water-saturated	(420)	300	0,085	1,15	2,78	4,05
		150	0,106	3,16	3,68	6,45
		Clay shale				
Air-dried	(400)	320	0,049	1,38	1,37	4,60
		250	0,067	1,08	2,16	7,20
Water-saturated	(180)	145	0,008	0,248	0,31	0,24
		100	0,014	0,286	3,45	4,30
		Limestone				
Water-saturated	(1600)	1270	4,3	—	57,0	177,0
		1000	4,5	—	41,0	131,0

*) Mean uniaxial compression strength perpendicular to the lamination.

The work of WAWERSIK and BROWN (1973) has shown that time-dependent deformations in granite and sandstone are augmented by increases in water content. In uniaxial compression the secondary creep rates in air-dry and water-saturated specimens differ by approximately two orders of magnitude. AFROUZ and HARVEY (1974) found that in saturated soft rocks, the creep rate increased by 3 fold in coal and 8 fold in shales (Table 41).

In triaxial tests, creep is related to the magnitude of the effective stresses (i.e. applied stresses minus pore pressure). It is anticipated, therefore, that time-dependent deformations in rock will be influenced by changes in pore pressure regardless of the nature of the pore fluid.

9.8.7. Structural Factors

LECOMTE (1965) studied the effect of grain size on the creep behaviour of rock salt and found that an increase in grain size decreased the creep rate. The results obtained by him are represented in Fig. 9-32. It is seen that increasing the grain size from 0.1 to 0.63 mm (0.004 to 0.248 in) reduced the creep rate by a factor of 2. LECOMTE further found that the addition of 2% of calcium sulphate to rock-salt reduced the creep rate by 3% while a specimen containing 5% calcium sulphate increased the strain-rate by a factor of 2 compared with pure rock-salt specimens.

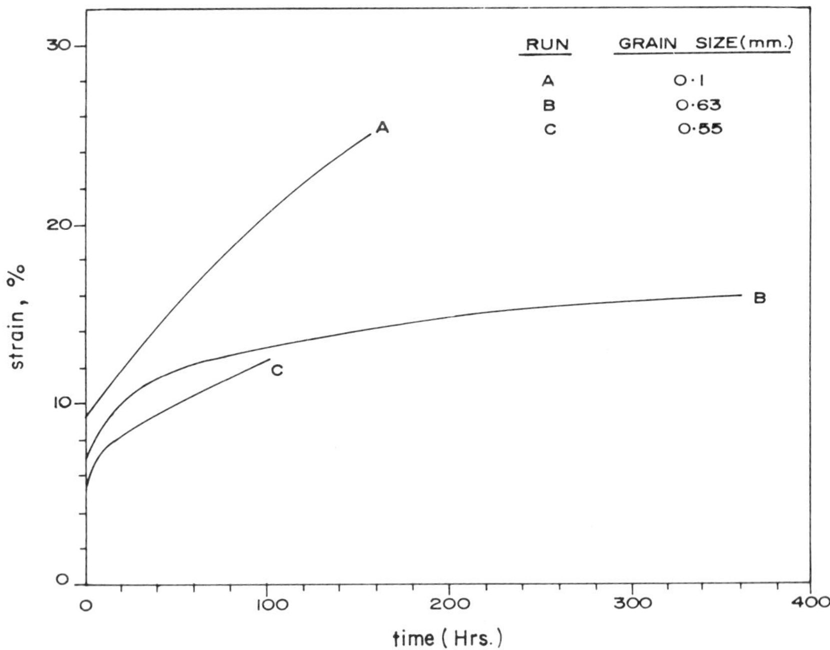

Fig. 9-32. Effect of grain size on the creep of rock salt (after LECOMTE, 1965).

WIID (1966) suggested that rock structure is an important influencing factor in the creep studies. The effect of orientation of grains was studied on Yule marble* (GRIGGS et al, 1953, 1960). It was found that stress required at the commencement of transient creep was much higher when the marble was compressed along the direction of predominant orientation of the optic axis than when the compression

* Yule marble is characterised by a well defined anisotropic structure with a clearly manifested preferred orientation of the optic axis.

was at right angles to the optic axis. Porosity, due to stress concentrations around the pores enhanced creep rate.

There are several other factors such as geometry of the grains, binding material and grain interaction, degree of interlocking, porosity, strain rate etc. which have not been much investigated. The effect of strain rate was investigated by HEARD (1961), at different temperatures (25 to 500 °C) and confining pressure of 5000 atm. They found that experimental data with lower strain rate can be defined by the equation

$$\dot{\varepsilon} = A.10^{k} \exp\left(-\frac{B}{RT}\right) Sh\left(\frac{\sigma}{C}\right) \tag{9.60}$$

where
A, B, C and k = constants of a given rock
R = gas constant
T = absolute temperature and
σ = strength.

9.9. Creep of Rock In Situ

Most of the work on in situ-creep has been done on salt and measurements on other rocks are very much limited.

REYNOLDS and GLOYNA (1961) measured time-dependent deformation of openings in salt mines (closure measurements) and found that creep rate could be represented by

$$\dot{\varepsilon} = 0.03 e^{-0.635t} \tag{9.61}$$

where t = time in years.

On integration of the Eq. 9.61, the total deformation of the openings becomes

$$\varepsilon = \int_{t=0}^{t=\infty} 0.03 e^{-0.635t} \cdot dt = 4.7\%$$

and the time required for 95% of this total deformation becomes

$$\varepsilon = 0.95 \times 4.7\%$$
$$= 0.03 \int_{0}^{t} e^{-0.635t} \cdot dt$$
$$t = 4.7 \text{ years}$$

They further found that creep rate decreased as the age of the tunnel increased. Cavities 5 years old had creep rates 35 times greater than those 10 years old.

It is not known whether this was due to some change in the properties of the salt (strain hardening) or due to decreased stresses around the excavation as a result of destressing. They also found that creep rate at the centre of the openings was about 14% greater than at the wall.

BARRON and TOEWS (1963) carried out measurements of displacements, relative to the shaft axis, of points on the surface of a shaft and within the solid surrounding the shaft in the unlined portion of a shaft in salt above the potash beds. A diametral extensometer and an extensometer to measure the longitudinal deformation of boreholes around the shaft were used. The radial displacement, U_r, for a point in the solid at a radius r from the shaft axis may be expressed in the form

$$U_r = -pB\,(a^2/r)\,\log_{10}(1+bt) \tag{9.62}$$

where B and b = constants
a = shaft radius
t = time and
p = pressure.

The values of the constants for the points 4 ft, 7 ft and 10 ft deep in the wall of the 18 ft diameter shaft were found to be
$b = 1/2.7$
$pB = -7.7 \times 10^{-3}$ (There is insufficient data to separate p and B.). U_r, a and r are in inches and t is the time in days. Surface points on the shaft did not conform to the $1/r$ proportionality, but the time function was the same. They argued that perhaps the change in the properties of the material at the surface affected the results.

HEDLEY's (1967) tests on creep in salt were conducted to determine the stability of pillars in 5 different mines in U.K., Canada and U.S.A. He correlated the results of convergence measurements with stress in pillars calculated from the principle of "extraction ratio proportionality". He found that convergence rate could be expressed by the relationship

$$\dot{\varepsilon} = A\sigma^n \tag{9.63}$$

where $\dot{\varepsilon}$ = convergence rate
σ = stress and
A and n = constants.
The values of the constants A and n from his data are
$\dot{\varepsilon} = 15 \times 10^{-8}\,\sigma^{2.7}$
where $\dot{\varepsilon}$ = convergence rate, in/in/day $\times 10^{-6}$ and
σ = pillar stress, lbf/in^2.

Some work on creep of schistose rocks in situ was conducted by KUBETSKY and UKHOV (1966) and KUBETSKY and ERISTOV (1969). They carried out creep investigations by plate loading of 1 m^2 (10.76 ft^2) area in chloritic mica-schists, graphitic schists, and schistose sandstones. They found that the deformation of the rock mass (S—mean settlement of the surface of the rock mass, under the mean pressure, p) was not linear with the increasing pressure at small values of load ($p = 2.5$ to 10.0 kgf/cm^2) (0.245 to 0.98 MPa) (35.5 to 142.0 lbf/in^2) but might be assumed to be linear at higher load values ($p = 40.0$ kgf/cm^2) (3.92 MPa) (568 lbf/in^2) (Fig. 9-33). They applied the classical BOLTZMANN-VOLTERRA theory of linear creep to describe the time-dependent behaviour. Assuming that the POISSON's ratio of the rock mass remained constant, the settlement of an arbitrary point on the surface of a rock mass $S(r,t)$ under the action of local loading at any time t could be given by the relationship

$$S(r,t) = A\,\frac{1-v^2}{E}\left[p(t) + \int_0^t K(t-\tau)\,p(\tau)\,d\tau\right] \tag{9.64}$$

where A = coefficient representing the shape of the loaded area and the position of the point where the settlement is to be determined. For a rigid rectangular plate of 1 m × 1 m or 0.9 m × 1.1 m (3.28 ft × 3.28 ft or 2.95 ft × 3.61 ft),

$A_{x,\,y=0} = \omega_x 2b;\; A_{x=0,\,y} = \omega_y 2a.$

For a circular plate, $A = \frac{\pi d}{4}$; outside the limit of the plate

$A = \frac{d}{2}\sin^{-1}\left(\frac{d}{2r}\right),$

where

ω_x, ω_y = shape coefficients for absolutely rigid rectangular plate

a, b = half-sides of a rectangular plate, $a > b$
d = diameter of circular plate
r = distance from the centre of the circular plate
E = instantaneous modulus of deformation
$K(T-\tau)$ = creep kernel—for the rocks investigated by KUBETSKY and ERISTOV, its value is given by

$$\frac{1}{E}K(t-\tau) = \lambda_1\theta_1 \exp\left[-\lambda_1(t-\tau)\right] + \lambda_2\theta_2 \exp\left[-\lambda_2(t-\tau)\right]$$

where $\lambda_1, \theta_1, \lambda_2, \theta_2$ are the creep parameters

v = POISSON's ratio of the rock mass.

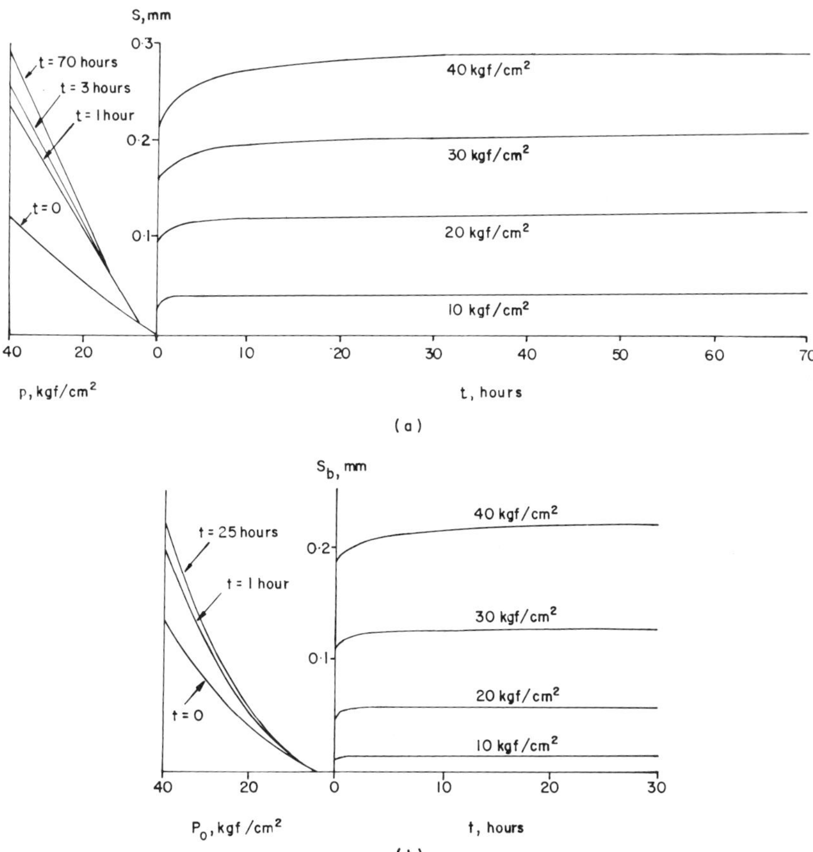

Fig. 9-33. Series of creep and isochronous settlement curves for graphite schist
(a) for increasing loads
(b) for unloading
(after Kubetsky and Eristov, 1969).

For a constant value of p, the settlement with time was given by

$$S(r,t) = [A(1-\nu^2)p/E] + [\theta_1(1-e^{-\lambda_1 t}) + \theta_2(1-e^{-\lambda_2 t})] \qquad (9.65)$$

The values of both E and the various creep parameters were different for both loading and unloading conditions.

HOFER and KNOLL (1971) correlated laboratory test results on small specimens of polycrystalline carnallitite with observations in salt mines in East Germany in areas liable to sudden failures of pillars and noted that if the creep behaviour of the pillars corresponded to the logarithmic law of the type

$$\varepsilon = c_1 + c_2 \ln t \tag{9.66}$$

the pillars could be supposed to be stable but if the creep law corresponded to the exponential law of the type given by Eq. 9.27 and the pillars were unstable.

BRADSHAW, BOEGLY and EMPSON (1964) have carried out creep tests on model salt pillars and found that the creep rate could be described by the general equation

$$\dot{\varepsilon} = 9 \times 10^{-8} \cdot \sigma^{3.1} \cdot t^{-0.6} \tag{9.67}$$

The predicted values of convergence of pillars using this equation in salt mines are in reasonable agreement with actual measurements in mines. They further found that vertical convergence rates in salt mines continued to decrease even after 12 years where the pillar stress was well below the ultimate strength. The transverse pillar expansion rate was considerably lower than the vertical closure rate at the initial stage but after 10 years, this tended to approach the vertical convergence rate.

Creep rate in pillars at different depths is not constant. It greatly decreases in the centre of the pillars reaching 1/100th of that at the edges depending upon the depth and pillar dimensions (MCCLAIN and BRADSHAW, 1970).

WINKEL, GERSTLE and KO (1972) have compared the results of laboratory studies with the time-dependent deformation of horizontal boreholes in potash mines (Carlsbad, New Mexico and Moab, Utah). They found that deformation could not be predicted using the laboratory calculated parameters. Modified parameters had to be used to conform to the observed in situ deformation values.

LANGER (1969, 1974) has given the suitable rheological models for the different types of rocks. According to him, a HOOKEAN model coupled in parallel with a series of KELVIN models can successfully represent the creep of most of the rocks. The rheological constants given in Table 43 are based mostly on in situ plate bearing tests with some obtained from borehole deformation tests.

9.10. Time-Dependent Strength of Rock

Time-dependent strength of a rock can be defined as the maximum stress sustained by the rock at which the failure will just not occur, or just occur no matter how long the force has been applied. This strength has been termed as

TABLE 43
(after LANGER, 1974)

(a) HOOKEAN model parallel with a KELVIN model

Rock	Description	Elastic elements E_H kp/cm²	Elastic elements E_K kp/cm²	Relaxation time T_K min	Plastic element $3\eta_K$ kp/cm²/s	Time of test min
Dolomite	coarse crystalline	200000	64000	14	$5 \cdot 10^7$	600
Dolomite	brecciated	104000	28000	150	$2{,}5 \cdot 10^8$	6000
Dolomite	fractured	96000	12200	27	$2 \cdot 10^7$	1000
Limestone	—	120000	13500	3,5	$3 \cdot 10^6$	60
Bunt sandstone	thick bedded, ⊥ to bedding	100000	25000	21	$3 \cdot 10^7$	300
Bunt sandstone	⊥ to bedding, clay intercalations	60000	6700	18	$7 \cdot 10^6$	300
Basalt tuff	—	7400	1600	5	$5 \cdot 10^5$	20
Metaphyr	tuffig	68000	3800	80	$2 \cdot 10^7$	600
Graywacke shale	clayey, // to cleavage	160000	2000	20	$2{,}5 \cdot 10^6$	600
Claystone	fest	6000	820	1,5	$7 \cdot 10^4$	10
Clay shale	// to cleavage	128000	14500	72	$6 \cdot 10^4$	3000
	⊥ to cleavage	64000	6250	60	$2 \cdot 10^7$	3000

(b) HOOKEAN model in parallel with 3 KELVIN models

Rock	Description	Elastic elements E_H kp/cm²	E_{K_1} kp/cm²	E_{K_2} kp/cm²	E_{K_3} kp/cm²	Relaxation time T_{K_1} min	T_{K_2} min	T_{K_3} min	Plastic elements $3\eta_{K_1}$ kp/cm²/s	$3\eta_{K_2}$ kp/cm²/s	$3\eta_{K_3}$ kp/cm²/s
Clay shale	along cleavage	140000	48000	73000	–	9	12	–	$2{,}5 \cdot 10^7$	$3 \cdot 10^9$	–
	slightly inclined	140000	64000	76000		9	6		$3{,}5 \cdot 10^7$	$1{,}6 \cdot 10^9$	
	highly inclined to cleavage	120000	80000	280000	200000	9	1	0,5	$4 \cdot 10^7$	$1 \cdot 10^9$	$8{,}5 \cdot 10^9$
Quartzite clay shale	highly inclined to cleavage	170000	11000	13000		40	12		$2{,}5 \cdot 10^7$	$5 \cdot 10^8$	
Gypsum/anhydrite	compact	240000	76000	40000		8		3	$4 \cdot 10^7$		$1 \cdot 10^{10}$
Rock salt	—	150000	130000	240000		10	10		$8 \cdot 10^7$	$9 \cdot 10^9$	

"fundamental strength" (GRIGGS, 1936), "true strength" (PHILLIPS, 1948), "time safe stress" (POTTS, 1964), "long-term strength" (PRICE, 1966) or "sustained load strength" (IYENGAR et al, 1967). There are many methods for the determination of this strength. They may be grouped under two main heads; direct methods and indirect methods.

In the direct method, specimens are subjected to different sustained loads and the highest value of the load at which no failure takes place with time is determined. The method is very cumbersome and involves much labour and time. As such, a number of indirect methods have been used. These indirect methods are dependent upon the understanding of the phenomenon of creep, microseismicity, volumetric strain, etc. and are named after the phenomenon utilised and described below.

Transient Creep Method

In the creep method (GRIGGS, 1940; POTTS, 1964 and PRICE, 1964) the assumption is that the stress at which steady state strain rate is zero represents the time-dependent strength. In other terms, it represents the stress level at which only transient creep is observed. The method is slightly simpler than the direct method, but its accuracy depends upon the sensitivity of the strain measuring equipment and is time consuming if not impossible to determine exactly the point where the secondary creep rate is zero. The method is very conservative.

Dilatancy Method

Dilatancy has been observed in specimens subjected to compression. This concept has been utilised by BIENIAWSKI (1967) to explain the mechanism of brittle fracture of rock. At lower stress, microcracks first close with elastic deformation of rock; at higher stress, deformation becomes inelastic and crack propagation takes place. At first this is stable and then tends to become unstable. The stress at which crack propagation becomes unstable represents the point where reversal in the volumetric strain curve (and pore pressure curve in undrained test) takes place. This is the point which represents the stress value corresponding to the long term strength (Fig. 9-34).

It is essential that axial and transverse strains are measured accurately along the whole length of the specimen for unstable points to be fully representative of the deformation characteristics of the specimen.

Based on the principle, DESAYI and VISWANATHA (1967) have suggested the use of log stress—log strain curve. They found that for concrete, the plot consists of 3 straight lines with different slopes so that two kinks are present in the log stress—log strain curve. The first kink (15–25% of short term strength) represents closure of crack and the second kink (70–90% of short term strength) represents unstable crack propagation.

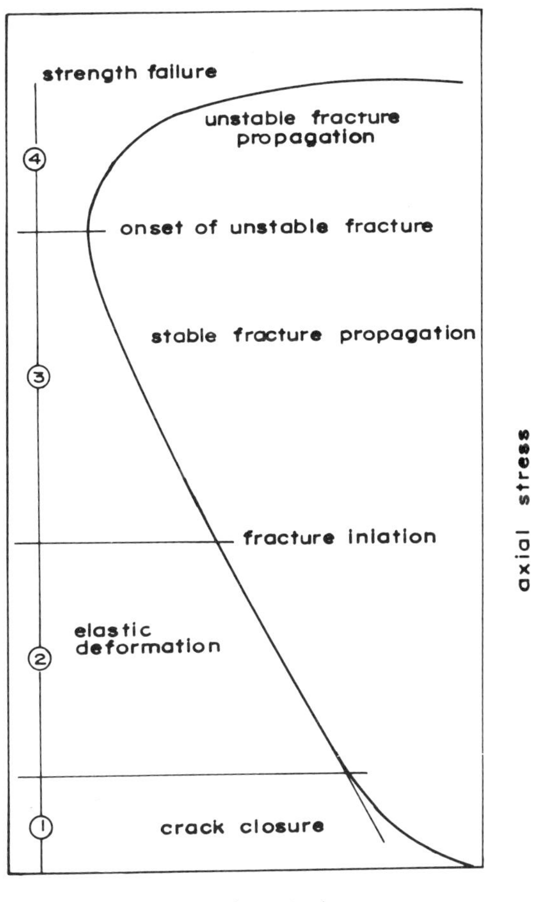

Fig. 9-34. Typical stress – volumetric strain curve for hard rock showing fracturing processes (after BIENIAWSKI, 1967).

WIID (1970) used the dilatancy concept to determine the influence of moisture on the time-dependent strength of dolerite and two sandstones and found some agreement between the time-dependent strength measured from dilatancy and actual strength tests. The agreement is better with dolerite which is of lower porosity (<1%) than sandstones (porosities 21 and 15%). From the results of WIID, Table 44, Fig. 9-35, it appears that the point of crack initiation is a better measure of the ultimate or long-term strength of dolerite than is the point of unstable crack growth. The region between crack initiation and unstable crack growth is the region of finite time-dependent strength.

TABLE 44
Influence of moisture on strength of dolerite
(after WILD, 1970)

Moisture Control	Mean Uniaxial Compressive Stress: Fracture Initiation σ_{in} (MPa)	Unstable Propagation σ_{un} (MPa)	Strength Failure σ_c (MPa)	Mean Tensile Strength σ_t (MPa)	Ratios: $\frac{\sigma_{in}}{\sigma_c}$	$\frac{\sigma_{un}}{\sigma_c}$	$\frac{\sigma_{in}}{\sigma_t}$
Dry	287.40	427.10	480.10	31.40	0.60	0.89	9.1
50 % R.H.	212.40	299.90	353.40	27.60	0.60	0.85	7.7
Sat.	181.60	266.30	310.30	25.50	0.58	0.86	7.1

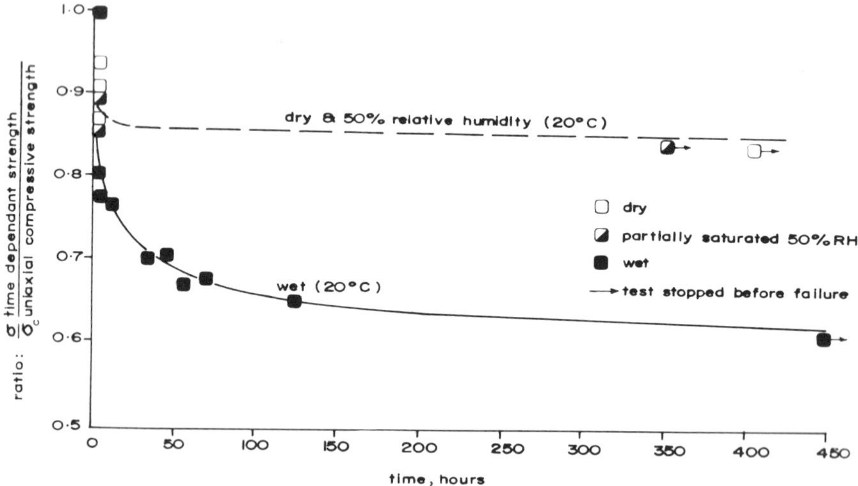

Fig. 9-35. Time-dependent uniaxial compressive strength tests on dolerite (after WIID, 1970).

Strain Rate Method

The method is based on determining the influence of strain rate on the strength. The asymptotic value (Fig. 9-36) gives the long-term strength of the material (SANGHA and DHIR, 1972). SANGHA and DHIR also concluded that unstable crack propagation, which determines finally the long-term strength, commences at the stress level at which the incremental Poisson's ratio becomes 0.50 and not when the material begins to dilate. They recommended this stress level be taken as the critical stress of a material.

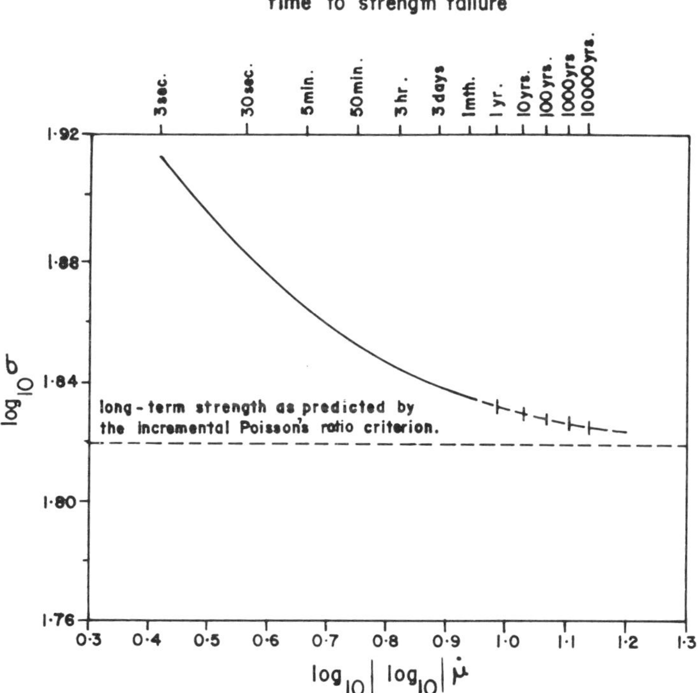

Fig. 9-36. Influence of strain rate on the strength of Laurencekirk sandstone (after SANGHA and DHIR, 1972).

WAWERSIK and BROWN (1971) used the creep properties and the complete quasi-static stress-strain curves of rock to predict its time-dependent strength. The hypothesis underlying this method is that time-dependent failure occurs when rock has been strained a critical amount. The allowable strain is provided by complete quasi-static stress strain-curves (ε_q' between the ascending and descending parts, Fig. 9-37) and the time-dependent strains are calculated from suitable constitutive equations.

The point is best illustrated by means of Figs. 9-37 and 9-38. For example, in Fig. 9-38 the uniaxial compressive strength is plotted versus the strain ε_c' which was observed for water-saturated Westerly granite at the onset of fracture under constant uniaxial compression (triangular points). Fig. 9-38 also shows a plot of the strain ε_q between the ascending and descending parts of the uniaxial stress-strain curve for water-saturated granite. At any stress $\sigma = \sigma'$ the strain ε_q is proportional to the distance from point A to B in Fig. 9-37. Clearly there is a strong correlation between the strains ε_c' and ε_q provided they are measured under the same conditions. Equally acceptable agreement was obtained between ε_c' and ε_q for Nugget sandstone and Tennessee marble. Because of this agreement it appears that the failure time of the rock used can now be estimated in the following manner. Firstly the time-dependent strain $\varepsilon(\sigma, t)$ is calculated by means of suitable constitutive relations. Then an estimate is obtained of the maximum strain ε_q which rock can undergo prior to failure. Finally the failure time of the rock or the strength of rock for a prescribed time interval is ascertained by solving the equation

$$\varepsilon(\sigma, t) - \varepsilon_q'(\sigma) = 0 \tag{9.67}$$

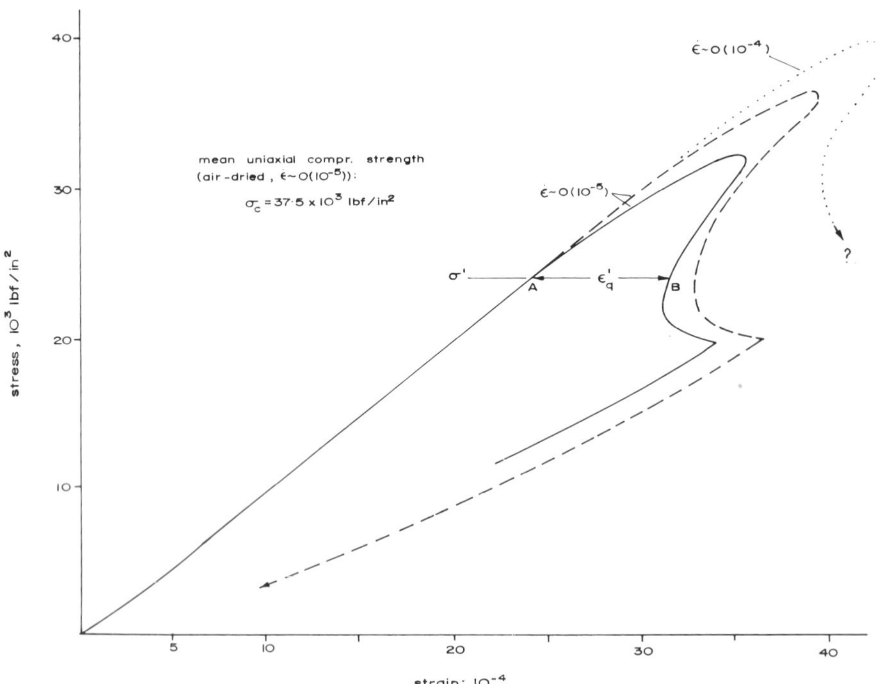

Fig. 9-37. Complete quasi-static stress-strain curve of Westerly granite under uniaxial compression (solid line: water saturated; dashed and dotted lines: air-dry) (after WAWERSIK and BROWN, 1971).

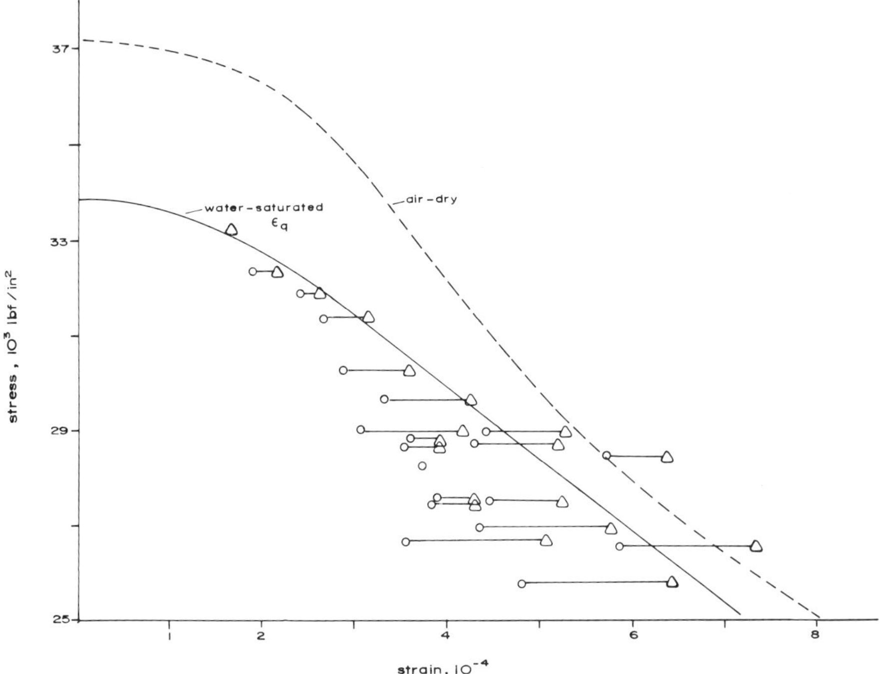

Fig. 9-38. Stress-failure strain data for Westerly granite (curve: strain between ascending and descending parts of complete stress-strain curve; circles: onset of tertiary creep; triangles: creep fracture) (after WAWERSIK and BROWN, 1971).

Micro-Seismic Method

One of the techniques which has been very commonly used to determine stability of underground openings is the micro-seismic technique which is also very often called the micro-acoustic technique. It has been found that in concrete amplitude of the acoustic emissions first increases with closure of cracks and thereafter it diminishes gradually and then more rapidly as the failure approaches (RUSCH, 1959). For concrete it has been found that amplitude drops below the initial value at about 75–80% of short-term strength of concrete which is quite in agreement with the long-term strength tests.

THILL (1972) discussed acoustic methods for monitoring failure in rock. The acoustic data were used to infer the stress level and deformation characteristics in rock under stress. Deformation conditions that were recognised are 1. crack closure 2. linear elasticity 3. fracture initiation and development 4. accelerated fracture development and 5. gross failure and post-peak behaviour.

Creep Rate Method

This method of determining the long term strength is based upon relating the time to failure to the secondary creep rate which is further a function of the applied stress and temperature. This concept has been adopted in predicting the creep rupture behaviour at high temperature of metals and alloys in a number of criteria (GAROFALO, 1966) and can be written as

$$T_f = \frac{C}{\dot{\varepsilon}} \tag{9.68}$$

where T_f = time of failure
C = constant and
$\dot{\varepsilon}$ = creep rate at a given stress level.

It has been found that at larger creep rates, C is not constant but is itself a function of the creep rate, strain and temperature and can be represented by general equation

$$T_f = \frac{F(\varepsilon, \dot{\varepsilon}, T)}{\dot{\varepsilon}} \tag{9.69}$$

For a large number of alloys, this equation can be approximated by (MONKMAN and GRANT, 1956)

$$T_f = \frac{C}{(\dot{\varepsilon})^m} \tag{9.70}$$

where C and m are constants for a given temperature. The value of m varies from 0.77 to 0.93 for a large variety of materials and C is the difference between the strain at failure or onset of tertiary creep (ε_f) and the total elastic and primary creep strain (i.e. total recoverable strain ε_r). In other words

$$T_f = \frac{(\varepsilon_f - \varepsilon_r)}{(\dot{\varepsilon})^m} \tag{9.71}$$

The value of ε_f does not seem to be constant for the rock for different stress levels and the method is only an approximation. GRIGGS (1940) has shown that for alabaster immersed in water, there is a critical strain before the onset of tertiary creep under uniaxial compression. Under biaxial or triaxial loading the critical strain is definitely different and depends upon the differential stress. However, this concept has been applied for frozen soils (LADANYI, 1972). In metals a straight line relationship between rupture time and tensile stress has been obtained (YOKOBORI, 1964). CRUDEN (1974) has analysed GRIGGS

(1940) data and shown that for alabaster immersed in water at 24° C, a linear relationship exists between the stress and fracture time (Fig. 9.39). Very similar results have been obtained by STAVROGIN and LODUS (1974) on potash salt and Cambrian clays. They found that the relationship between the compressive stress and the time to failure can be represented by a relationship of the type

$$t = t_0 e^{-\alpha_1 \sigma} \tag{9.72}$$

where t = time for failure
σ = stress level and
t_0 and α_1 = constants of the material.

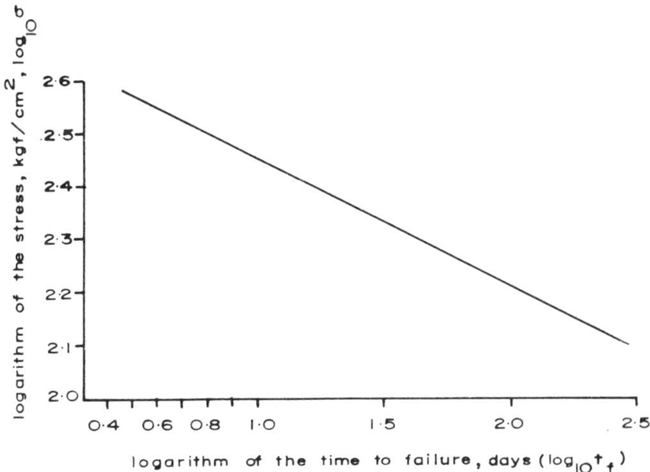

Fig. 9-39. The dependence of time to failure on stress in GRIGG's experiments on alabaster (after CRUDEN, 1974).

Similarly, the creep rate has the same relationship $\dot{\varepsilon} = \dot{\varepsilon}_0 e^{\alpha_2 \sigma}$ (Eq. 9.54) except that the sign of the coefficient is positive. These relationships are represented in Fig. 9.40. The values of α for both the time and strain rate functions for the 3 different materials are given in Table 45. The approximate equality of the coefficients for the two salt deposits permits to determine the long term strength by knowing only one long term strength value of the rock (say at 80 to 95% of the short term strength) and measuring the creep rate of the pillar in actual practice and following the method shown in Fig. 9-40. Knowing the value of α and one value t at a certain σ value, a line is drawn on the log $t-\sigma$ graph with an inclination α through the one long term strength point. The slope of $\dot{\varepsilon}-\sigma$ line being the same, and if the $\dot{\varepsilon}$ at any stage is known, the time t at which failure of pillar will occur can be calculated. To overcome certain uncertainties, it is advisable to determine the creep rate at two stress levels at least.

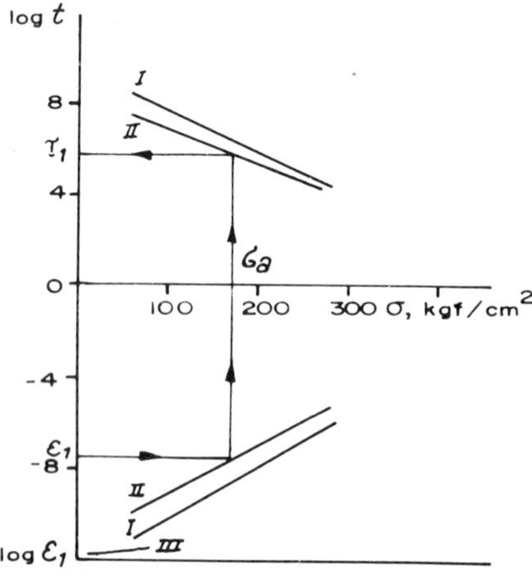

Fig. 9-40. Long-term strength and creep rate versus stress for sylvinites and Cambrian clay (after STAVROGIN and LODUS, 1974).

TABLE 45

Values of α_1 and α_2 for various rocks in Eqs. 9.72 and 9.54

(after STAVROGIN and LODUS, 1974)

Rock	α_1 (time function) Eq. 9.72	α_2 (strain function) Eq. 9.54
Potash salt from Verkhnekamskaya deposit	0.047	0.045
Potash salt from Starobin deposit	0.0143	0.0133
Cambrian clay	—	0.0061

WAWERSIK (1972) used this method to calculate the time dependent strength of water saturated Westerly granite and Nugget sandstone. The predited and measured values are given in Table 46.

It may be pointed out that this technique is useful if the straight line relationship for the rock has been proved in the laboratory and mechanical similarity exists between field and laboratory conditions.

TABLE 46
(after WAWERSIK, 1972)

Rocks Water-saturated	Stress lbf/in² (MPa)	Failure time, hours	
		Predicted	Measured
Westerly granite	31000 (218)	0.5	0.8
	29000 (200)	10	10
	25000 (172)	555	542
Nugget sandstone	25000 (172)	0.05	0.07
	20000 (138)	197	150 (range 90 to 300)
	19000 (131)	1480	1300 (range 1250 to 1475)

Relaxation Method

Under the influence of an external load, the sample creeps and if strain is maintained constant, there is a relaxation of load value till an equilibrium stage is reached between the applied stress and the internal stress. This value of the load is called the "steady stress". If at this stage the load is taken off and the sample allowed to recover for a long time, a certain value of the strain

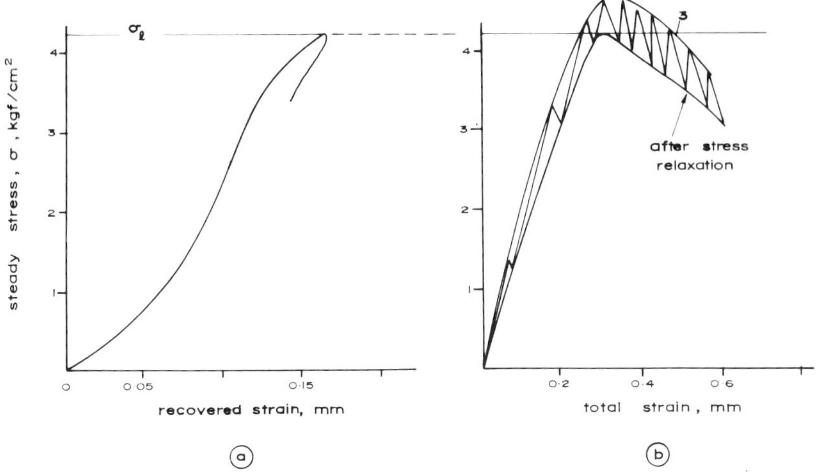

Fig. 9-41. Determination of long-term strength by relaxation method.

introduced in the sample is recoverable. This recoverable strain is dependent upon the stress level. PUSHKAREV and AFANASEV (1973) carried out a series of tests on weak rocks such as sandstones, siltstones, argillites, clays, etc. and found that the recoverable strain decreases after a certain load value (Fig. 9-41), σ_1 which corresponds to the point where a large scale fracturing starts. If the loading (stress-total strain) curve is plotted along with the relaxed stress-total strain curve (Fig. 9-41 b), the curve so obtained represents the one obtained at infinite small strain rates and hence the peak value represents the long-term strength of the rock. Some examples of this are given in Fig. 9-42.

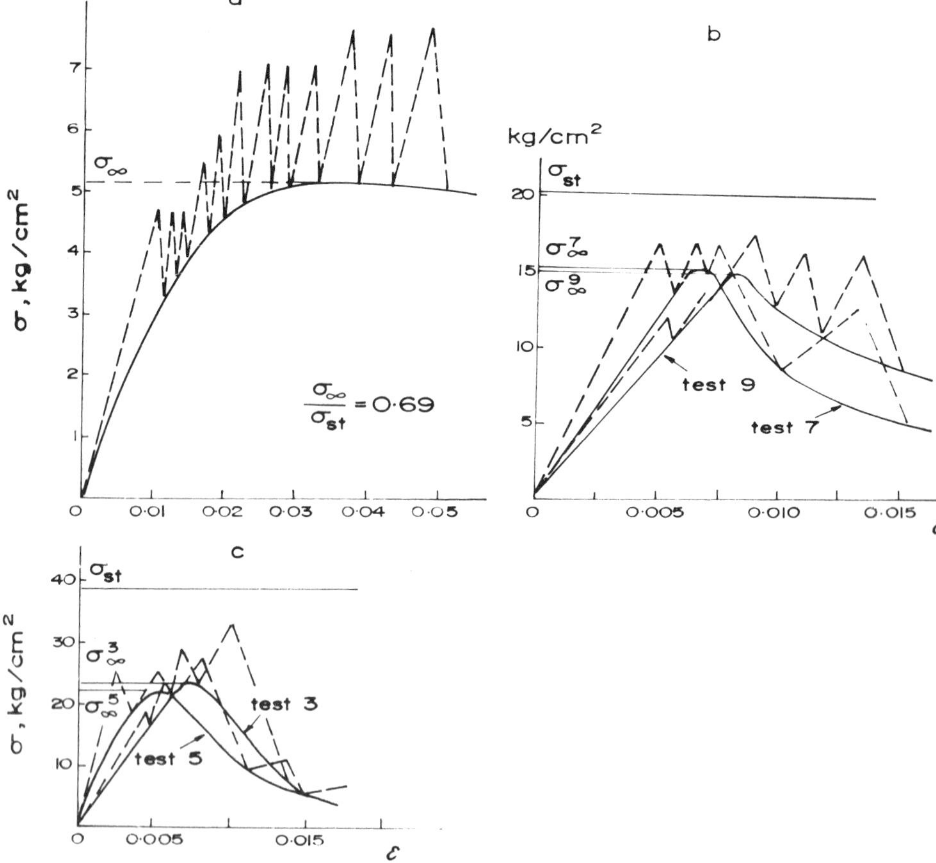

Fig. 9-42. Examples of determination of long-term strength
(a) sandy clay (Tyulgan deposit);
(b) argillite (Bogoslovo deposit);
(c) siltstone (Bogoslov deposit)
σ_∞ = long-term strength: σ_{st} = short-term strength
(after PUSHKAREV and AFANASEV, 1973).

9.11. Rock Fatigue

Very little attention has been paid to rock fatigue under cyclic or dynamic loads though the phenomenon is of importance in bridge abutments, foundations, dams and road pavements, open cut benches, rock drilling and structures subjected to earthquake. Studies undertaken on other materials such as metals, concrete and soil have shown that the effect of repeated loads causes failure at stresses lower than the normally determined strength and this phenomenon is called fatigue.

Fatigue characteristics are usually presented in the form of *S-N* curves (stress level versus the number of cycles required to bring about failure) Fig. 9-43. This curve indicates clearly that at high maximum stresses the number of cycles required to cause failure is small and increases as the maximum stress is lowered. The maximum applied stress level at which the material can stand an infinite number of cycles is called fatigue stress.

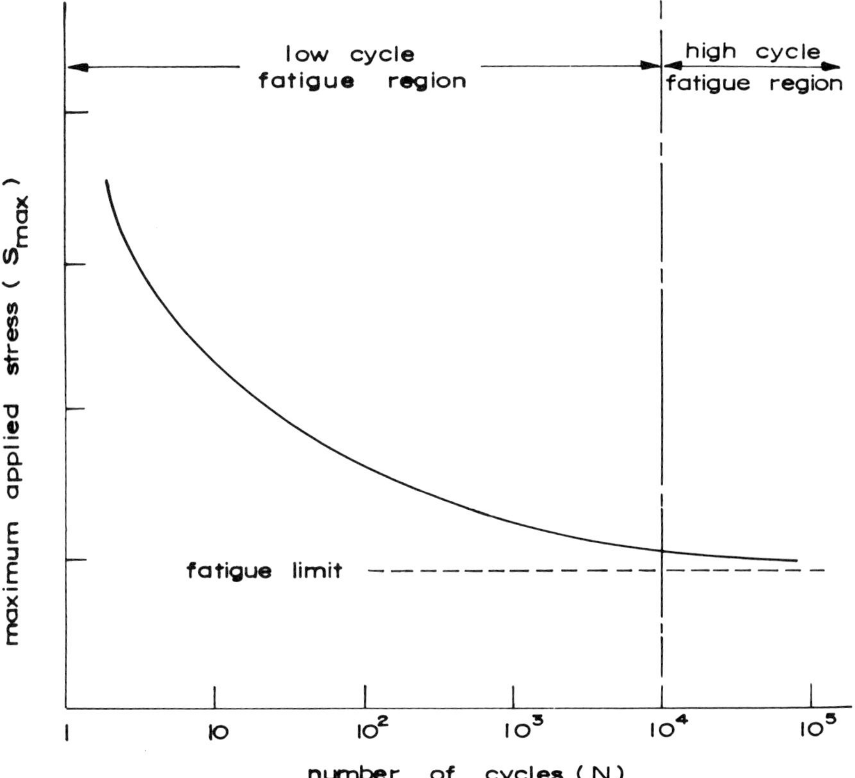

Fig. 9-43. General form of a typical S-N curve.

In metals, two regions of fatigue failure limits have been recognised. The first refers to the region of high stress where the failure occurs due to plastic deformation and is called as low cycle fatigue region. The second is called the high cycle fatigue region where failure occurs due to elastic deformation. In metals, the dividing region lies somewhere at 10^4 cycles.

The testing of rock for fatigue requires the preparation of the specimens as in usual uniaxial compression or direct tension tests and subjecting them to alternating cycles of loading and unloading in a suitable testing machine. Servo-hydraulic testing machines are best suited for the purpose and are capable of giving frequencies up to and in excess of 100 Hz. The tests till date have been conducted in the compressive range i.e. the minimum stress always has a certain positive compressive value, on alternating tension-compression cycles or only tension cycles.

The system for loading of the specimens for compression fatigue of rocks as adopted by HARDY and CHUGH (1970) is given in Fig. 9-44. In direct tension fatigue testing, it is important to use a gluing material with an adhesive fatigue strength more than that of rock. This is difficult to achieve. In such cases it is more suitable to use special-shape specimens (Chapter 1, Volume I) and grip the specimens using suitable grips. In other cases indirect tension tests may be

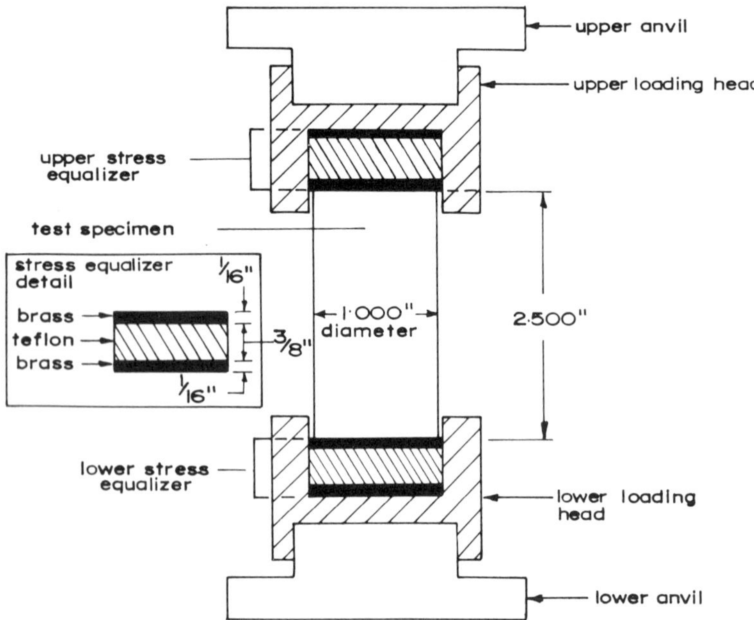

Fig. 9-44. Specimen loading jig used in fatigue studies (after HARDY and CHUGH, 1970).

used to determine fatigue characteristics under repeated tensional loads. The analysis however becomes more complicated due to stress gradients of which the influence has not been fully understood as yet.

One of the earliest studies on rock fatigue was undertaken by GROVER, DEHLINGER and MCCLURE (1950) on limestone but due to the limited number of tests undertaken, the results are inconclusive. BURDIN (1963) conclusively demonstrated that the influence of repeated loading on Berea sandstone was to decrease its strength. HARDY and CHUGH (1970) conducted tests on Barre granite, Tennessee sandstone and Indiana limestone in the low cycle fatigue range at testing frequency of 0.25 Hz. The dynamic compressive strengths (determined at loading rate of 14,000 lbf/in^2 equal to the load cycling frequency) of the rocks were in the range as follows:

	Mean dynamic compressive strength	
	lbf/in^2	(MPa)
Barre granite	32,081	(221)
Tennessee sandstone	21,153	(146)
Indiana limestone	8,194	(56)

The *S-N* data for these rocks fits to a general equation of the type $S = A - B \log N$ with the fatigue limit defined as the percentage of the dynamic strength varying 65 to 80%. The lower value is for Barre granite and the higher for Tennessee sandstone. From similar tests, HAIMSON and KIM (1971) found a similar relationship for Tennessee marble [dynamic compressive strength = 23,285 lbf/in^2 (161 MPa)] and the samples at 75% of this value did not fail at 10^6 cycles. Since in practice it is unlikely that rock shall be subjected to higher load cycles during its life time, these values could be taken as the compressive fatigue limit of rock.

Tests by CAIN, PENG and PODNIEKS (1975) on charcoal granite $\sigma_c = 230$ MPa, $\sigma_t = 12.2$ MPa, $E = 67.6$ GPa), Tennessee marble ($\sigma_c = 116$ MPa, $\sigma_t = 8.4$ MPa, $E = 62.0$ GPa) and quartzite (Jasper, Minn., $\sigma_c = 389$ MPa, $\sigma_t = 18.3$ MPa, $E = 69.9$ GPa) showed that fatigue life was related to the strain amplitude to which the specimens of these rocks were subjected. At high frequency cycles (10 kHz) (Fig. 9-45), the fatigue limit was higher for higher tensile strength rocks. The failure was associated with tensile cracking and specimen temperature rose linearly depending upon the energy input and energy dissipated in the specimen. Temperatures up to 160° C above ambient temperature were recorded depending upon the strain amplitude and rock type.

HAIMSON and KIM (1971) have observed the stress-strain curve produced as a result of cyclic loading with the complete stress-strain curve obtained under monotonically increased strain (Figs. 9-46 a & b). Surprisingly, there is a sudden failure as the strain reaches close to the failure envelope of rock both in the pre-failure and post-failure tests. The summary of the results on failed Georgia marble in the form of *S-N* curve are given in Fig. 9-47. The rock in the post-failure region shows surprisingly large capacity to withstand cyclic loading.

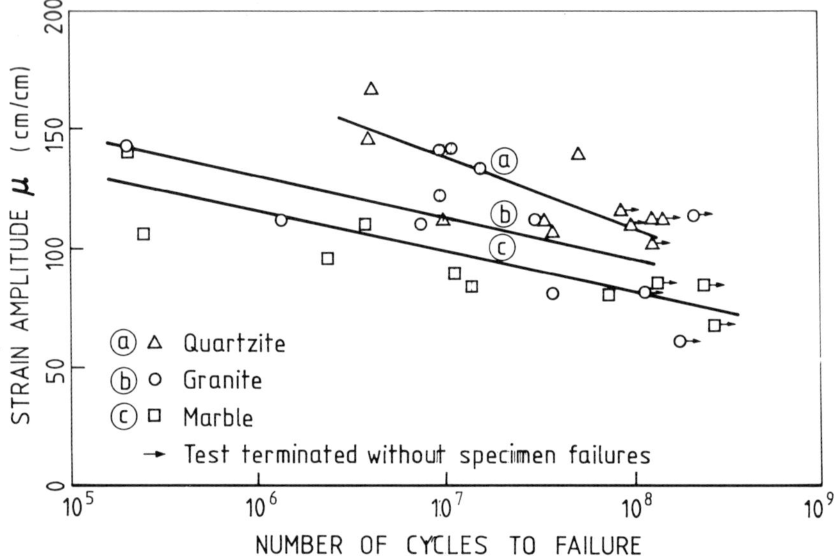

Fig. 9-45. Variation of fatigue life with strain amplitude (after CAIN, PENG and PODNIEKS, 1975).

ATTEWELL and FARMER (1973) have studied the dynamic creep under cyclic loading. Maximum and minimum strains were measured at different test frequencies; the dynamic modulus values were calculated at different stress cycles (Fig. 9-48). The modulus indicates a drop in the values with increase in number of cycles, the decrease to be larger for higher excitation frequencies. The interpretation is that the fatigue test data for rocks are more strain than stress dependent. Failure possibly occurs when sufficient strain energy can be stored to be released at a critical energy level by sub-failure cyclic stresses. If the rate of recovery of the material is elastically unretarded, deformation under cyclic loading is accumulative where the deformation from succeeding cycles is superimposed on residual deformations from earlier cycles. This deformation becomes more important at higher frequencies due to less chance of recovery on unloading.

Dynamic creep strain-stress cycle (*N*) curves at different maximum stresses are very much similar to the creep curves obtained on constant loading (Fig. 9-49) showing the 3 stages of creep.

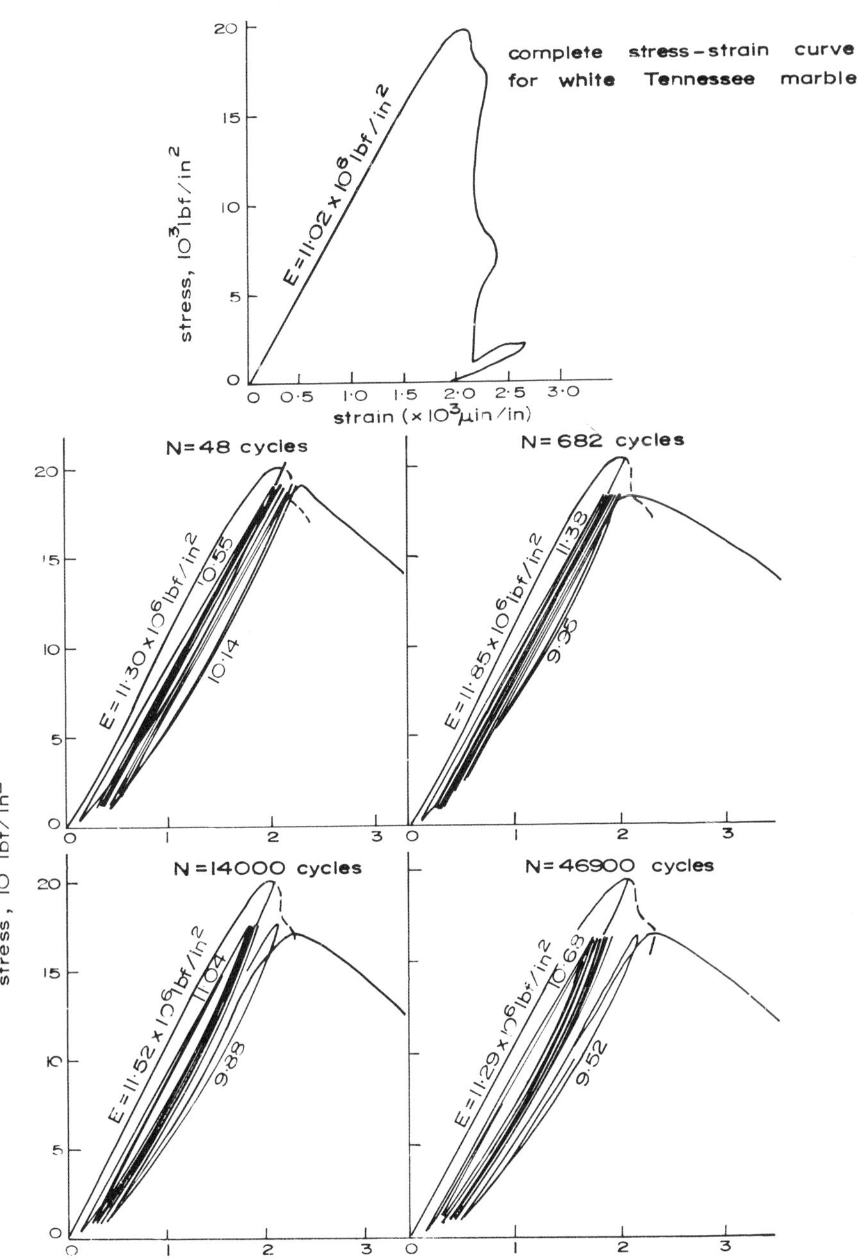

Fig. 9-46a. Stress-strain curves for failed White Tennessee marble (Dashed lines are expected to complete stress-strain curves) (after HAIMSON and KIM, 1971).

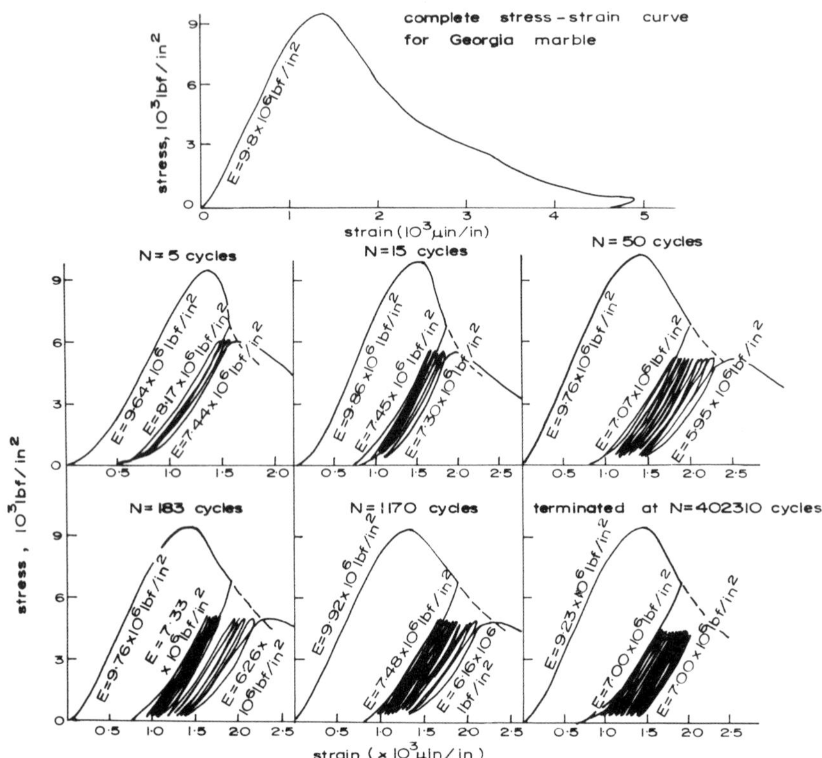

Fig. 9-46b. Stress-strain curves for failed Georgia marble
(Dashed lines are expected to complete stress-strain curves) (after HAIMSON and KIM, 1971).

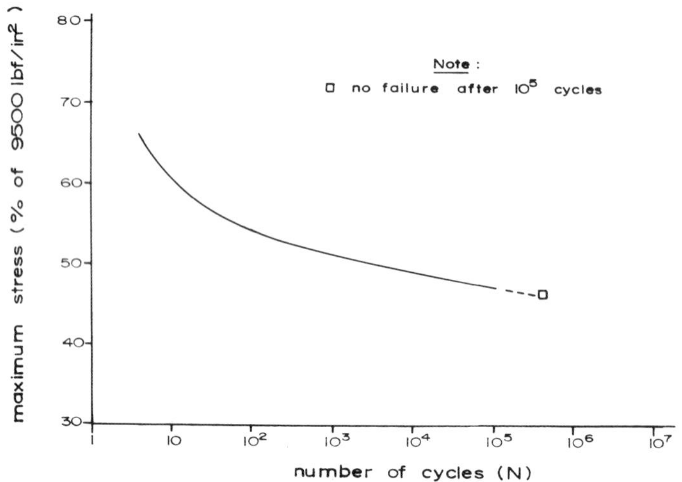

Fig. 9-47. S-N curve for failed Georgia marble (after HAIMSON and KIM, 1971).

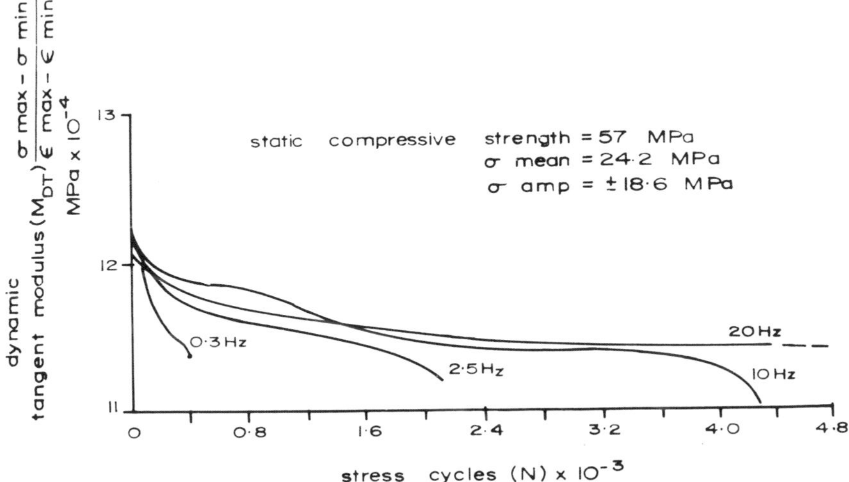

Fig. 9-48. Change of dynamic modulus with stress frequency and loading history (after Attewell and Farmer, 1973).

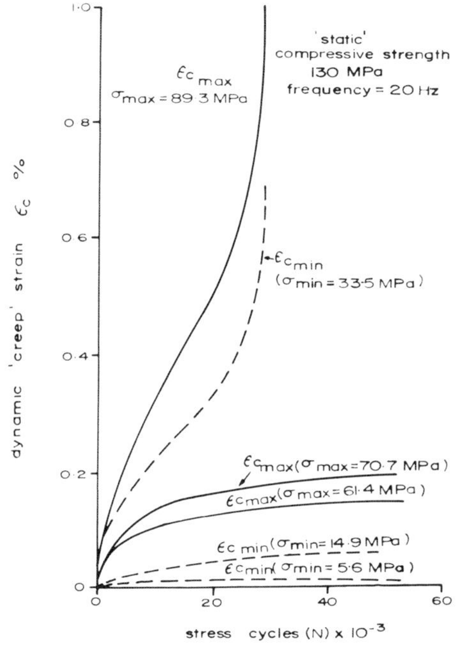

Fig. 9-49. Influence of cyclic stress variation at constant frequency on the strain variation as a function of time (initial elastic strains discounted; both maximum and minimum initial creep strain set at zero base line) (after Attewell and Farmer, 1973).

PENG, PODNIEKS and CAIN (1974) have studied fatigue of Salem limestone under cyclic compression loading, cyclic tensile loading and cyclic compression-tensile loading. The cycles were applied at a certain percentage of compressive and tensile strengths. They found that fatigue strength under cyclic compressive loading to be about 78% of the compressive strength. The area of the loop sharply decreases during the first few cycles and then reaches a constant level until a few cycles before failure when it increases sharply.

In cyclic tensile loading, the fatigue limit is about 70% of the tensile strength. There is no sign of increase in area as specimen approaches failure and the fracture of specimen is instantaneous.

In compressive-tensile cyclic loading, the endurance is much smaller and specimens fail predominantly at compressive cycle. The area of the loop decreases rapidly in the first few cycles and then increases exponentially. The maximum deformation for each tensile half cycle remain constant but for each compressive half cycle increased exponentially.

PENG and ORTIZ (1972) have explained the mechanism of rapid failure of rock under compressive-tensile loading as a development of en-echlon cracks parallel and perpendicular to compressive loading which extend and coalesce much earlier under cyclic compressive-tensile loading than under cyclic compressive loading.

The damping capacity (ψ_s) defined as the ratio of the specific damping energy of the rock E_s represented by the energy absorbed per unit volume per cycle of loading and the total strain energy E_t at maximum displacement

$$\psi_s = E_s/E_t \tag{9.73}$$

for the rock under different systems is given in Figs. 7-50 to 7-52. The decrease in damping capacity in the first few cycles is probably associated with stabilisation

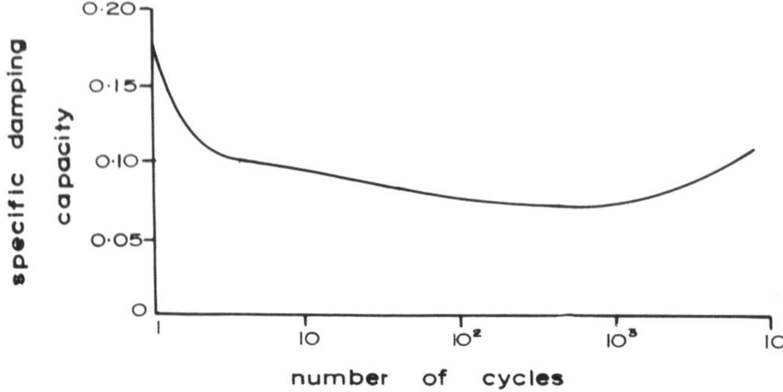

Fig. 9-50. Specific damping of Salem limestone under cyclic compressive loading (after PENG, PODNIEKS and CAIN, 1974).

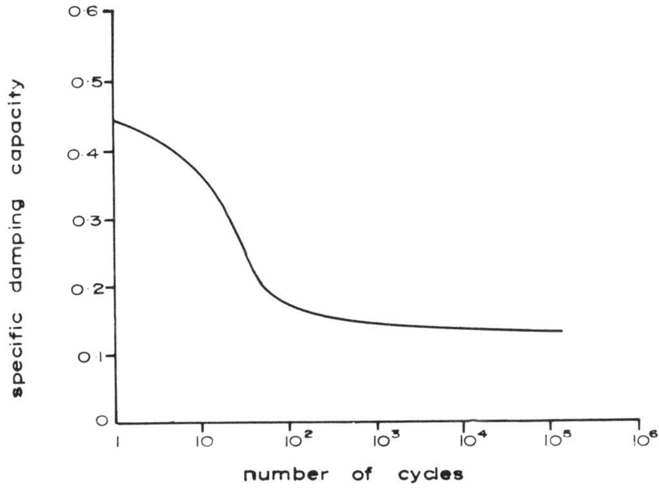

Fig. 9-51. Specific damping of Salem limestone under cyclic tensile loading (after PENG, PODNIEKS and CAIN, 1974).

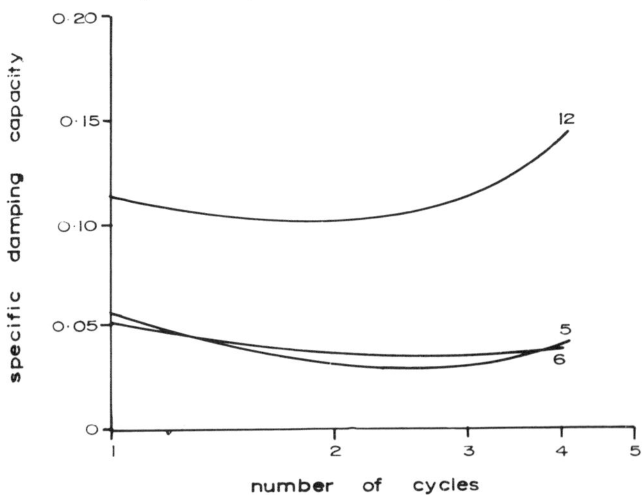

Fig. 9-52. Specific damping of Salem limestone under cyclic compressive-tensile loading. The numbers in the curves represent specimen numbers (after PENG, PODNIEKS and CAIN, 1974).

of the internal structure which is also observed in the change in secant modulus values. With approach of failure, cracks develop, and the friction between crack surfaces accounts for increase of specific damping capacity. This behaviour however is not observed in cyclic tensile loading because of sudden failure of specimen and there is hardly any chance of friction between new surfaces developed.

HARDY and CHUGH (1970) observed changes in the internal structure of the rock under the influence of fatigue strain. The data on static strength and porosity for undeformed and deformed (subjected to a fatigue cycling of 10,000 cycles or below their fatigue limits) rocks is given in Table 47. The mean static strength increased by 21.0%, 10.9% and 3.4% for Tennessee sandstone, Indiana limestone and Barre granite whilst porosity increased appreciably (28.8%) in Barre granite. Though dilatancy has been reported (HARDY et al, 1969; SAINT-LEU and SIRIEYS, 1971), this increase in strength is not understandable and could only be a stray event.

TABLE 47
Comparison of static strength and porosity for deformed and undeformed rocks
(after HARDY and CHUGH, 1970)

Rock type	Before Cyclic Loading		After 10,000 Cycles	
	Mean Static Strength (lbf/in^2)	Mean Porosity (%)	Mean Static Strength (lbf/in^2)	Mean Porosity (%)
Barre Granite	28,132 (10)*	0.66 (1)	29,100 (2)	0.85 (1)
Tennessee Sandstone Crab Orchard)	19,662 (10)	5.5 (1)	23,800 (2)	5.6 (1)
Indiana Limestone	7,327 (10)	17.0 (1)	8,130 (2)	16.9 (1)

* Numbers in parenthesis indicate the number of specimens tested.

Rate of change in lateral strain under repeated cycling is more prominent than longitudinal strain at no confining pressures for a number of rocks (granite, marble, etc.) as the number of cycles increase for a given maximum stress value, but under confined conditions, a stabilisation effect has been observed (SAINT-LEU and SIRIEYS, 1971).

9.12. Creep of Fractured Rock

Very limited work has been done on the time-dependent behaviour of fractured rock. The creep behaviour of a jointed rock can be idealised into two categories:
1. The creep of joints.
2. The creep of the solid material.

The mechanism of shear failure along joints is not that of pure friction but is associated with the failure of the asperities (failure of the solid material). Influence of strain (stress) rates has shown that the rock strength is highly dependent upon higher stress rates but at lower stress rates, the influence is small. The drop in strength as the stress rate is decreased from 0.1 MPa/s to 0.001 MPa/s is only 1%. It is only at higher stress rates that the influence on strength becomes appreciable. It will therefore be expected that at higher strain rates, the frictional coefficients shall be higher. Test results have not confirmed this. It is seen on the contrary that the frictional coefficient tends to slightly fall with high strain rates. In general, the influence of strain rate is so small that it could, for all practical purposes, be neglected. A detailed discussion of this aspect is given in Chapter 10.

The creep of the solid material at low confining pressures and temperatures which is mostly associated with changes in the structure (fracturing) causing increased deformation has been studied using stiff machines. BIENIAWSKI (1970) conducted tests on sandstone at different strain rates and obtained the post-failure curve (Fig. 9-53). These curves were obtained on specimens of 21.6 mm (0.86 in) diameter and 10.8 mm (0.43 in) height. The strain rates were varied from 33×10^{-6}/s to 0.43×10^{-6}/s. The curves are similar in shape to those obtained by RUSCH (1960) on concrete. These curves will tend to show that the influence of time is very much marked on the post-failure part of the curves. These results are however not in agreement with results of more recently conducted tests by PENG and PODNIEKS (1972) and LAMA (1973). PENG and PODNIEKS (1972) conducted tests on tuff using 1.25 in (31.5 mm) diameter specimens with height/diameter ratio of 2 and found no influence on the general

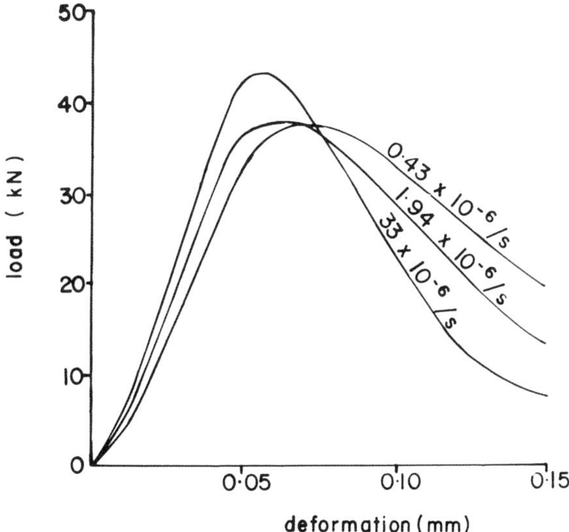

Fig. 9-53. Influence of strain rate on the complete load-deformation curve of sandstone in uniaxial compression (after BIENIAWSKI, 1970).

shape of the pre- and post-failure curve in the strain rate regions of 10^{-2} to 10^{-7}/s. Tests conducted by LAMA on marble using 61.5 mm (2.45 in) diameter specimens with height/diameter ratio of 2 showed that the influence of strain rate on the pre-failure region is more marked than on the post-failure region (Fig. 9-54). This clearly indicates the difference in the test conditions conducted by BIENIAWSKI (1970) where the stresses were very non-uniform resulting in rather triaxial conditions which permitted the material to flow at lower strain rates and behave less plastically at higher strain rates thereby giving a steeper post-failure curve.

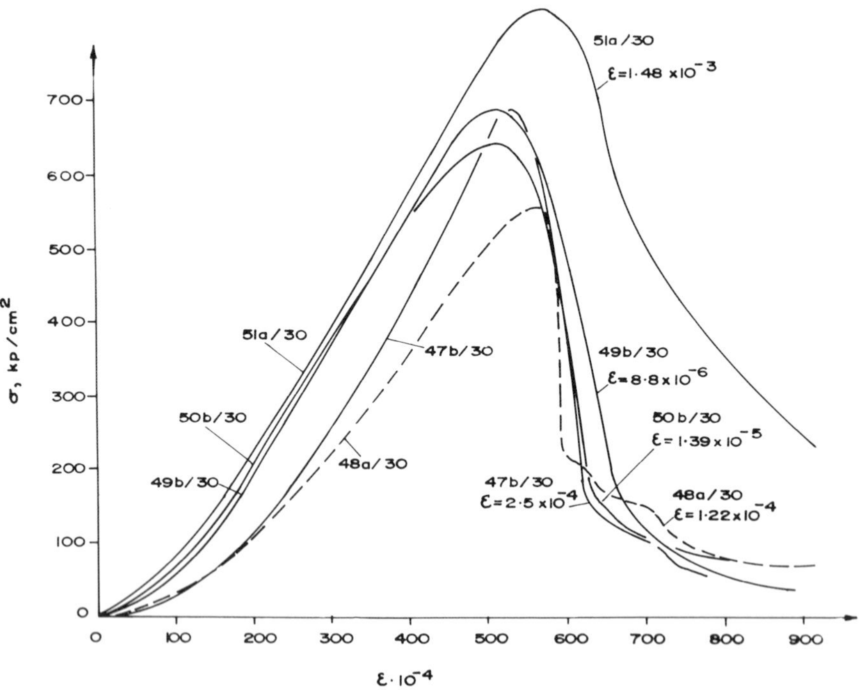

Fig. 9-54. Influence of strain rate on pre-and-post-failure behaviour of marble (after LAMA, 1973).

LAMA (1973) tested marble specimens at constant load at different stress levels. The loads were kept constant using servo-controlled machine and continued till failure. The results are shown in Fig. 9-55. Constant stress creep tests were conducted on specimens in the post-failure range obtained by unloading the specimens at different points on the post-failure curve. These results are shown in Fig. 9-56 along with average complete stress-strain curve.

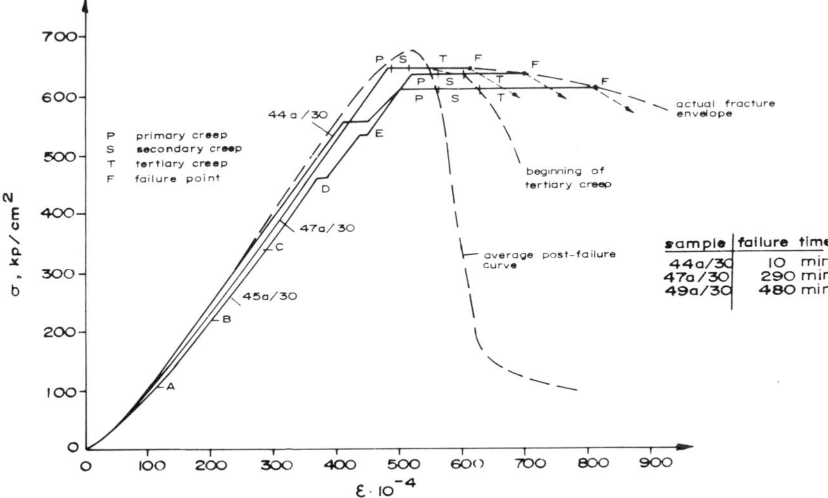

Fig. 9-55. Creep of marble specimens (pre-failure range) and projected (expected) failure envelope (after LAMA, 1973).

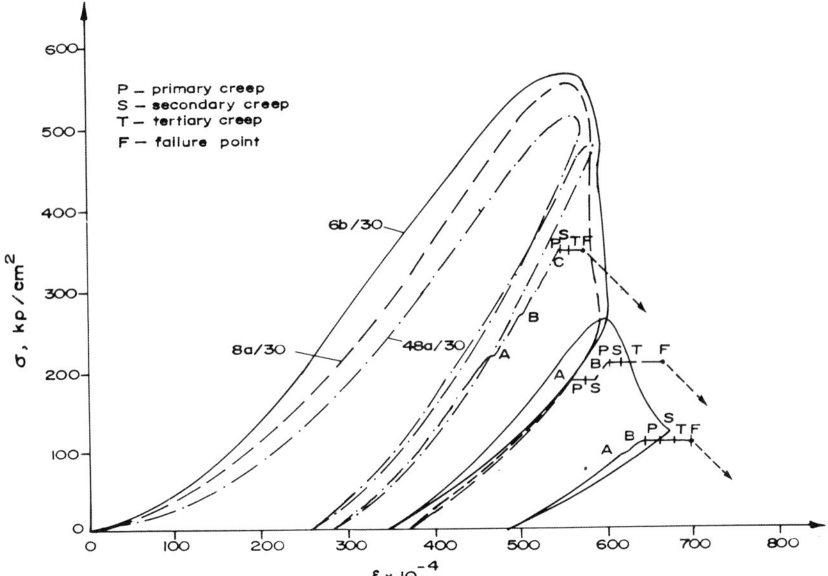

Fig. 9-56. Creep of marble (post-failure range) (after LAMA, 1973).

From both these studies, it can be seen that the creep failure envelope for the pre-failed rock lies beyond the usual stress-strain curve irrespective of whether the failure envelope is drawn using the point where the secondary creep ends and tertiary creep starts or where actual failure takes place. The creep failure envelope of the specimens in the post-failure range lies very much close to the usual post-failure curve. It, therefore, permits the conclusion that whilst deformation at failure with time for a fractured rock can be predicted by its usual complete stress-strain curve, for an unfractured rock it is not possible to predict using this test.

The reason for this abnormality lies in the total number of cracks that can propagate in the specimen before final failure occurs. At a certain load value, the probable number of cracks in the prefailure rock is very large. The effective deformation modulus that a specimen shows with a crack length 'c' is given by (COOK, 1965)

$$E_{eff} = \frac{\sigma}{\varepsilon_{eff}} \times \frac{E}{1 + \left[\dfrac{\pi c^2 (\mu^2+1)^{1/2} - \mu}{2bl\,(\mu^2+1)^{1/2}}\right]} \tag{9.74}$$

where E = Young's modulus of rock substance containing no cracks
σ = stress
ε_{eff} = effective strain
c = equivalent or total crack length
b = width of specimen
l = length of specimen and
μ = coefficient of friction between crack surfaces.

Hence, larger the value of c, the larger the total strain at a given stress level.

In the post-failure range, most of the cracks have already propagated and the final failure is occurring along a well defined path with a small change in the total crack length.

9.13. Transverse Creep and Microfracturing

The work of EVANS and WOOD (1937) on slate, granite and marble showed that transverse creep is always accompanied by a corresponding longitudinal creep. They found that there exists a stress level at which transverse creep rate may exceed the longitudinal creep rate. The cleavage planes present in slate specimens influence transverse creep. The magnitude of transverse creep in the direction perpendicular to the cleavage planes is twice that of parallel to the cleavage planes.

MATSUSHIMA (1960) and RUMMEL (1969) reported that transverse creep rate increases more rapidly with stress than does longitudinal creep rate. At low stresses, transverse creep is extremely small, but at high stress (0.5 – 0.66 of the compressive strength) it increases rapidly. SINGH (1970) found that lateral and longitudinal creep curves follow the same pattern, but at stress near to the strength of rock, lateral creep rate is far greater than the axial creep rate (Fig. 9-57).

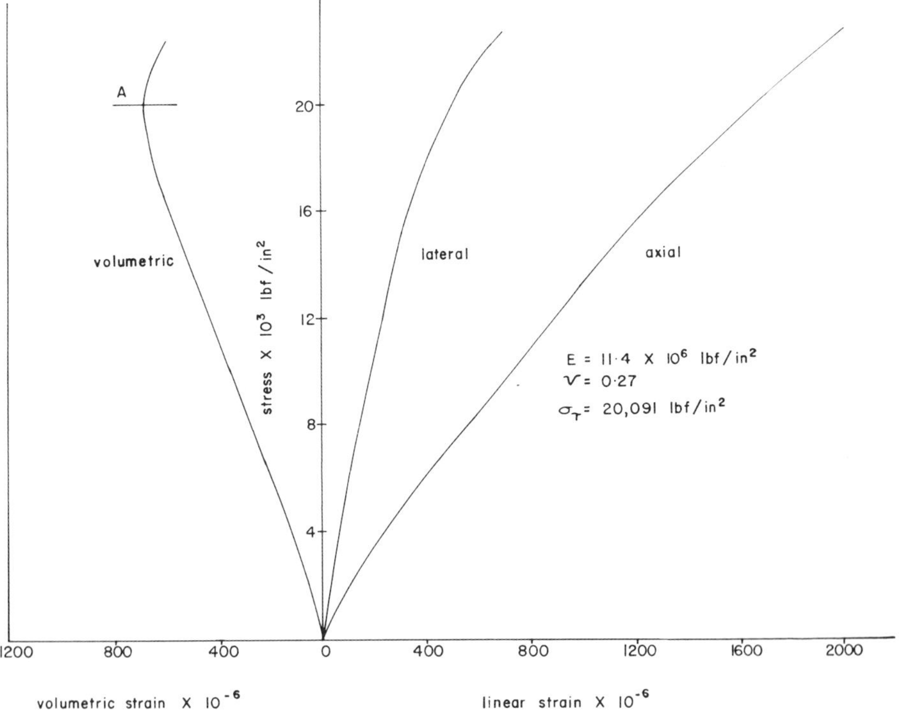

Fig. 9-57. Stress-strain curves for Lilydale limestone tested in uniaxial compression
A = long-term strength
(after SINGH, 1970).

Transverse creep and microfracturing are a manifestation of one and the same process occurring in the specimen. SCHOLZ (1968a) found that cracking in rocks is proportional to creep and the creep in rocks at low temperature and pressure is due to time-dependent microfracturing. Work of GOLD (1960) on ice has shown that cracking activity is directly proportional to creep rate. Whether cracking is due to creep or creep due to cracking is a debatable question. SAVAGE and MOHANTY (1969) are of the opinion that creep is the cause of internal cracking and not the result of cracking as proposed by SCHOLZ (1968b).

Dilatancy* of rocks has been observed by several investigators (BRIDGMAN, 1949; BRACE, 1963; PAULDING, 1965; BIENIAWSKI, 1967; HARDY et al, 1969; SINGH, 1970) which is an indication of opening and propagation of cracks in rocks under stresses. BRACE et al (1966) found that rock becomes dilatant when the stress difference was of the order of 1/3–2/3 the compressive strength under triaxial conditions. Under uniaxial conditions, this value is about 50–70% depending upon rock type. Studies on pore pressure changes have shown that firstly there is an increase in pore pressure and then a decrease. This state is reached at about half the short term strength of the rock.

HARDY et al (1969) found that both creep strain and accumulated micro-seismic activity data against time fit the generalised BURGER's model with two KELVIN units. They further found that near linear relationships exist between these two parameters (Fig. 9-58).

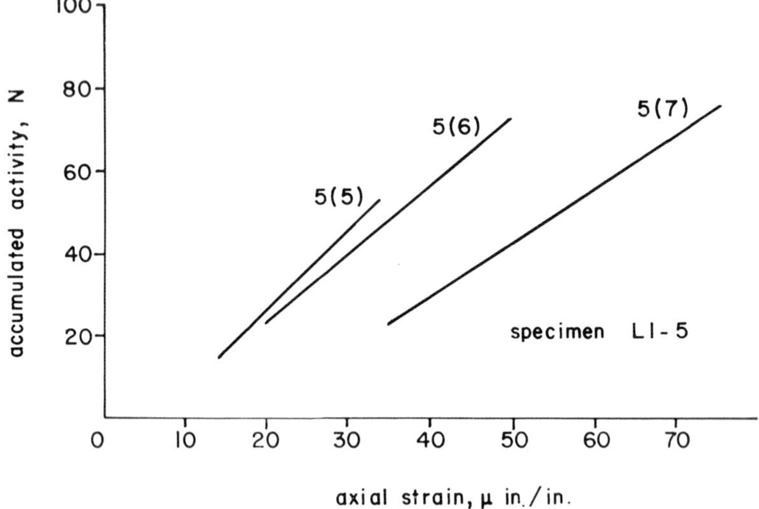

Fig. 9-58. Variation of accumulated activity with axial creep strain for Indiana limestone. (Numbers in brackets refer to the level of the load increment, 5 – sample number.) (after HARDY et al, 1969).

WU and THOMSEN (1975) report that accumulated microseismicity has a general exponential trend (Fig. 9-59) which establishes after an initial transient and lasts as much as 95% of the total time-to-failure. There are, however, deviations from the trend where relatively active periods are followed by quiet periods and the general trend is recovered. The slope of the curves is higher in the

* The term dilatancy is referred to the phenomenon in rocks exhibiting a marked increase in volume when stressed near their load-bearing capacity (see Section 6.11.).

active periods though immediately before the active period, there may or may not be any changes in the slope, but immediately after the event there is a slow transient decrease in slope.

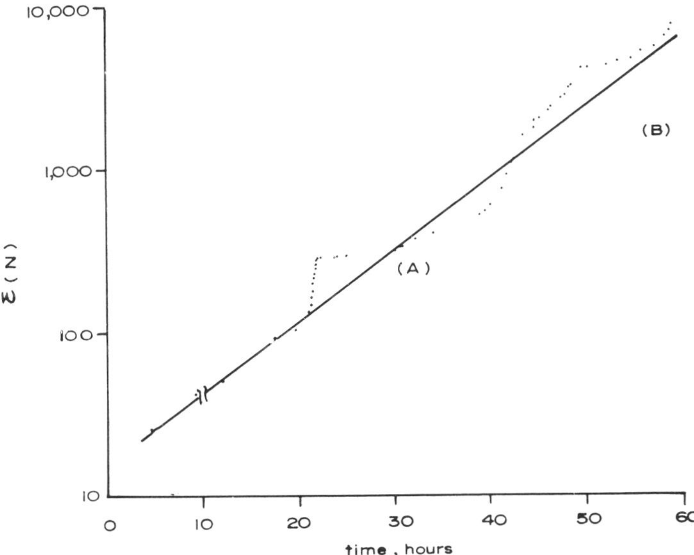

Fig. 9-59. $\Sigma(N)$ versus time at 2.1 kbar uniaxial load for Westerly granite (after WU and THOMSEN, 1975).

There is consistent qualitative agreement between axial strain (ε_z), radial strain (ε_r), volumetric strain ($\Delta V/V_0$) and the integrated microseismic activity (ΣN) (Fig. 9-60). The presence of moisture increased both the longitudinal creep rate as well as the integrated microseismic activity at the same load (Fig. 9-61). In the initial stage of injection of water, though the ε_z curve had a slope increase, the $\Sigma(N)$ curve had a slope decrease. About half an hour after the addition of water, there was a clear acceleration of $\Sigma(N)$ activity. The decrease of seismicity in the initial stage was perhaps due to formation of a film in the cracks facilitating dislocation without acoustic emission. After a certain time, the corrosion effect of water decreased the strength causing enhanced seismicity. Heating the specimen from 25° to 50° caused a sudden increase in seismicity (though it had got established earlier—Fig. 9-62a) and resulted in sudden failure. But when a sample had been previously heated to a higher temperature (150° C) a change in temperature of $\pm 25°$ altered only transient seismicity and the sample stabilised again without causing fracture (Fig. 9-62b). It shows that the major part of the weakening effect as a result of change in temperature was due to presence of water.

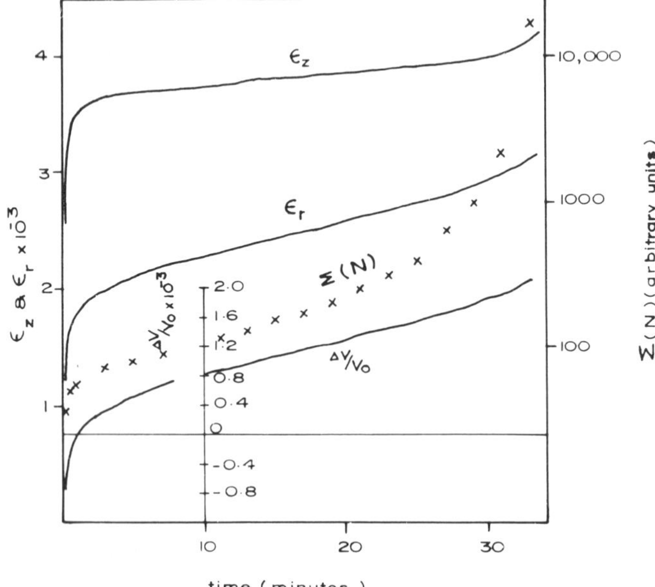

Fig. 9-60. Microfracturing $\Sigma(N)$, ε_z, ε_r and $\Delta V/V$ at 2.25 kbar for Westerly granite (after Wu and Thomsen, 1975).

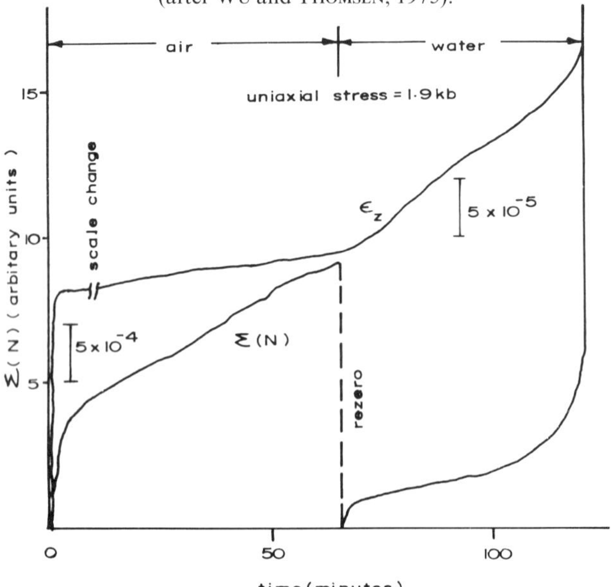

Fig. 9-61. Microfracturing $\Sigma(N)$ and axial strain before and after immersion at 1.9 kbar for Westerly granite (after Wu and Thomsen, 1975).

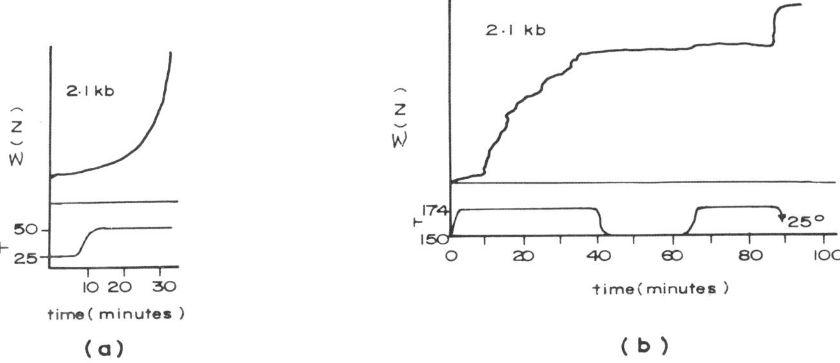

Fig. 9-62. (a) Westerly granite failed catastrophically after heating from 25 to 50° C under 2.1 kbar.
(b) Effects on $\Sigma(N)$ curve of raising and lowering temperature after the specimen has been under compression at 2.1 kbar and 150° C for Westerly granite (after WU and THOMSEN, 1975).

The frequency content of micro-fracturing for Westerly granite, the dominant period of the P-wave at 150° C was found to be about 2μs. After lowering the temperature to 25°–28° C, this mode persisted though other periods started appearing more frequently. In general, the period varied from 1 to 4.5μs.

These results allow to calculate the crack radius. WYSS and HANKS (1972) have utilised seismic source theory according to which the corner frequency of the radiation spectrum depends upon the crack size and is related by

$$r = \frac{1.17v}{\pi f_0(P)} \tag{9.75}$$

where r = radius of a circular crack
v = velocity and
$f_0(P)$ = corner frequency.

Using the above dominant peak frequency the value of r works out to be between 5 and 9 mm. Actual observations using optical microscopes have given the value to be of the order of 1 mm only.

9.14. Theories of Rock Creep

Experimental results on the time-dependent behaviour of rocks discussed earlier make it amply clear that creep in rocks can be described by a logarithmic or power law at lower temperature (room temperature). At higher temperatures, it is found that creep is a function of activation energy and that melting temperatures of the material (T_m) plays an important role.

Work on creep in metals is very much advanced and many theories of creep applicable to metals have been applied to explain creep in rocks.

Most of the theories of creep are based upon 3 main concepts (COTTRELL, 1963):

1. Every material has a certain defined yield stress which must be achieved for flow to take place without the need of any local fluctuation in the stress field. In creep conditions, where the applied stress is smaller than the yield stress, creep takes place due to stress fluctuations and the creep rate is dependent upon the frequency of fluctuations in the stress field.
2. At the initial stages of creep, creep rate is very high because of the applied stress equal to the yield stress.
3. With increase in creep strain, strain hardening takes place and yield stress increases progressively requiring increasingly large fluctuations to bridge the gap between yield stress and applied stress resulting in drop in creep rate. The rate of flow decreases until it is equal to the rate of recovery from strain hardening. Steady-state creep is then observed.

Based upon these basic concepts a number of theories have been developed and are discussed here in brief.

9.14.1. Strain Hardening or Dislocation Theory

This is based upon the concept of existence of dislocations in crystal lattices (MOTT, 1952, 1953, 1956; COTTRELL, 1963). According to this theory, inelastic deformation is first concentrated within the crystal boundaries which happens as long as the slip resistance of the crystals exceeds the resolved shear stresses. The material responds as a visco-elastic body which deforms by creep of the boundary material and by relative motion and rotation of the grains along the boundaries. This character of creep is changed when either the slip resistance of the crystal regions is reduced (at high temperatures) or stress increases such that rotation of the crystals within the boundaries is followed by slip in a certain number of crystals. With further deformation more and more crystals are involved resulting in increasing volume of disordered material produced by slip and fragmentation. This instability promotes local recovery and recrystallisation. This grain boundary deformation, rotation of neighbouring crystal grains, slip and recrystallisation cannot go on indefinitely without the opening up of a large number of cracks within the material. Hence, the third stage creep does not represent a pure deformation process but a process of progressive rupture.

This concept developed by MOTT was basically developed for metals but has been applied to rock also (MISRA and MURRELL, 1965).

According to MOTT (1953), the movement of the dislocation is obstructed and the strain rate $\dot{e}$ can be given by

$$\dot{e} = NA \exp(-U/KT) \tag{9.76}$$

where N = number of obstacles per unit volume
A = parameter describing the size and spacing of dislocations and
U = energy required to overcome an obstacle.

Assuming

$$U = Bhe/F_y \tag{9.77}$$

where B = constant
h = strain-hardening parameter
e = strain and
F_y = yield stress of the material

$$\dot{e} = NA \exp(-Bhe/KTF_y) \tag{9.78}$$

Integrating the Eq. 9.78 from zero to a time, t, we get

$$e_t = Q \log(1 + Pt) \tag{9.79}$$

where $Q = KTF_y/Bh$ and
$P = NA/Q$.

Differentiating Eq. 9.79 we get the creep rate at any given time

$$\dot{e}_t = QP/(1 + Pt) \tag{9.80}$$

or

$$\dot{e}_t = Q/t \tag{9.81}$$

when Pt is large compared to one.

When stress on a test specimen which is creeping is raised after sometime to a higher value and creep test continued (single-increment creep tests) the above equations can be modified as follows (CRUDEN, 1971b):

If the specimen has creeped to a certain strain e_0 in a time t_0 under a stress F_0 and the stress is then raised to F_1, the strain e_t at a time t_1 can be given by (from Eq. 9.78)

$$\int_{e_t}^{e_t+e_0} \exp(e/Q_1)\, de = NA \int_{t_0}^{t_0+t} dt$$

or

$$e_t + e_0 = Q_1 \log(P_1 t + k) \tag{9.82}$$

where $k = \exp(e_0/Q_1)$.

From Eq. 9.82

$$\dot{e}_t = Q_1/(t + k/P_1) \tag{9.83}$$

$$k/P_1 = (P_0 t_0/P_1)^{(Q_0/Q_1)} \tag{9.84}$$

as $e_0 = Q_0 \log(P_0 t_0)$.

Assuming $P_0/P_1 = Q_1/Q_0$, Q_1 and Q_0 proportional to F_1 and F_0, Eq. 9.84 gives

$$k/P_1 = (F_1 t_0/F_0)^{(F_0/F_1)} \tag{9.85}$$

So, from Eqs. 9.83 and 9.85, the form of the curve obtained in the second stage of the single increment creep test (daughter curve) is determined by the form of the first curve (parent curve), the duration of the parent experiment and the stress increment. The strain rate $\dot{e}_{tP}$ in the parent experiment in terms of the strain rate is $\dot{e}_{tD}$ in the daughter experiment can be given by

$$\dot{e}_{tP} = \dot{e}_{tD}(F_0/F_1)(t + (k/P_1))/t \tag{9.86}$$

9.14.2. Exhaustion Hypothesis

This hypothesis assumes the existence of a number of soft spots where a relatively low activation energy is sufficient to cause deformation. These soft spots are brought into existence by the elastic deformation on loading and decelerating creep is due to exhaustion of these soft spots. These soft spots have the following properties (CRUDEN, 1971 b):

1. The characteristic yield stress F_y of each is greater than the applied stress F and their deformation takes place only when the difference between the yield stress and the applied stress is augmented by the activation stress F_a brought about by local fluctuations.
2. The activation energy of all these elements $U(F_a)$ is assumed to be the same function of F_a for all the elements. COTTRELL (1963) suggested

$$U(F_a) = AF_a = B(1 - F/F_y) \tag{9.87}$$

3. When an element is activated, a displacement takes place resulting an increment equal to v to the strain, after which the activation stress of the element becomes so large that the element does not displace again.
4. The displacements in creep are stochastic events.

With these assumptions, if $N(F_a, t)dF_a$ are the number of elements with activation stresses between F_a and $F_a + dF_a$ at a time t after the start of creep and if n is the frequency of the fluctuations, the chance that an element with activation energy F_a will deform in the time interval dt is

$$ndt\, N(F_a, t)\, dFa \exp(-U(F_a)/KT) = f(F_a)\, dt \tag{9.88}$$

The creep rate $\dot{e}_t$ is then

$$\dot{e}_t = v \int_0^\infty N(F_a, t) f(F_a) dF_a \tag{9.89}$$

Since each element is displaced only once, the number of elements available for deformation at each activation stress decreases at a rate

$$d(N(F_a, t))/dt = -N(F_a, t) f(F_a) \tag{9.90}$$

After integration from zero time to t, Eq. 9.90 gives

$$N(F_a, t) = N(F_a, 0) \exp(-f(F_a)t) \tag{9.91}$$

where $N(F_a, 0)$ is the distribution of the elements at the start of creep ($t = 0$).

Substituting Eq. 9.91 into Eq. 9.89

$$\dot{e}_t = v \int_0^\infty N(F_a, 0) \exp(-f(F_a)t) f(F_a) dF_a \tag{9.92}$$

Cottrell (1963) suggested that the distribution $N(F_a, 0)$ might be considered uniform in the interval zero to F_m and zero outside this interval. Thus $N = N(F_a, 0)$ where N is a constant.

Substituting Eq. 9.87 into Eq. 9.88, we get

$$f(F_a) = n \exp(-AF_a/KT) \tag{9.93}$$

Differentiation of the above Eq. 9.93 with respect to F_a, we get

$$\begin{aligned} df(F_a) &= n(-A/KT) \exp(-AF_a/KT) dF_a \\ (-KT/A) df(F_a) &= f(F_a) dF_a \end{aligned} \tag{9.94}$$

Substitution of Eq. 9.94 into Eq. 9.92 gives

$$\dot{e}_t = (vNKT/A) \int_{(F_m)}^{(F_0)} \exp(-f(F_a)t) df(F_a) \tag{9.95}$$

$$\dot{e}_t = (vNKT/At)(\exp(-f(F_a)t))_{f(0)}^{f(F_m)} \tag{9.96}$$

Misra and Murrell (1965) have discussed the evaluation of Eq. 9.96. They suggested that at small values of nt the first two terms of the exponential series are adequate. Then $\exp(-f(0)t) = 1 - nt$ and $\exp(-f(F_m)t) = 1 - nt \exp(-AF_m/KT)$.

Eq. 9.96 can then be written as follows:

$$\dot{e}_t = vNKTn(1-\exp(-AF_m/KT))/A \qquad (9.97)$$

At longer times, they suggested $\exp(-f(0)t)=0$, $\exp(-f(F_m)t)=1$. These conditions represent a situation in which all the lowest activation energy elements have been exhausted while those with much higher activation energies are unaffected. Then Eq. 9.96 can be evaluated as

$$\dot{e}_t = vNKT/At \qquad (9.98)$$

Integrating Eq. 9.98 we get

$$e_t - e_1 = (vNKT/A) \log t \qquad (9.99)$$

where e_1 is the strain at one time unit.

The basis of a simpler method of deriving Eq. 9.99 is due to COTTRELL (1963). When $f(F_a)$ is small, exp $(-f(F_a)t)$ is practically a step function of F_a, being nearly zero or nearly one and changing sharply from one value to the other. The process of exhaustion is equivalent to the advance of the step function across the distribution, $N(F_a)$, towards the higher values of F_a.

If an element with zero activation energy is defined as probability of displacement in the time interval dt equal to one, and as there can be no probability higher than one, the thermodynamic probability of this event can range only from n to zero. The advance of the step function can be measured by the value F_t of the activation stress of those elements which are certain to displace in time t. Thus

$$\begin{aligned} n &= nt \exp(-AF_t/KT) \\ F_t &= (KT/A) \log t \\ e_t &= vNF_t \\ e_t &= (vNKT/A) \log t \end{aligned} \qquad (9.100)$$

COTTRELL's step approximation has been used to extend his theory to increment-creep experiments (CRUDEN, 1971 b). Suppose that after a strain e_0 in time t_0 at stess F_0, the stress is increased to F_1. After a time t at F_1 the position of the step is given by F_t and

$$1 = t \exp(-A_1 F_t/KT)$$

The strain in time t is given as $e_t = vN(F_t-(F_x-F_i))$. F_x is the position of the step after t_0. The strain increment $F_i(=F_1-F_0)$ displaces the step F_i towards the origin

$$e_t = (vNKT/A_1) \log t - vN(F_x - F_i)$$

Negative values of e_t have no physical significance but no strain is recorded till after the elapse of a waiting time t_w which is defined by

$$(vNKT/A_1)\log t_w = vN(F_x - F_i)$$
$$e_t = (vNKT/A_1)\log t - \log t_w \tag{9.101}$$

At times greater than t_w, the strain rate curve takes the form

$$\dot{e}_t = vNKT/A_1 t \tag{9.102}$$

The strain rate in the experiment on 1st loading (parent experiment) $\dot{e}_{tp}$ is then given in terms of the strain rate $\dot{e}_{tD}$ in the subsequent loading experiment by

$$\dot{e}_{tP} = \dot{e}_{tD}(F_0/F_1) \tag{9.103}$$

9.14.3. Structural Theory of Brittle Creep

The concept of the stress-aided corrosion at the tips of the microcracks present in the material has been applied to describe the creep in rock. Corrosion causes the cracks to lengthen and after a certain period of time under condition of sustained tensile stresses at the crack tips, the crack reaches a critical limit and then starts propagating unstably causing complete failure of the specimen. The influence of the tensile stresses is to stretch the bonds between the atoms causing ease of the diffusion of the corroding ions while the influence of compressive stresses is to prohibit corrosion. The influence of temperature on the crack growth is explained by assuming that corrosion is a rate process with an activation energy.

A sub-critical crack under uniaxial compression extends in its own plane by stress corrosion due to the tensile stresses near the crack tips. When it reaches a critical length (L_{cr}) it propagates in the manner described by BRACE and BOMBOLAKIS (1963) causing fracture of the specimen. Then, the principal contribution to creep strain comes from strain about propagating cracks. Once these cracks have propagated, they are "stable" or "crack hardened".

Considering a highly elliptical crack with a major axis L placed in a plate subjected to a uniform stress S_y in the direction perpendicular to the major axis, then crack extension rate can be given by (CHARLES, 1958)

$$(dL_x/dt) = C(S_x/S_{cr})^n + k \tag{9.104}$$

where S_x = tensile stress at the tip of the crack tangential to the crack surface
S_{cr} = tensile strength of the atomic bond at the crack tip
k = corrosion rate at zero stress
n = constant (positive) and
C = maximum velocity of crack.

Two different aspects of stress corrosion which oppose each other occur simultaneously. Stress relaxation helps retard crack growth and stress activation helps accelerate it. If static fatigue is to occur, stress activated corrosion must occur at a much faster rate. In such a case, crack would grow at a constant curvature till it reaches its critical length.

The temperature dependence of crack growth in glasses has been translated that corrosion is a rate process with an activation energy. Under the circumstance (valid for glass), Eq. 9.104 can be written as

$$\dot{L}_x = B(L/L_{cr})^{n/2} \exp(-A/KT) \tag{9.105}$$

Integrating Eq. 9.105 with respect to time and assuming n and t_f are large, we get

$$\log t_f = -n \log S_y - \log D \tag{9.106}$$
$$D = (rS_{cr}^2/4)^{-n/2} L_0^{(n-2)/2} (B/2)(n-2) \exp(A/KT)$$

Eq. 9.106 is a static-fatigue law for thin glass rods under a uniaxial tension S_y. The parameter n can be determined from the slope of a $\log t_f - \log S_y$ plot. CHARLES (1958) has given its value about 16.

The growth of sub-critical cracks under tension has been directly observed in glass (WIEDERHORN, 1967) and in sapphire (WIEDERHORN, 1968). There is reasonable agreement between values of n estimated from this data (CRUDEN, 1969) and CHARLES' estimate and this theory seems plausible.

The extension of CHARLES' theory to the growth of cracks under uniaxial compression requires analysis of the stresses at the crack margin. JAEGER and COOK (1969) have discussed the general problem of stress concentrations at crack tips in a perfectly elastic medium. The maximum tensile stress parallel to the crack margin is at the crack tip when the crack major axis is parallel or perpendicular to the principal stress. When the crack axes are not parallel to the stress axes the maximum tensile stress is not at the crack tip; but it will be close to it if the crack has a high aspect ratio, and this is assumed to be the general case.

Let there be $M(L,\alpha)dL$ cracks in a specimen with lengths at zero time between L_0 and $L_0 + dL$ at angles to the principal stress between α and $\alpha + d\alpha$. Each crack on propagating causes a strain increment v. The total strain de due to those cracks is $M(L,\alpha)v dL$. The time t_f for a crack length L_0 to grow to its critical length L_{cr} is given by (SCHOLZ, 1968b)

$$\begin{aligned} t_f &= \exp(A/KT) L_{cr}^{n/2} (2/B(n-2)) L_0^{-((n-2)/2)} \\ &= E L_0 \end{aligned} \tag{9.107}$$

Similarly, the time $(t_f - dt)$ required for a crack of length $(L_0 + dL)$ to grow to L_{cr} can be given as

$$(t_f - dt) = E(L_0 + dL)^{-((n-2)/2)} \quad (9.108)$$

Subtracting Eq. 9.108 from Eq. 9.107, we get

$$dt = EL_0^{-((n-2)/2)}(1-(1+dL/L_0))^{-((n-2)/2)} \quad (9.109)$$

If dt and dL are small and n is large, Eq. 9.109 can be written

$$dt = ((n-2)/2)EL_0^{-n/2}dL \quad (9.110)$$

Therefore the strain rate at t_f due to the propagating cracks is

$$de/dt_f = M(L,\alpha)\, v dL/dt$$
$$de/dt_f = (2/E(n-2))L_0^{n/2}M(L,\alpha)v \quad (9.111)$$

As it is reasonable to expect more short cracks than long ones, the $M(L, \alpha)$ is unlikely to be independent of L_0. In the absence of any direct way to determine crack length distribution, an indirect method based on GILVARRY's analysis of the size distribution of the fragments in the single fracture of an infinitely extensive, brittle body due to the propagation of internal flaws can be used (OROWAN, 1967). It leads to the suggestion (CRUDEN, 1969)

$$M(\alpha)L^{-m} = M(L,\alpha) \quad (9.112)$$

and thus

$$de/dt_f = (2/(n-2)E)L_0^{(n-2m)/2}M(\alpha)v \quad (9.113)$$

As $L_0 = (t_f/E)^{-(2/(n-2))}$, from Eq. 9.107

$$de/dt_f = ((B\exp(-A/KT))^{(2m-2)}(2/(n-2)t)^{(n-2m)}$$
$$L_{cr}^{-n(m-1)})^{1/(n-2)}M(\alpha)v \quad (9.114)$$

As $S_{cr} = S_y 2(L_{cr}/r)^{1/2}(\sin^2\alpha - \sin\alpha)$ (HOEK, 1965) using CHARLES' notation for cracks inclined at α to the principal compressive stress S_y

$$de/dt_f = ((B\exp(-A/KT))^{(2m-2)}(2/(n-2)t)^{(n-2m)}$$
$$(S_y(\sin^2\alpha - \sin\alpha)/S_{cr}r^{1/2})^{2n(m-1)})^{1/(n-2)}M(\alpha)v \quad (9.115)$$

The creep rate of the whole specimen is the sum of the values of Eq. 9.115 for all the appropriate values of α. Unfortunately, the values of A, B, v and $M(\alpha)$ are uncertain. However, as Eq. 9.115 does predict the time, temperature and stress dependence of the transient-creep rate in the specimen, the theory can still be tested.

The theory has been extended to describe the creep behaviour of the specimen in increment tests (CRUDEN, 1969).

Suppose a crack of length L, at zero time, has a length L_0, after a duration t_0.

From Eq. 9.107

$$t_0 = E_0(L^{-(n-2)/2} - L_0{}^{-(n-2)/2})$$

$$L_0 = L(1 - t_0 L^{(n-2)/2}/E_0)^{-2/(n-2)} \tag{9.116}$$

Similarly, a crack length $(L + dL)$ will have length L_0' after time t_0.

$$L_0' = (L + dL)(1 - t_0(L + dL)^{(n-2)/2}/E_0)^{-2/(n-2)} \tag{9.117}$$

The difference $L_0' - L_0$ can be given (approximately) by

$$L_0' - L_0 = (1 - t_0 L^{(n-2)/2}/E_0)^{-2/(n-2)} dL \tag{9.118}$$

Eq. 9.118 is an increasingly coarse approximation as t_0 increases. Because there were $M(L, \alpha)dL$ cracks between L and $L + dL$, the crack density after time t_0, M_0 is given by

$$M(L, \alpha)dL = M_0(1 - t_0 L^{(n-2)/2)}/E_0)^{-2/(n-2)} dL \tag{9.119}$$

If after a duration t_0 at S_y, the stress is raised to S_{y1}, then the creep rate at a time t after this event can bė given by

$$de/dt = (2/(n-2)E_1)L_0^{(n-2m)/2} M_0 v dL \tag{9.120}$$

Eq. 9.120 can be written in terms of the original length of the cracks at zero time using Eqs. 9.116 and 9.119

$$de/dt = (2/(n-2)E_1)\,(L(1 - t_0 L^{(n-2)/2}/E_0)^{-2/(n-2)})^{(n-2m)/2}$$

$$M(L_0, \alpha)/(1 - t L_0{}^{(n-2)/2}/E_0)^{-2/(n-2)} \tag{9.121}$$

The creep rate at some time t_x (greater than t_0) under stress S_y is given by

$$de/dt_x = (2/(n-2)E_0)L^{(n-2m)/2} M(L, \alpha) v$$

where $t_x = E_0 L^{-(n-2)/2}$ from Eq. 9.107. Therefore,

$$de/dt = de/dt_x \cdot E_0/E_1 \cdot (1 - t_0/t_x)^{-(n-2m-2)/(n-2)} \tag{9.122}$$

Since $t_x = t_0 + (E_0/E_1)t$, Eq. 9.122 gives the increment curve in terms of the parent curve and is thus a suitable form for predicting the increment curve.

9.14.4. General Mechanism of Creep

The mechanism of deformation in rocks and minerals has been studied by a number of investigators at room temperatures, at elevated temperatures and high pressures (GRIGGS and MILLER, 1951; HANDIN and GRIGGS, 1951; TURNER et al, 1956; BREDTHAUER, 1957; HANDIN and HAGER, 1957, 1958; BORG et al, 1960; GRIGGS et al, 1960; WHITE, 1971). Some of the important findings are cited below (LAMA, 1969).

HUGHES and MAURETTE (1957) found that in calcite crystals twin gliding along certain planes is a dominant mechanism of deformation at all temperatures while translation gliding at low temperatures is rather subordinate. TURNER et al (1956) observed in Yule marble that external rotation (rotation of the optic axis of the grain through some angle depending upon the stress direction) invariably accompanies plastic deformation. The higher the temperature or greater the plastic deformation, the more distinct is the development of orientation. GRIGGS et al (1960) reported that at temperatures above 400°C intragranular and intergranular recrystallisation is still associated along with orientation of grains and twinning. The greatest development of recrystallisation takes place at 600°C but the process declines at higher temperatures because of the two opposing effects, namely increase in the rate of recrystallisation activity because of decrease in strength of marble with rise in temperature.

GRIGGS and MILLER (1951) and HANDIN and GRIGGS (1951) found that deformation of calcite crystals in Yule marble at room temperature and high pressures is due to intercrystalline gliding, the important glide mechanism being translation gliding and mechanical twinning.

The work of HIGGS and HANDIN (1959) on the deformation of single crystals of dolomite at temperatures up to 300°C and confining pressures up to 5000 atm, showed that translation gliding or twinning is prominent in strained specimens. HANDIN and HAGER (1958) conducted tests on Hasmark dolomite up to 2000 atm. and up to 500°C and found that beginning with a temperature of 300°C, the yield point declines sharply. They explain this as due to intracrystalline gliding. GRIGGS, TURNER and HEARD (1960) conducted microstructural studies and found that plastic deformation of dolomite takes place by translation gliding while twinning develops extensively only at higher temperatures.

Work of HANDIN and HAGER (1958) on single crystals of rock salt showed that temperature is the basic factor affecting the strain process. In their work on Blain anhydrite, HANDIN and HAGER (1957) showed that its ultimate strength increases with temperature and is associated with strain hardening and that intracrystalline gliding is facilitated with rise in temperature.

BORG et al (1960) conducted tests on St. Peter sandstone and concluded that confining pressure leads to compression of sand grains by forces applied to the point of contact with neighbouring grains and these forces cause sharp concentrations leading to brittle fracture. GRIGGS, TURNER and HEARD (1960) did not detect any plastic deformation in microstructure of deformed specimens.

BREDTHAUER (1957) conducted tests on sandy shales (70% sand, and rest cement containing half clay and half siliceous material) and Rush Spring polymictic sandstones (with 60% grains and rest dolomite cement) and found that while the capacity for plastic deformation of the former is due to clay the brittle failure of the latter is due to dolomite cement. This indicates that deformability of sandstones at pressures is governed chiefly by cementing material.

Work on silicate rocks has been limited to room temperatures and not much of systematic work on the changes in microstructure has been done. CINZBURG and ROZANOV (BAIDYUK, 1967) conducted tests on granite (syenite) by enclosing them in steel jackets and found difference in the deformability of various minerals present in granite (biotite, aegerine, arfvedsonite etc). Biotite deformed most easily followed by aegerine and arfvedsonite and even filled in the fractures formed. They further found that individual parts of the crystals reoriented by crushing and displacement.

GRIGGS, TURNER and HEARD (1960) tested dunite at high pressures and temperatures and found that plastic deformation with crushing of pyroxene and olivine takes place at smaller pressures while at higher pressures and deformation there is further segregation of subgrains and disorientation within individual grains. In pyroxenite, they found crushing and bending of grains with twinning. In granite, they found kink bands in biotites. These bands were found to be zones of strong concentration of plastic deformation with some sort of gliding.

WHITE (1971) studied deformation effects in quartzites from various regions of metamorphic environments. He examined petrological slides optically and from these discs were selected for electron microscopic studies. He found that creep in the deformation of quartzites is not a simple process of diffusion controlled dislocation movement but encompasses grain boundary sliding and migration, micro-fracturing, recrystallisation and complex dislocation movements involving not only the movement of individual dislocations but also walls of dislocations as usually happens in the natural creep deformation of olivines.

MURRELL and CHAKRAVARTY (1971) conducted creep tests on dolerite, microgranodiorite and dunite up to 1000° C but at lower stress levels and found that creep rate follows ANDRADE's law and never reached a constant value. These rocks at this temperature and at low pressure behave in a brittle manner.

Boland and Hobbs (1973) deformed 7 mm diameter, 15 mm long specimens of peridotite at 0.5 to 5 kb confining pressures at temperatures from 780 to 950° C and studied the specimens using transmission electron microscope with ion bombardment technique. Three broad categories of cataclastic microstructure were distinguished by them.

(a) Clean microcracks with very little plastic deformation.
(b) Micro-fractures with large amount of displacement (up to 70 μm) along the fracture surface.
(c) Blunted microcracks by severe plastic deformation.

Even at small deformations to which the specimens were subjected (under 8 %), the specimens showed a considerable cataclastic strain component.

All these studies indicate that creep is associated with twinning, intercrystalline gliding, slippage along grain boundaries, translation gliding, strain hardening etc., but all these changes in the micro-structure of rocks take place at high pressures and temperatures or both. At ordinary temperatures (room temperature) and low pressures, it is microfracturing that plays a dominant role in the creep behaviour of rocks, particularly in such rocks which have low strength compared to their grain strength. A general theory of creep of polycrystalline rock may thus be described as follows:

Creep in polycrystalline materials such as rocks may be due to interaction of constituent phases during deformation and due to the difference in the responses of the phases to changes, stress and temperature with time. The character of the interaction between the phases depends upon relative rigidity of the phases. In the earlier stages, deformation may be concentrated in one phase (e.g. intercrystalline boundaries) with the transfer of the response to the other phase (e.g. crystalline phase). In other terms, this means at earlier stages, flow shall be essentially like a relatively simple visco-elastic material in which the response of some phases is elastic and the other phase inelastic. At lower stresses, the creep rate will reduce with time as the skeleton becomes more and more rigid with greater contact of the more rigid phase. This is what happens in the transient and secondary stages of the creep curve, where the creep rate slowly reduces and may become zero at sufficiently low stresses. This concept has been supported by Voropinov (1964) and Dreyer (1972).

At higher stresses, this character may be changed where the resistance of the second, more rigid, phase is overcome resulting in fragmentation. This gives rise to what is termed as secondary creep. With deformation, greater and greater number of particles become aligned followed by intercrystalline gliding and slip; this process may continue for a much longer time at constant rate.

As indicated, tertiary creep which is not purely a process of deformation but progressive damage is also involved. It is obvious that its rate will increase with the passage of time till final fracture.

Microfractures will enhance the effect of these stages of creep. Some suitably oriented cracks will propagate first as the critical (minimum) stress for these is reached, followed by other cracks, which are less favourably oriented and less severe giving rise to secondary creep. At final stages, because of large scale destruction of the specimen, crack propagation may become unstable giving tertiary creep.

Single increment creep tests conducted by CRUDEN (1971 b) on Carrara marble and Pennant sandstone have shown that the structural theory of creep which suggests that creep is caused by stress aided corrosion gives a satisfactory fit to the test data. The other theories such as strain hardening theory and exhaustion theory do not fit well. WAWERSIK (1974) showed that secondary creep for water saturated Westerly granite decreases markedly as the octahedral normal stress is increased relative to the octahedral shear stress. Increase in the ratio of these stresses inhibits crack initiation and propagation indicating rock creep as structural phenomenon. However, the test conditions in which creep occurs are important and it is possible that under higher pressures and temperatures, the other two theories may equally fit well. There definitely occurs a transition zone due to the difference in the mechanism governing deformation process. GOETZE (1971) found poor agreement between the high temperature and low temperature creep results on Westerly granite. Work of RENZHIGLOV and PAVLISHCHEVA (1970) on softer rocks (clay shale and sandstone) showed that creep rate and viscosity of a rock are not constant values and perhaps dependent upon the different phases of the deformation process. In soils it has been well demonstrated by TER-STEPANIAN (1975) that there are sudden changes in the creep rate which he refers to as reorganisation of the soil fabric. Fabric changes in rock with creep have been noted by HOSHINO and KOIDE (1970) for sandstone. Tests were conducted in a triaxial cell. The rose pattern of fractures obtained varied at different stages of deformation at almost constant load (peak strength). At lower deformation, the crack pattern was within the bands of maximum shear, while at higher deformation, there was an increasing number of axial cleavage fractures. It is, therefore, obvious that no one theory could be applicable.

9.15. Summary and Conclusions

Even the ideal solids due to the very nature of the forces holding the atoms together do not behave according to HOOKE's law for linear elasticity when stressed beyond a certain limit, much less in the case of rocks due to their non-homogeneity and differential response of the different constituents forming them. The mechanical behaviour of rocks is governed by the mutual response of the different constituents depending upon which of the constituents under a given condition starts yielding and plays a dominant role.

The large deformation that rocks exhibit under constraint at low confining pressures and low temperatures (fraction of the melting temperature) is not an indication of plasticity of rocks in the true sense but a result of propagation of a large number of cracks. Constraint and friction between the fractured parts do not allow them to fall apart. It is only at higher pressures and temperatures that deformation is associated with twinning, intercrystalline gliding, slippage along grain boundaries etc. (In certain e.g. calcite rocks, this may happen at lower temperatures and pressures too.)

The mechanical behaviour of rock under stress as a function of time can be represented by a combination of springs and dashpots and the most suitable type of models are the generalised KELVIN's model and BINGHAM's model. These models can represent sufficiently accurately the creep behaviour of rocks.

The lever operated dead-weight loaded or spring-loaded machines are quite satisfactory in creep work.

The idealised creep curve for rock at constant stress can be represented by Fig. 9-20 and consists of the following 4 sections.

1. the instantaneous elastic strain (AB),
2. primary or transient creep (BC),
3. secondary or steady-state creep (CD) and
4. tertiary or accelerated creep (DE).

A large number of factors influence the creep properties of rocks. The most important of these are stress level, temperature, type of stress and humidity. In certain rocks such as rock salt, grain size plays an important role.

The mechanism of creep of small specimens of rocks in the laboratory can be interpreted in terms of the structural theory of brittle creep. Certain indications are that the creep behaviour of rock-mass in situ can be interpreted in terms of the BOLTZMANN-VOLTERRA theory.

The determination of time-dependent strength of rocks by short term methods is quite difficult and no definite reliable method is as yet available. Many investigators have used dilatancy method (BIENIAWSKI, 1967; SINGH, 1970) but indications are that this gives higher values than reality. The stress level at which crack initiation with time just takes place represents the long-term strength. SANGHA and DHIR (1972) concluded that this occurs at the stress level at which incremental POISSON's ratio becomes 0.5. Micro-seismic technique has been used to assess stability of workings and this may be useful to determine time-dependent strength. Stress relaxation method of PUSHKAREV and AFANASEV (1973) seems very promising.

The creep failure envelope of rock specimen in the post-failure envelope lies very close to the normal post-failure curve, but far apart for the solid rock specimen. Under the dynamic loads, the failure envelope could be represented very closely with the post-failure curve and the fatigue strength is about 65 to 75 % of the quasi-static strength.

References to Chapter 9

1. ADAMS, F.D. and NICOLSON, J.T.: An experimental investigation into the flow of marble. Phil. Trans. Roy. Soc. London, Series A, Vol. 195, 1901, pp. 363–401.
2. AFROUZ, A. and HARVEY, J.M.: Rheology of rocks within the soft to medium strength range. Int. J. Rock Mech. Min. Sci. & Geomech. Abstr., Vol. 11, No. 7, July, 1974, pp. 281–290.
3. ALBRECHT, H. and LANGER, M.: The rheological behaviour of rock salt and related stability problems of storage caverns. Proc. 3rd Cong. Int. Soc. Rock Mech., Denver, 1974, Vol. 2, Part B, pp. 967–974.
4. ANDRADE, E.N. da C.: On the viscous flow in metals, and allied phenomena. Proc. Roy. Soc. London, Series A, Vol. 84, 1910, pp. 1–12.
5. ANDRADE, E.N. da C.: The flow in metals under large constant stresses. Proc. Roy. Soc. London, Series A, Vol. 90, 1914, pp. 329–342.
6. ANTONIDES, L.E.: How you can predict rock falls. Eng. Min. J., Vol. 156, No. 12, Dec., 1955, pp. 75–77, 103.
7. ATTEWELL, P.B. and FARMER, I.W.: Fatigue behaviour of rock. Int. J. Rock Mech. Min. Sci., Vol. 10, No. 1, Jan., 1973, pp. 1–9.
8. BAIDYUK, B.V.: Mechanical properties of rocks at high temperatures and pressures. Translated from Russian by J.P. Fitzsimmons. New York, Consultants Bureau, 1967, 75 p.
9. BARRON, K. and TOEWS, N.A.: Deformation around a mine shaft in salt. Proc. 2nd Can. Rock Mech. Symp., Kingston, Ontario, 1963, pp. 115–134.
10. BIENIAWSKI, Z.T.: Mechanism of brittle fracture of rock. Parts 1, 2 and 3. Int. J. Rock Mech. Min. Sci., Vol. 4, No. 4, Oct., 1967, pp. 395–430.
11. BIENIAWSKI, Z.T.: Time-dependent behaviour of fractured rock. Rock Mech., Vol. 2, No. 3, Sept., 1970, pp. 123–137.
12. BOBROV, G.F.: Anisotropy and rheological properties of Kuzbass coal. Sov. Min. Sci., No. 2, March-April, 1970, pp. 159–163.
13. BOLAND, J.N. and HOBBS, B.E.: Microfracturing process in experimentally deformed peridotite. Int. J. Rock Mech. Min. Sci. & Geomech. Abstr., Vol. 10, No. 6, Nov., 1973, pp. 623–626.
14. BORG, I., FRIEDMAN, M., HANDIN, J. and HIGGS, D.V.: Experimental deformation of St. Peter sand: A study of cataclastic flow. Geol. Soc. Am. Mem. 79, 1960, pp. 133–191.
15. BRACE, W.F.: Brittle fracture of rocks. Proc. Int. Conf. State of Stress in the Earth's Crust, Santa Monica, California, 1963, pp. 110–174.
16. BRACE, W.F. and BOMBOLAKIS, E.G.: A note on brittle crack growth in compression. J. Geophys. Res., Vol. 68, No. 12, June 15, 1963, pp. 3709–3713.
17. BRACE, W.F., PAULDING, B.W. and SCHOLZ, C.: Dilatancy in the fracture of crystalline rocks. J. Geophys. Res., Vol. 71, No. 16, Aug. 15, 1966, pp. 3939–3953.
18. BRADSHAW, R.L., BOEGLY, W.J. and EMPSON, F.M.: Correlation of convergence measurements in salt mines with laboratory creep-test data. Proc. 6th Symp. Rock Mech., Rolla, Missouri, 1964, pp. 501–514.
19. BREDTHAUER, R.O.: Strength characteristics of rock samples under hydrostatic pressure. Trans. A.S.M.E., Vol. 79, 1957, pp. 695–706.
20. BRIDGMAN, P.W.: Volume changes in the plastic stages of simple compression. J. Appl. Phys., Vol. 20, 1949, pp. 1241–1251.

21. BURDIN, N.T.: Rock failure under dynamic loading conditions. Soc. Pet. Eng. J., Vol. 3, 1963, pp. 1–24.

22. BUTCHER, B.M., CORNISH, R.H. and ROUFF, A.L.: Techniques for measuring creep strains at high hydrostatic pressures. Am. Soc. Mech. Eng., Paper 64-WA-PT-30 for Meeting in Nov., 1964, 7 p.

23. CAIN, P.J., PENG, S.S. and PODNIEKS, E.R.: Rock fragmentation by high frequency fatigue, U.S.B.M., Rep. Invest. 8020, 1975, 21 p.

24. CAREY, S.W.: The rheid concept in geotectonics. J. Geol. Soc. Aust., Vol. 1, 1953, pp. 67–117.

25. CHARLES, R.J.: Static fatigue of glass. J. Appl. Phys., Vol. 29, 1958, pp. 1554–1567.

26. CHUGH, Y.P.: Viscoelastic behaviour of geologic materials under tensile stress. Trans. A.I.M.E., Vol. 256, 1974, pp. 259–264.

27. CHUGH, Y.P., HARDY, H.R. and STEFANKO, R.: Microseismic activity in rocks under uniaxial tension. Trans. Am. Geophys. Union, Vol. 48, 1967, pp. 204–205.

28. COOK, N.G.W.: The failure of rock. Int. J. Rock Mech. Min. Sci., Vol. 2, No. 4, Dec., 1965, pp. 389–403.

29. COTTRELL, A.H.: The time laws of creep. J. Mech. Phys. Solids, Vol. 1, 1952, pp. 53–63.

30. COTTRELL, A.H.: Dislocations and plastic flow in crystals. Oxford, Clarendon Press, 1963, 223 p.

31. CROUCH, S.L.: Experimental determination of volumetric strains in failed rock. Int. J. Rock Mech. Min. Sci., Vol. 7, No. 6, Nov., 1970, pp. 589–603.

32. CRUDEN, D.M.: A laboratory study of time-strain behaviour and acoustic emission of stressed rock. Ph. D. Thesis, Univ. London, London, 1969.

33. CRUDEN, D.M.: The form of the creep law for rock under uniaxial compression. Int. J. Rock Mech. Min. Sci., Vol. 8, No. 2, March, 1971a, pp. 105–126.

34. CRUDEN, D.M.: Single-increment creep experiments on rock under uniaxial compression. Int. J. Rock Mech. Min. Sci., Vol. 8, No. 2, March, 1971b, pp. 127–142.

35. CRUDEN, D.M.: The static fatigue of brittle rock under uniaxial compression. Int. J. Rock Mech. Min. Sci. & Geomech. Abstr., Vol. 11, No. 2, Feb., 1974, pp. 67–73.

36. DESAYI, P. and VISWANATHA, C. S.: True ultimate strength of plain concrete. Bull., RILEM, No. 36, Sept., 1967, pp. 163–173.

37. DREYER, W.: The science of rock mechanics—Part 1—The strength properties of rocks. Clausthal-Zellerfeld, Trans Tech Publications, 1972.

38. EVANS, R.H.: The elasticity and plasticity of rocks and artificial stone. Proc. Leeds Phil Lit. Soc., Vol. 3, 1936, pp. 145–158.

39. EVANS, R.H. and WOOD, R.H.: Transverse elasticity of natural stones. Proc. Leeds Phil. Lit. Soc., Vol. 3, 1937, pp. 340–352.

40. GAROFALO, F.: Fundamentals of creep and creep rupture in metals. New York, Macmillan, 1966.

41. GOETZE, C.: High temperature rheology of Westerly granite. J. Geophys. Res., Vol. 76, No. 5, Feb. 10, 1971, pp. 1223–1230.

42. GOLD, L.W.: The cracking activity in ice during creep. Can. J. Phys., Vol. 38, No. 9, Sept., 1960, pp. 1137–1148.

43. GORANSON, R.W.: "Flow" in stressed solids: An interpretation. Geol. Soc. Am. Bull., Vol. 51, 1940, pp. 1023–1033.

44. GRIGGS, D.T.: Deformation of rocks under high confining pressures —1: Ex-

periments at room temperature. J. Geol., Vol. 44, No. 5, July-Aug., 1936, pp. 541–577.

45. GRIGGS, D.: Creep of rocks. J. Geol., Vol. 47, No. 3, April-May, 1939, pp. 225–251.

46. GRIGGS, D.: Experimental flow of rocks under conditions favouring recrystallisation. Geol. Soc. Am. Bull., Vol. 51, 1940, pp. 1001–1022.

47. GRIGGS, D.T.: Hydrolytic weakening of quartz and other silicates. Geophys. J. Roy. Astr. Soc., Vol. 14, 1967, pp. 19–31.

48. GRIGGS, D. and MILLER, W.B.: Deformation of Yule marble: Part 1. Geol. Soc. Am. Bull., Vol. 62, 1951, pp. 853–862.

49. GRIGGS, D., TURNER, F.J., BORG, I. and SOSOKA, J.: Deformation of Yule marble: Part 4. Geol. Soc. Am. Bull., Vol. 62, 1951, pp. 1385–1405.

50. GRIGGS, D., TURNER, F.J., BORG, I. and SOSOKA, J.: Deformation of Yule marble: Part 5. Geol. Soc. Am. Bull., Vol. 64, 1953, pp. 1327–1342.

51. GRIGGS, D.T., TURNER, F.J. and HEARD, H.C.: Deformation of rocks at 500° to 800° C. Geol. Soc. Am. Mem. 79. 1960, pp. 39–104.

52. GROVER, H.J., DEHLINGER, P. and MCCLURE, G.M.: Investigation of fatigue characteristics of rocks. Rep. Battelle Mem. Inst. Drilling Res. Inc., Nov. 30, 1950.

53. HAIMSON, B.C. and KIM, C.M.: Mechanical behaviour of rock under cyclic fatigue. Proc. 13th Symp. Rock Mech., Urbana, Illinois, 1971, pp. 845–863.

54. HANDIN, J.W. and GRIGGS, D.: Deformation of Yule marble: Part 2. Predicted fabric changes. Geol. Soc. Am. Bull., Vol. 62, 1951, pp. 863–885.

55. HANDIN, J. and HAGER, R.V.: Experimental deformation of sedimentary rocks under confining pressure: Tests at room temperature on dry samples. Bull. Am. Assoc. Pet. Geol., Vol. 41, No. 1, Jan., 1957, pp. 1–50.

56. HANDIN, J. and HAGER, R.V.: Experimental deformation of sedimentary rocks under confining pressure: Tests at high temperature. Bull. Am. Assoc. Pet. Geol., Vol. 42, No. 12, Dec., 1958, pp. 2892–2934.

57. HARDY, H.R.: Time-dependent deformation and failure of geologic materials. Proc. 3rd Symp. Rock Mech., Golden, Colo., 1959, pp. 135–175.

58. HARDY, H.R.: Determination of the inelastic parameters of geologic material from incremental creep experiments. Proc. 3rd Conf. Drilling and Rock Mech., Austin, Texas, 1967a.

59. HARDY, H.R.: Analysis of the inelastic deformation of geologic materials in terms of mechanical models. Paper presented at Soc. for Exp. Stress Analysis, 1967 Spring Meeting, Ottawa, Canada, 1967b.

60. HARDY, H.R. and CHUGH, Y.P.: Failure of geologic materials under low cycle fatigue. Proc. 6th Can. Rock Mech. Symp., Montreal, 1970, pp. 33–47.

61. HARDY, H.R., KIM, R.Y., STEFANKO, R. and WANG, Y.J.: Creep and microseismic activity in geologic materials. Proc. 11th Symp. Rock Mech., Berkeley, California, 1969, pp. 377–413.

62. HEARD, H.C.: The effect of time on the experimental deformation of rocks. J. Geophys. Res., Vol. 66, 1961, p. 2534.

63. HEARD, H.C. and CARTER, N.L.: Experimentally induced natural intragranular flow in quartz and quartzite. Am. J. Sci., Vol. 266, 1968, pp. 1–42.

64. HEDLEY, D.G.F.: An appraisal of convergence measurements in salt mines. Proc. 4th Can. Rock Mech. Symp., Ottawa, 1967, pp. 117–135.

65. HIGGS, D.V. and HANDIN, J.: Experimental deformation of dolomite single crystals. Geol. Soc. Am. Bull., Vol. 70, 1959, pp. 245–278.

66. Hobbs, D.W.: Stress-strain-time behaviour of a number of coal measure rocks. Int. J. Rock Mech. Min. Sci., Vol. 7, No. 2, March, 1970, pp. 149–170.

67. Hoek, E.: Rock fracture under static stress conditions. Rep. S. African C.S.I.R., No. MEG 383, 1965, 228 p.

68. Hofer, K.H.: The principles of creep in rock salts and their general significance to mining engineering. Proc. Int. Strata Control Cong., Leipzig, 1958, pp. 49–63.

69. Hofer, K.H.: The results of rheological studies in potash mines. Proc. 4th Int. Conf. Strata Control and Rock Mech., New York, 1964.

70. Hofer, K.H. and Knoll, P.: Investigations into the mechanism of creep deformation in carnallitite, and practical applications. Int. J. Rock Mech. Min. Sci., Vol. 8, No. 1, Jan., 1971, pp. 61–73.

71. Hooper, J.A.: Apparatus for applying sustained loads to large specimens. Int. J. Rock Mech. Min. Sci., Vol. 4, No. 3, July, 1967, pp. 353–361.

72. Hoshino, K. and Koide, N.: Process of deformation of the sedimentary rocks. Proc. 2nd Cong. Int. Soc. Rock Mech., Belgrade, 1970, Vol. I, pp. 353–359.

73. Hughes, D. and Maurette, C.: Determination des vitesses d'onde elastique dans diverses roches en fonction de la pression et de la temperature. Rev. Inst. Franc. Petr. et Ann. Comb. liquides, Vol. 7, No. 6, 1957.

74. Iyengar, K.T.S.R., Desayi, P. and Viswanatha, C.S.: A new approach for the prediction of true ultimate strength of concrete. Mat. Res. Stand., Vol. 7, No. 11, Nov., 1967, pp. 486–493.

75. Jaeger, J.C. and Cook, N.J.W.: Fundamentals of rock mechanics. London, Methuen, 1969, 513 p.

76. Jones, R.: The failure of concrete test specimens in compression and flexure. Proc. Conf. Mech. Prop. Non-metallic Brittle Materials, London, 1958, pp. 29–32.

77. Kanagawa, T. and Nakaarai, K.: Restraint of swelling creep and effect of absorption of water on triaxial strength and deformability of rocks. Rock Mechanics in Japan, Vol. 1, 1970, pp. 74–76.

78. Kidybinski, A.: Rheological models of upper Silesian carboniferous rocks. Int. J. Rock Mech. Min. Sci., Vol. 3, No. 4, Nov., 1966, pp. 279–306.

79. Kubetsky, V.L. and Eristov, V.S.: In situ investigations of creep in rock for the design of pressure tunnel linings. Proc. Conf. In situ Invest. Soils and Rocks, London, 1969, pp. 83–91.

80. Kubetsky, V.L. and Ukhov, S.B.: Towards the investigation of deformation of fissured rock masses based on the theory of creep. Donau-Europäische Konferenz für Bodenmechanik in Strassenbau, Vienna, 1966.

81. Kuznetsov, Y.U.F. and Vashchilin, V.A.: Rock creep at high temperature. Sov. Min. Sci., No. 5, Sept.-Oct., 1970, pp. 586–588.

82. Ladanyi, B.: An engineering theory of creep of frozen soils. Can. Geotechnical J., Vol. 9, 1972, pp. 63–80.

83. Lama, R.D.: Effect of nonhomogeneities and discontinuities on deformational behaviour and strength of rocks, Metals and Minerals Review, Vol. VIII, 1969, No. 7, pp. 3–10.

84. Lama, R.D.: SFB-77, Jahresbericht 1973, Inst. Soil Mech. and Rock Mech., Univ. Karlsruhe, Karlsruhe.

85. Lama, R.D.: Rheological properties of rocks, SFB-77, Report No. K-128, University of Karlsruhe, August 1972.

86. Langer, M.: Grundlagen einer theoretischen Gebirgskörpermechanik, Proc. Ist

Cong. Int. Soc. Rock Mech., Lisbon, 1966, vol. I, pp. 277–282.

87. LANGER, M.: Rheologie der Gesteine. Deut. Geol. Ges. Z., Vol. 119, Aug., 1969, pp. 313–425.

88. LANGER, M.: The determination of rheological numerical characteristics of rocks. Proc. 2nd Cong. Int. Soc. Rock Mech., Belgrade, 1970, Vol. 1, pp. 405–414.

89. LANGER, M.: Mechanical and rheological characteristics of rock as natural building material. Proc. 2nd Int. Cong. Int. Assoc. Eng. Geol., Sao Paulo, 1974, Paper IV-PC-3.1.

90. LECOMTE, P.: Creep in rock salt. J. Geol., Vol. 73, No. 3, May, 1965, pp. 469–484.

91. LOMENICK, T.F. and BRADSHAW, R.L.: Acelerated deformation of rock salt at elevated temperature. Nature, Vol. 207, No. 4993, July 10, 1965, pp. 158–159.

92. LOMENICK, T.F. and BRADSHAW, R.L.: Deformation of rock salt in openings mined for the disposal of radioactive wastes. Rock Mech., Vol. 1, 1969, pp. 5–29.

93. LOMNITZ, C.: Creep measurements in igneous rocks. J. Geol., Vol. 64, 1956, pp. 473–479.

94. LOONEN, H.E. and HOFER, K.H.: Proc. 6th Meeting Int. Bur. Rock Mech., Leipzig, 1964.

95. MATSUSHIMA, S.: On the flow and fracture of igneous rocks. Kyoto Univ. Disaster Prevent. Res. Inst. Bull., Vol. 10, No. 36, 1960, pp. 2–9.

96. MCCLAIN, W.C.: Time-dependent behaviour of pillars in the Alsace potash mines. Proc. 6th Symp. Rock Mech., Rolla, Missouri, 1964, pp. 489–500.

97. MCCLAIN, W.C.: The effect of nonelastic behaviour of rocks. Proc. 8th Symp. Rock Mech., Minneapolis, Minn., 1966, pp. 204–216.

98. MCCLAIN, W.C. and BRADSHAW, R.L.: Deformation and stress transference in a salt mine resulting from the application of heat. Proc. 2nd Cong. Int. Soc. Rock Mech., Belgrade, 1970, Vol. 1, pp. 559–565.

99. MENZEL, W. and SCHREINER, W.: Das Festigkeits- und Verformungsverhalten von Carnallitit als Grundlage für die Standsicherheitsbewertung von Grubenbauen. Neue Bergbautechnik, Vol. 5, No. 6, 1975, pp. 451–457.

100. MICHELSON, A.A.: The laws of elastico-viscous flow. J. Geol., Vol. 25, No. 5, July-Aug., 1917, pp. 405–410.

101. MICHELSON, A.A.: The laws of elastico-viscous flow. II. J. Geol., Vol. 28, 1920, pp. 18–24.

102. MISRA, A.K. and MURRELL, S.A.F.: An experimental study of the effect of temperature and stress on the creep of rocks. Geophys. J. Roy. Astr. Soc., Vol. 9, No. 5, July, 1965, pp. 509–535.

103. MONKMAN, F.C. and GRANT, N.J.: An empirical relationship between rupture life and minimum creep rate in creep rupture tests. A.S.T.M. Proc., Vol. 56, 1956, pp. 593–605.

104. MONTOTO, M.: Fatigue in rocks: Failure and internal fissuration of Barre granite under loads cyclically applied, 3rd Cong. Int. Soc. Rock Mech., Denver, Colorado, Vol. II-A, 1974, pp. 379–84.

105. MOTT, N.F.: A theory of work-hardening of metal crystals. Phil. Mag., Vol. 43, 1952, pp. 1151–1178.

106. MOTT, N.F.: Theory of strain hardening of metals—II. Phil. Mag., Vol. 44, 1953, pp. 742–765.

107. MOTT, N.F.: A discussion of some models of the rate-determining process in creep. Proc. Nat. Phys. Lab. Symp. on the Creep and Fracture of Metals at High Tem-

perature, London, H. M. Stationery Office, 1956, pp. 21–24.

108. MTS Systems Corporation, Grips and Tension Fixtures, Catalogue, 1975.

109. MURRELL, S. A. F.: Micromechanical basis of the deformation and fracture of rocks. Proc. Int. Conf. Struct., Solid Mech. and Eng. Design in Civ. Eng. Mater., Southhampton, 1969, Vol. 1, pp. 239–248.

110. MURRELL, S. A. F. and CHAKRAVARTY, S.: Some new rheological experiments on igneous rocks at temperatures up to 1000° C and their implications for mechanical processes in the mantle. Report, Dept. Geólogy, Univ. Coll., London, 1971, 48 p.

111. MURRELL, S. A. F. and MISRA, A. K.: Time-dependent strain or 'creep' in rocks and similar non-metallic materials. Bull. Inst. Min. Metall., Vol. 71, Part 7, April, 1962, pp. 353–378.

112. NAKAMURA, S. T.: On visco-elastic medium. Sci. Rep. Tokyo Univ., 5th Series, Geophysics, Vol. I, No. 2, 1949, pp. 91–95.

113. NISHIHARA, M.: Stress-strain relation of rocks. Doshisha Eng. Rev., Vol. 8, No. 2, Aug., 1957a, pp. 13–55.

114. NISHIHARA, M.: Stress-strain-time relation of rocks. Doshisha Eng. Rev., Vol. 8, No. 3, Nov., 1957b, pp. 85–115.

115. NISHIMATSU, Y. and HEROESEWOJO, R.: Rheological properties of rocks under pulsating loads, 3rd Cong. Int. Soc. Rock Mech., Denver, Colorado, Vol. II-A, 1974, pp. 385–389.

116. OBERT, L. and DUVALL, W.: Use of subaudible noises for the prediction of rock-bursts, Part II. U. S. B. M. R. I. 3654, 1942, 22 p.

117. OBERT, L. and DUVALL, W.: Microseismic method of predicting rock failure in underground mining. Part 1. U. S. B. M. R. I. 3797, 1945a, 7 p.

118. OBERT, L. and DUVALL, W.: The microseismic method of predicting rock failure in underground mining. Part 2. U. S. B. M. R. I. 3803, 1945b, 14p.

119. OBERT, L. and DUVALL, W. I.: Rock mechanics and the design of structures in rock. New York, Wiley, 1967, 650 p.

120. OKSENKRUG, E. and SAFARENKO, E.: Creeping and long term strength of rock salt. Soil Mech. & Found. Eng. (Translated from Russian), May, 1975, pp. 387–389.

121. OROWAN, E.: Seismic damping and creep in the mantle. Geophys. J. Roy. Astr. Soc., Vol. 14, 1967, pp. 191–218.

122. PAULDING, B. W.: Techniques used in studying the fracture mechanics of rock. Proc. Symp. Testing Techniques for Rock Mech., Seattle, Wash., 1965, pp. 73–86.

123. PENG, S. and PODNIEKS, E. R.: Relaxation and the behaviour of failed rock. Int. J. Rock Mech. Min. Sci., Vol. 9, 1972, pp. 699–712.

124. PENG, S. S., PODNIEKS, E. R. and CAIN, P. J.: The behaviour of Salem limestone in cyclic loading. Soc. Pet. Eng. J., Feb., 1974, pp. 19–24.

125. PENG, S. S. and ORTIZ, C. A.: Crack propagation and fracture of rock specimens loaded in compression. Proc. Int. Conf. Dynamic Crack Propagation, Lehigh Univ., Bethlehem, Pa, July, 1972.

126. PHILLIPS, D. W.: The nature and physical properties of some coal-measure strata. Trans. Inst. Min. Eng., Vol. 80, Part 4, Jan., 1931, pp. 212–239.

127. PHILLIPS, D. W.: Further investigation of the physical properties of coal-measure. rocks and experimental work on the development of fractures. Trans. Inst. Min. Eng., Vol. 82, Part 5, Feb., 1932, pp. 432–446.

128. PHILLIPS, D. W.: Tectonics of mining. Coll. Eng., Vol. 25, 1948, pp. 199–202; 206; 278–282; 312–316; 349–352.

129. Pomeroy, C.D.: Creep in coal at room temperature. Nature, Vol. 178, No. 4527, Aug. 4, 1956, pp. 279–280.

130. Potts, E.L.J.: An investigation into the design of room and pillar workings in rock salt. Min. Eng., No. 49, Oct., 1964, pp. 27–44.

131. Price, N.J.: A study of the time strain behaviour of coal measure rocks. Int. J. Rock Mech. Min. Sci., Vol. 1, No. 2, March, 1964, pp. 277–303.

132. Price, N.J.: Fault and joint development in brittle and semi-brittle rock. Oxford, Pergamon, 1966, 176 p.

133. Price, N.J.: Laws of rock behaviour in the earth's crust. Proc. 11th Symp. Rock Mech., Berkeley, California, 1969, pp. 3–23.

134. Pushkarev, V.I. and Afanasev, B.G.: A rapid method of determining the long-term strengths of weak rocks. Sov. Min. Sci., Vol. 9, No. 5, Sept.-Oct., 1973, pp. 558–560.

135. Renzhiglov, N.F. and Pavlishcheva, T.V.: On the viscosity of rocks. Sov. Min. Sci., No. 5, Sept.-Oct., 1970, pp. 582–585.

136. Reynolds, T.D. and Gloyna, E.F.: Creep measurements in salt mines. Proc. 4th Symp. Rock Mech., Univ. Park, Penn., 1961, pp. 11–17.

137. Robertson, E.C.: Experimental study of the strength of rocks. Geol. Soc. Am. Bull., Vol. 66, 1955, pp. 1275–1314.

138. Robertson, E.C.: Creep of Solenhofen limestone under moderate hydrostatic pressure. Geol. Soc. Am. Mem. 79, 1960, pp. 227–244.

139. Robertson, E.C.: Viscoelasticity of rocks. Proc. Int. Conf. State of Stress in the Earth's Crust, Santa Monica, California, 1963, pp. 180–224.

140. Robinson, L.H.: The effect of pore and confining pressure on the failure process in sedimentary rock. Proc. 3rd Symp. Rock Mech., Golden, Colo., 1959, pp. 177–199

141. Rummel, F.: The rheological behaviour of some quartz-phyllite and limestone-jura specimens under uniaxial, static pressure. Boll. die Geofisica Teorica ed Applicata, Vol. 7, 1965, pp. 165–174.

142. Rummel, F.: Studies of time-dependent deformation of some granite and eclogite rock samples under uniaxial, constant compressive stress and temperatures up to 400° C. Zeitschr. Geophysik, Vol. 35, 1969, pp. 17–42.

143. Ruppeneit, K.V. and Libermann, Y.M.: Einführung in die Gebirgsmechanik, 1960.

144. Rusch, H.: Physical problems in testing of concrete. Zement-Kalk-Gips, Vol. 12, No. 1, Jan., 1959, pp. 1–9.

145. Rusch, H.: Researches toward a general flexural theory for structural concrete. Proc. Am. Conc. Inst., Vol. 57, 1960, pp. 1–28.

146. Rutter, E.H.: On the creep testing of rocks at constant stress and constant force. Int. J. Rock Mech. Min. Sci., Vol. 9, No. 2, March, 1972, pp. 191–195.

147. Saint-Leu, C. and Sirieys, P.: Rock fatigue. Proc. Symp. Rock Fracture, Nancy, 1971, Paper II-18.

148. Salustowicz, A.: Rock as an elastic viscous medium. (In Polish). Archiwum Gornictwa, Vol. 3, 1958, pp. 141–172.

149. Sangha, C.M. and Dhir, R.K.: Influence of time on the strength, deformation and fracture properties of a Lower Devonian sandstone. Int. J. Rock Mech. Min. Sci., Vol. 9, No. 3, May, 1972, pp. 343–354.

150. Savage, J.C. and Mohanty, B.B.: Does creep cause fracture in brittle rocks? J. Geophys. Res., Vol. 74, No. 17, Aug. 15, 1969, pp. 4329–4332.

151. SCHOLZ, C.H.: Microfracturing and the inelastic deformation of rock in compression. J. Geophys. Res., Vol. 73, No. 4, Feb. 15, 1968a, pp. 1417–1432.

152. SCHOLZ, C.H.: Mechanism of creep in brittle rock. J. Geophys. Res., Vol. 73, No. 10, May 15, 1968b, pp. 3295–3302.

153. SERATA, S.: Application of continuum mechanics to design of deep potash mines in Canada. Int. J. Rock Mech. Min. Sci., Vol. 5, No. 4, July, 1968, pp. 293–314.

154. SINGH, D.P.: A study of time-dependent properties and other physical properties of rocks. Ph. D. Thesis, Univ. Melbourne, Melbourne, 1970, 219 p.

155. SINGH, D.P.: A study of creep of rocks. Int. J. Rock Mech. Min. Sci. & Geomech. Abstr., Vol. 12, No. 9, Sept., 1975, pp. 271–275.

156. STAVROGIN A.N. and LODUS, E.V.: Creep and the time dependence of the strength in rocks. Sov. Min. Sci., Vol. 10, No. 6, Nov.-Dec., 1974, pp. 653–658.

157. TERRY, N.B. and MORGANS, W.T.A.: Studies of the rheological behaviour of coal. Proc. Conf. Mech. Prop. Non-metallic Brittle Materials, London, 1958, pp. 239–256.

158. TER-STEPANIAN, G.: Creep of a clay during shear and its rheological model. Geotechnique, Vol. 25, No. 2, June, 1975, pp. 299–320.

159. THILL, R.E.: Acoustic methods for monitoring failure in rock. Proc. 14th Symp. Rock Mech., Univ. Park, Penn., June, 1972, pp. 649–687.

160. THOMPSON, E. and RIPPERGER, E.A.: An experimental technique for the investigation of the flow of halite and sylvinite. Proc. 6th Symp. Rock Mech., Rolla, Missouri, 1964, pp. 467–488.

161. TURNER, F.J.: Preferred orientation of calcite in Yule marble. Am. J. Sci., Vol. 247, 1949, pp. 593–621.

162. TURNER, F.J., GRIGGS, D.T., CLARK, R.H. and DIXON, R.H.: Deformation of Yule marble. Part 7. Geol. Soc. Am. Bull., Vol. 67, 1956, pp. 1259–1293.

163. VON KARMAN, TH.: Festigkeitsversuche unter allseitigem Druck. Z. Ver. dt. Ing., Vol. 55, 1911, pp. 1749–57.

164. VOROPINOV, J.: The rheological coaction of rocks and supports and the indirect tensometry of the rock massif as an application of physics of rocks. Proc. 4th Int. Conf. Strata Control and Rock Mech., New York, 1964, pp. 349–359.

165. WACHTMAN, J.B. and MAXWELL, L.H.: A bend test method of determining the stress required to cause creep in tension. A.S.T.M. Bull., Jan., 1956, pp. 38–39.

166. WAWERSIK, W.R.: Time-dependendent rock behaviour in uniaxial compression, 14th Symp. Rock Mech., University Park, Pennsylvania, 1972, pp. 85–106.

167. WAWERSIK, W.R.: Time-dependent behaviour of rock in compression. Proc. 3rd Cong. Int. Soc. Rock Mech., Denver, 1974, Vol. 2, Part A, pp. 357–363.

168. WAWERSIK, W.R. and BROWN, W.S.: Creep fracture in rock in uniaxial compression. Report No. UTEC-ME-71-242, Mech. Eng. Dept., Univ. Utah, Salt Lake City, Utah, Dec., 1971.

169. WAWERSIK, W.R. and BROWN, W.S.: Creep fracture of rock. Report No. UTEC-ME-73-197, Mech. Eng. Dept., Univ. Utah, Salt Lake City, Utah, July, 1973.

170. WHITE, S.: Natural creep deformation of quartzites. Nature, Physical Science. Vol. 234, No. 52, Dec. 27, 1971, pp. 175–177.

171. WIEDERHORN, S.M.: Influence of water vapour on crack propagation in sodalime glass. J. Am. Cer. Soc., Vol. 50, No. 8, Aug., 1967, pp. 407–414.

172. WIEDERHORN, S.M.: Moisture assisted crack growth in ceramics. Int. J. Fracture Mech., Vol. 4, 1968, pp. 171–178.

173. WIID, B.L.: The time dependent behaviour of rock; considerations with regard

to a research programme. Rep. S. African C.S.I.R., No. MEG 514, 1966, 32 p.

174. WIID, B.L.: The influence of moisture on the pre-rupture fracturing of two rock types. Proc. 2nd Cong. Int. Soc. Rock Mech., Belgrade, 1970, Vol. 2, pp. 239–245.

175. WINKEL, B.V., GERSTLE, K.H. and KO, H.Y.: Analysis of time-dependent deformations of openings in salt media. Int. J. Rock Mech. Min. Sci., Vol. 9, No. 2, March, 1972, pp. 249–260.

176. WU, F.T. and THOMSEN, L.: Microfracturing and deformation of Westerly granite under creep conditions. Int. J. Rock Mech. Min. Sci. & Geomech. Abstr., Vol. 12, No. 5/6, June, 1975, pp. 159–166.

177. WYSS, M. and HANKS, T.C.: The source parameters of the San Fernando earthquake inferred from teleseismic body waves. Bull. Seism. Soc. Am., Vol. 62, No. 2, April, 1972, pp. 591–602.

178. YOKOBORI, T.: Strength, fracture and fatigue of materials. Groningen, Noordhoff, 1964, 324 p.

Uncited References (Chapter 9)

1. AFANASIEV, V.B. and ABRAMOV, B.: Determination of the longterm shear strength on the contact surface of rocks, Fiziko Tech. probl. rozrobtki, Vol. 11, No. 5, 1975, pp. 131–134.

2. AIYER, A.K.: An analytical study of the time-dependent behaviour of underground openings. Ph. D. Thesis, Univ. Illinois, 1969, 239 p.

3. AKLONIS, J.J. and TOBOLSKY, A.V.: Relationship between creep and stress relaxation. J. Appl. Phys., Vol. 37, No. 5, April, 1966, pp. 1949–1952.

4. BARTENEV, G.M. and RAZUMOVSKAYA, I.V.: Time dependence of the strength of brittle materials in surface active media. Sov. Phys., Vol. 8, No. 6, Dec., 1963, pp. 602–604.

5. BOYEL, J.M.: Transition from brittle to ductile behaviour in alabaster under confining pressure to 5 kilobars. M.S. Thesis, Univ. Toronto, Toronto, 1967, 67 p.

6. BOZHINOV, G.M.: On some aspects of application of the rheological models. Proc. Int. Conf. Struct., Solid Mech. and Eng. Design in Civ. Eng. Mater., Southampton, 1969, Vol. 1, pp. 527–532.

7. BRANINSKII, V.G. and RAZUMOVSKAYA, I.V.: Safe stress level and creep. Sov. Mat. Sci., Vol. 2, No. 4, 1966, pp. 291–296.

8. BRUECKL, E. and SCHEIDEGGER, A.E.: The rheology of spacially continuous mass creep in rock. Rock Mech., Vol. 4, No. 4, 1972, pp. 237–250.

9. BUCHHEIM, W.: The time dependence in the theory of the mechanical behaviour of rock masses. In German. Geol.-u. Bauwesen, Vol. 26, No. 4, 1960, pp. 218–233.

10. CAIN, P.J., PENG, S.S. and PODNIEKS, E.R.: Fatigue damage in rock by sonic energy. Paper presented at 84th Meeting of Acoustic Soc. Am., Miami, Florida, Nov. 28 to Dec. 1, 1972.

11. CLARKE, J.M.: Model for uniaxial creep based on internal stress redistribution. J. Strain Analysis, Vol. 4, No. 2, 1969, pp. 95–104.

12. Doring, T., Heinrich, F. and Pforr, H.: Deformation and strength behaviour of static isotropic and homogeneous rock with inelastic deformations properties. In German. Proc. 6th Meeting Int. Bur. Rock Mech., Leipzig, 1964, pp. 68–80.

13. Douglas, G. and McDougall, D.J.: Strain energy build up in fatigue cycling, 14th Symp. Rock Mech., University Park, Pennsylvania, 1972, pp. 121–126.

14. Gossick, B.: Design of extensometer for creep studies. Rev. Sci. Inst., Vol. 25, 1954, pp. 907–909.

15. Hardy, H.R.: Physical properties of mine rock and minerals. Time strain studies. Development of new apparatus for maintaining constant stress on rock specimens for time strain tests. Can. Dept. Mines. Tech. Surveys, Mines Branch, T.M. 18, March, 1958, 22 p.

16. Hardy, H.R.: Physical properties of mine rock and minerals. Time strain studies. Initial creep experiments using a hydraulic loading system. Can. Dept. Mines. Tech. Surveys, Mines Branch, T.M. 19, March, 1958, 14 p.

17. Hardy, H.R.: Design of instrumentation for the measurement of time-dependent strain in stressed rock specimens, Winter Instrument Automation Conf. Instrument Soc. Am., Houston, Texas, 1960. Paper No. 1–460.

18. Hardy, H.R.: New rock mechanics program underway by Canadian Government, Mining Engineering, Jan. 1965, pp. 62–67.

19. Hardy, H.R.: A loading system for the investigation of the inelastic properties of geologic materials. Proc. Symp. Testing Techniques for Rock Mech., Seattle, Wash., 1965, pp. 232–265.

20. Hoagland, R.G., Hahn, G.T. and Rosenfield, A.R.: Influence of microstructure on fracture propagation in rock. Rock Mech., Vol. 5, No. 2, 1973, pp. 77–106.

21. Hoagland, R.G., Hahn, G.T., Rosenfield, A.R., Simon, R. and Nicholson, G.D.: Influence of microstructure on fracture propagation in rock. Final Report, ARPA Contract H0210006, Battle Mem. Inst., Columbus, Ohio, Jan., 1972, 56 p.

22. Iida, K., Wade, T., Aida, Y. and Shichi, R.: Measurements of creep in igneous rocks. J. Earth Sciences, Nagoya Univ., Vol. 8, No. 1, 1960, pp. 1–16.

23. Inouye, K. and Tani, H.: Rheology of coal. Strength and fracture of coal. Bull. Chem. Soc. Japan, Vol. 20, No. 4, 1953, pp. 200–204.

24. Jeffreys, H.: A modification of Lomnitzs law of creep in rocks. Geophys. J. Roy. Astr. Soc., Vol. 1, 1958, pp. 92–95.

25. King, M.S.: A method for determining the long term strength of evaporites. Can. Min. Metall. Bull., Vol. 64, No. 711, July, 1971, pp. 48–51.

26. Langer, M.: Zur Theorie der Deformationsvorgänge in Tonen. Geol. Jb, Vol. 79, 1961, pp. 1–22.

27. Langer, M.: Principles of a theoretical rock mechanic. Proc. 1st Cong. Int. Soc. Rock Mech., Lisbon, 1966, Vol. 1, pp. 277–282.

28. Moore, J.F.A.: A longterm test on Bunter sandstone, Building Research Establishment, CP 23/75, February, 1975, 14p.

29. Murayama, S. and Shibata, T.: Rheological properties of clays. Proc. 5th Int. Conf. Soil Mech. Found. Eng., Paris, 1961, Vol. 1, pp. 269–273.

30. Nair, K. and Deere, D.U.: Creep behaviour of salt in triaxial extension tests. Proc. 3rd Symp. on Salt, Northern Ohio Geol. Soc., Cleveland, Ohio, 1970, Vol. 2, pp. 208–215.

31. Obert, L.: Creep in model pillars. Salt, trona and potash ore. U.S.B.M., R.I. 6703, 1965, 23 p.

32. ODQUIST, F.K. and MELLGREN, A.: Influence of nonhomogeneity of the material on the results of creep tests. Proc. Symp. Non-homogeneity in Elasticity and Plasticity, Warsaw, 1958.

33. OROWAN, E.: Creep in metallic and non-metallic materials. Proc. 1st U.S. Nat. Cong. Appl. Mech., Chicago, Illinois, 1951, pp. 453–472.

34. PARACHKEVOV, R.: Deformation and strength of rocks under longlasting stress. In French. Proc. 2nd Cong. Int. Soc. Rock Mech., Belgrade, 1970, Vol. 1, pp. 449–455.

35. POTTS, E.L.J. and HEDLEY, D.G.F.: The influence of time dependent phenomena on the dimensional design of mine pillars. In German. Proc. 6th Meeting Int. Bur. Rock Mech., Leipzig, 1964, pp. 35–41.

36. PRICE, N.J.: A study of the time-strain behaviour of coal-measure rocks. Int. J. Rock Mech. Min. Sci., Vol. 1, No. 2, March, 1964, pp. 277–303.

37. ROBERTSON, E.C.: Creep of Solenhofen limestone under moderate hydrostatic pressure. Geol. Soc. Am. Mem. 79, 1960, pp. 227–244.

38. RODRIGUEZ-CUEVAS, N.: Viscoelastic constants for a model representing the mechanical behaviour of materials. Proc. Int. Conf. Struct., Solid Mech. and Eng. Design in Civ. Eng. Mater., Southampton, 1969, Vol. 1, pp. 533–544.

39. RUMMEL, F.: The rheological behaviour of some quartz phyllite and limestone/jura specimen under uniaxial static pressure. Bull. die Geofisica Teorica ed Applicata, Vol. 7, No. 26, 1965.

40. RUTTER, E.H.: The influence of interstitial water on the rheological behaviour of calcite rocks. Tectonophysics, Vol. 14, 1972, pp. 13–33.

41. RUTTER, E.H. and SCHMID, S.M.: Experimental study of unconfined flow of Solenhofen limestone at 500° to 600° C. Geol. Soc. Am. Bull., Vol. 86, No. 2, Feb., 1975, pp. 145–152.

42. SAGINOV, A.S., ERZHANOV, Zh. S. and VEKSLER, Yu. A.: Rock breaking under bulk compression. Sov. Min. Sci., No. 6, Nov.-Dec., 1969, pp. 611–617.

43. SAGINOV, A.S., ERZHANOV, Zh. S. and VEKSLER, Yu. A.: The creep and destruction of rocks under conditions of all-round compression. In German. Proc. 2nd Cong. Int. Soc. Rock Mech., Belgrade, 1970, Vol. 1, pp. 463–466.

44. SCHEIDEGGER, A.E.: On the rheology of rock creep. Rock Mech., Vol. 2, No. 3, 1970, pp. 138–145.

45. SIEMES, H.: Experimental deformation of galena ores. In Experimental and Natural Rock Deformation, Berlin, Springer-Verlag, 1970.

46. SMITH, E.M., GRANT, C. and BOOTH, J.M.: Equipment for creep testing at variable load and temperature. J. Strain Analysis, Vol. 5, No. 2, 1970, pp. 145–154.

47. SOBOTKA, Z.: Two-dimensional and three-dimensional rheological models for orthotropic bodies with non-symmetrical shear effects. Proc. Int. Conf. Struct., Solid Mech. and Eng. Design in Civil Eng. Mater., Southampton, 1969, pp. 509–520.

48. STACEY, F.D.: Rock creep and convection in the mantle. Icarus. Int. J. Solar System, Vol. 6, 1963, pp. 305–312.

49. WACHTMAN, J.B. and MAXWELL, L.H.: A bend test method of determining the stress required to cause creep in tension. A.S.T.M. Bull., Jan., 1956.

50. WEIBULL, W.: Fatigue testing and analysis of results. Pergamon, 1961.

51. VALSANGKAR, A.J. and GOKHALE, K.V.: Stress-strain relationship for empirical equations of creep in rocks. Eng. Geol., Vol. 6, No. 1, 1972, pp. 49–53.

52. Vyalov, S.S.: Creep in rock. Proc. 2nd Cong. Int. Soc. Rock Mech., Belgrade, 1970, Vol. 1, pp. 307–312.
53. Yerzhanov, Z.H.S.: Theory of creep and its applications. In Russian. Alma Ata, U.S.S.R., Science Publishers, 1964.

APPENDIX III

In Situ Mechanical Properties of Rock

Abbreviations

C	– cohesion, MPa
cg	– coarse grained
E_d	– deformation modulus, MPa
E_e	– modulus of elasticity, MPa
E*	– dynamic modulus, MPa
fg	– fine grained
mg	– medium grained
σ_t	– tensile strength, MPa
P	– Pressure, bars
σ_c	– compressive strength, MPa
$\parallel$	– parallel to stratification
$\perp$	– right angle to stratification
σ_n	– normal pressure, bars
ϕ	– angle of friction, degrees
ν	– POISSON's ratio
horiz.	– horizontal
vert.	– vertical
av	– average
ρ	– density, g/cm^3

APPENDIX III

In Situ Mechanical Properties of Rocks

Location	Rock Type	Test Details		Field Properties (before grouting)					
		Type	P	E_d	E_c	E^*	C	ϕ	Remarks
Amagase Dam	paleozoic sandstone	shear test (60 × 60 cm concrete – rock surface)							
		⊥ bedding					1.52	51	
		∥ bedding					1.66	49	
		∥ bedding					1.76	54	
Ananaigawa Dam, Japan	Schalstein (sandstone + chert)	jack seismic			3500–4000	22000–35000			
Ambiesta, Italy	limestone	hydraulic chamber	12						
		right slope			12500	32500			
		left slope				25000			
Agri River, Italy	conglomerate	jack	293		11445	11031			
	sandstone	pressure-			1240	11031			
		chamber			1790				
Alvito Sussidenga, Portugal	quartzite	jack		6343					
				7308					
				1447					
				7308					
Andermatt, Switzerland	granite	jack	60						
	∥ bedding				40000				
	⊥ bedding				25000				
Adigüzel Dam, Turkey	schist	plate load (580 mm ∅)	1000	353–374 av = 364	1014–2362 av = 1811	10900			
	meta-limestone			335–3360 av = 1642	2749–8400 av = 8257	20900			
	meta-limestone + marble			387–1580 av = 1112	3436–12600 av = 8089	26500			
Altazar, Spain	paleozoic schist	seismic							
	middle					8500–9500			
	bottom					2800–3000			
Alpe Gera, Italy	serpentine	hydraulic	18		6750				
	schist	chamber	36						
		hydraulic	100	23090	25600				∥
		jack	100	7365	8900				⊥
			40	15989	16900				∥
			60	8880	14800				⊥
		from displacement measurement of dam			32000				
Achensee pressure shaft, Austria	Wetlerstein limestone, massive				10000–30000				
Afourer Tunnel, Morroco	marl	plate		4000–5000	6000–7500				
Abda Donthnala Tunnel, Morroco	conglomerate (medium compact)	plate			400–2600				
Amsteg Press. Tunnel, Switzerland	biotite gneiss (compact, unweathered)	pressure chamber	4		31700				
	seriţic shale		4.5	2700	3500–3200				
Andermatt, Switzerland	biotite shale	plate	60						
	∥ bedding			28000					
	⊥ bedding			8000					

Field Properties (after grouting)						Laboratory Properties				References
E_d	E_c	E^*	C	ϕ	Remarks	E_d or E_c	E^*	σ_t	σ_c	
										Ishii et al, 1966
						30300	15500			Kawabuchi, 1964
										Link, 1964
						18408				Lotti & Beamonte, 1964
					4688					Rocha, 1964
					5377					
					6825					
					12755					
										Müller, 1960
						47550	26500		40.5	Timur, 1970
									613	
						4300–	25900–		613	
						5840	29700		–	
						4130			808	
						43680				
										Serafim & Bollo, 1966
2625	4500									DEH, 1966
										Müller, 1960
										Müller, 1960
										Müller, 1960
										Müller, 1960
										Müller, 1960

(continued)

Location	Rock Type	Test Details		Field Properties (before grouting)					
		Type	P	E_d	E_c	E^*	C	ϕ	Remarks
Alto Rabagao Dam, Portugal	granite (weathered)				6600				
	granite (highly weathered)				1000				
Arrens Tunnel, France	limestone	jack		13500	40000				
	schist	jack			19200				
	schist	jack			25500				
	schist	jack		5000	12500				
	schist	jack		5000	18000				
Belasar, Spain	granite crystalline	seismic							
	top					1300–17500			
	middle					5400–20500			
	bottom					34000–51000			
Barbellino Dam, Italy	schist alternated	hydraulic chamber	50		48000				
	with quartz sandstone	,, ,,	50		47000				
Barbellino Dam, Italy	sandstone	overall structure			46200				
Beuregard, Italy	mica schist	hydraulic chamber	30		18000	85000			
	mylonite	,, ,,	30			3000–15000			
Bajina Basta Dam, Yugoslavia	shales with sandstone lentiles	pressure meter							
	depth 14.5 m		16.8	65	250				
	depth 15.3 m		16.8	110	800				
	depth 20.0 m		5.4	420					
Bor Copper Mine, Yugoslavia	pyroclastic	shear test					0.3	60	peak
								55–60	residual
	andesite						0.05–	30–65	peak
							0.4	30–55	residual
	conglomerate						0.4	70	peak
								63	residual
Born, Jura Mountains, Switzerland	jurassic limestone	shear test (3 m^2)							
	bedding plane							50–54	
	joint							48–54	
Beauregard, Italy	mica-schist & gneissic schist	hydraulic chamber	30	14000	24000				site a
	mica schist & gneissic schist locally mylonitised	hydraulic chamber	7	697	1000				site b (horiz.)
		,, ,,	20						site b (horiz.)
		,, ,,	11	2509	6000				site c (vert.)
		,, ,,	24						site c (vert.)
		,, ,,	24		17000				site d
									site d
Beauregard, Italy	schist gneiss	pressure chamber	30		23500	46130			
	vertical		7		965	16700			
	horizontal		11		5860	16700			

Field Properties (after grouting)						Laboratory Properties				References
E_d	E_c	E^*	C	ϕ	Remarks	E_d or E_c	E^*	σ_t	σ_c	
										WEYERMANN, 1970
										GTCNF, 1964a
						18000				
						22000				
										SERAFIM & BOLLO, 1966
										DEH, 1966
										DEH, 1966
										LINK, 1964
										KUJUNDZIC & STOJAKOVIC, 1964
										RADOSAVLJEVIĆ et al, 1970
(field values about 5° higher than laboratory values)										LOCHER & RIEDER, 1970
		4000			seismic					DEH, 1966
3808	5500									
5852	12500									
		42000			seismic					
		44000			seismic					
										DEH, 1966
	5380									
	12300	41000								

(continued)

Location	Rock Type	Test Details	Field Properties (before grouting)						
		Type	P	E_d	E_e	E^*	C	ϕ	Remarks
Benposta & Miranda, Portugal	shale	jack		3240–29372					
Bhakra Dam, India	sandstone	jack		2461–16892					
	calcareous siltstone			2737–7584					
	claystone			29716–51090					
Barbellino, Italy	schist sandstone	pressure meter			46900				
Bort Dam, France	gneiss	pressure chamber			15000–20000				
	biotite shale				5000				
Besserve Dam, France	gneiss	jack		1070	2160				
Burgin Mine, USA	shale, hard	jack	2.4 k		3839				
	shale, soft	jack	2.4 k		3712				
	quartzite (shaly)	jack	2.4 k		3094				
Cerny Vali, Czechoslovakia	triassic dolomite	dilatometer (112 mm)			30–140				
		radial jack (2 m)			140–250				
		plate load			40–65				
		flat jack			102–120				
Cachi Dam	agglomerate & tuff with andesite at places	hydraulic pressure chamber	20						
	chamber 1				2770				horiz.
					4800				vert.
	chamber 2				2100				horiz.
	chamber 3				8140				horiz.
	chamber 4				3030				horiz.
					8835				horiz.
	chamber 5				15760				vert.
Caprile Dam, Italy	compact sandstone	hydraulic chamber	40	7840	9800	35000			
		pressure chamber	25		10600	36500			
Careser Dam, Italy	quartziferrous phyllitic	hydraulic chamber	50		24000				
	gneiss	,, ,,	50		35000				
	,, ,,	pressure meter			23400				
		overall meter			34500				
Cabril Dam Portugal	granitic	jack pos. I			10900				vert.
					12800				horiz.
		pos. 2			4500				vert.
					5500				horiz.
		pos. 3			20000				vert.
					11500				horiz.
		pos. 4			950				vert.
					3000				horiz.
		pos. 5			13000				vert.
					12000				horiz.
Canicada Dam	granite (highly altered)	jack loading							
		gallery I			580				vert.
					1600				horiz.
		gallery II			1300				vert.
					14000				horiz.
		gallery V			10000				vert.
					1900				horiz.

Field Properties (after grouting)						Laboratory Properties				References
E_d	E_c	E^*	C	ϕ	Remarks	E_d or E_c	E^*	σ_t	σ_c	
						8343– 44057				ROCHA, 1964
18547– 44471						17582– 39300				BHATNAGAR & SHAH, 1964
										DEH, 1966
										MÜLLER, 1960
										GTCNF, 1964a
										COGAN, 1975
										BUKOVANSKY, 1970
								19000– 20200		SERAFIM et al, 1966
										DEH, 1966
										DEH, 1966
										DEH, 1966
										DEH, 1966
	18500				17300–					ROCHA et al, 1955
	18500				32100					
					16600–					
					19300					
	21000				13700–					
	20400				30800					
	4500				9000–					
	7200				18000					
	18000				13200–					
	21000				16800					
										ROCHA et al, 1955
	1100									
	2200									
	2900									
	3700									
	3000									
	5000									

APPENDIX III

(continued)

Location	Rock Type	Test Details		Field Properties (before grouting)					
		Type	P	E_d	E_c	E^*	C	ϕ	Remarks
Cambambe Dam, Angola	arkose sand-stone,	jack							
	(with shale	vert.			12343				
	layers, fg–cg)	horiz.			24016		0.08	50	
	clayshale	jack			1726		0.2	45	
Carillon Hydro-Power, Canada	shale	shear					0.04	15	
Czechoslovakian Dams	biotite	jack		15580	35900	49850			
	gneiss and	(0.71 m ∅)		4760	16500	20130			
	amphibolite			7170	24890	25850			
				2480	13450	17240			
				5240	24500	30700			
California, mine near Antioch, USA	sandstone	jack			7320	5170			
Cecita Dam, Italy	granite	pressure chamber	25						
Cabril Dam, Portugal	granite (massive unweathered)	jack	30		11500–23000				
Canicada Dam, Portugal	granite (lightly weathered)	plate	30		10900–12800				
Cancano Dam, Italy	limestone	pressure chamber	24		15000–6000				
Candes Dam, France	granite	jack		1510	3050				
Clear Creek Tunnel, USA	quartz-diorite					86500			
	metarhyolite blocky					73000			
	metarhyolite foliated					55000			
	metarhyolite jointed					49000			
	metarhyolite partly crushed					31600			
	metarhyolite weathered					22500			
Dloube Strane, Czechoslovakia	crystalline schist, gneiss	dilatometer			25–34				
		plate load (0.5 m^2)			58–98				
		jack			190				
D'Avene Dam, France	quartzite schist								
	site B–C					12500			
	site DE	jacking	48		5400	5500			
	∥-foliation								
	⊥-foliation				2850	5500			
	diorite								
	site A					6000			
	site B					12000			
	site C					15000			
Dubrovnik Power Plant, Yugoslavia	limestone & dolomite	pressure meter	49	4302 6000	6200				

Field Properties (after grouting)						Laboratory Properties				References
E_d	E_e	E^*	C	ϕ	Remarks	E_d or E_e	E^*	σ_t	σ_c	
	15213					63500–68000		88.5–135.0		SARAMENTO & VAZ, 1964
	26378									
	6231					10000		20.2		
										PIGOT & MACKENZIE, 1964
										DROZD & LOUMA, 1966
						5380	3720			GOODMAN & EWOLDSEN, 1965
	3000									MÜLLER, 1960
										MÜLLER, 1960
										MÜLLER, 1960
	17200–9800									MÜLLER, 1960
										GTCNF, 1964a
										WANTLAND, 1963
						240				BUKOVANSKY, 1970
						240				
						240				
										GTCNF, 1964
		15500								
	18100									
	11300									
		12500								
		18000								
		20000								
										KUJUNDZIC & STOJAKOVIC, 1964

(continued)

Location	Rock Type	Test Details		Field Properties (before grouting)					
		Type	P	E_d	E_e	E^*	C	ϕ	Remarks
Davis Dam,	rhyolite &	jack	50	345					
USA	gneiss			758					
			14	138					
				276					
		overall		2400–					
		structure		8300					
Dworshak Dam,	granite	jack	69						
USA	gneiss	(surface gauges)							
	vert.			18616	22753	63432			
	vert.			17237	26200	83432			
	horiz.			14479	19305	63432			
	horiz.			5516	8274	63432			
	horiz.			38611	50332	63432			
	horiz.			9997	11721	63432			
	vert.			5516	10342	48953			
	vert.			23442	43437	48953			
	horiz.			7584	12411	48953			
	horiz.			4826	8963	48953			
	vert.			15858	22753	53090			
	vert.			2758	5171	53090			
	horiz.			12411	15858	53090			
	horiz.			7584	11032	53090			
	vert.			26200	33784	48953			
	vert.			3792	689	49900			
	horiz.			15858	24800	48953			
	horiz.			29647	3500	48950			
	vert.			8963	15168	40679			
	vert.			4826	9653	40679			
	horiz.			13790	20340	40679			
	horiz.			8274	13100	40679			
	granite	jack	69						
	(burried gauges)								
	vert.			42747	52400	63432			
	vert.			28958	35163	63432			
	horiz.			37232	51711	63432			
	horiz.			28958	38600	63400			
	vert.			4130	8963	48950			
	horiz.			4130	7584	48900			
	vert.			50300	57900	53000			
	vert.			5170	5860	53000			
	horiz.			10680	15850	53000			
	horiz.			24132	30300	53000			
	vert.			34470	42700	48900			
	vert.			34470	42060	49000			
	horiz.			32400	38600	48900			
	horiz.			46200	55100	48900			
	vert.			41400	68900	40700			
	vert.			29600	34500	40700			
	horiz.			11000	18600	40700			
				6205	9700	40700			
		pressure	4	13800		4900			
		chamber	12	63400		63400			
Dez Dam,	conglomerate	press	18	1100		50000			
Iran		chamber							
		seismic							
	conglomerate	press			6700	50000			
		chamber			16200				
	horiz.				1800–	50000			
	vert.				5700				
	conglomerate	plate							
	horiz.				1400–	50000			
					57000				
	vert.				2800–	50000			
					50000				
	conglomerate	plate			9650–				
					21370				
		press			4895–				
		chamber			6653				

Field Properties (after grouting)						Laboratory Properties				References
E_d	E_c	E^*	C	ϕ	Remarks	E_d or E_c	E^*	σ_t	σ_c	
										USBR, 1948
										Shannon & Wilson, 1964
						59295	—			
						59294	80668			
						55847	77221			
						51711	79290			
						49642	77221			
						54469	82737			
						60674	83426			
						58605	81358			
						55158	86184			
						55158	76532			
						48953	77911			
						42747	89632			
						42747	89632			
						57226	86874			
						55158	79290			
						37232	79979			
						56537	74463			
						43437	75153			
						45505	84805			
						45505	81358			
						37232	79979			
						31716	82737			
						59295	—			
						52295	8069			
						49642	77221			
						54400	82700			
						60670	83400			
						55100	76500			
						48900	77900			
						51700	81300			
						42700	89600			
						57200	86800			
						55100	79200			
						37200	79979			
						56500	74400			
						43400	75200			
						45500	84800			
						45500	81400			
						37200	81400			
						31700	82700			
1930										
2410										
										Oberti & Fumagalli, 1964
										Müller, 1960
										Müller, 1960
										Müller, 1960
										Müller, 1960
										Dodds, 1965

(continued)

Location	Rock Type	Test Details	Field Properties (before grouting)						
		Type	P	E_d	E_e	E^*	C	ϕ	Remarks
Germany	hematite ore	uniaxial compression ($2\ m^2$ square)			70000				
Germany	shale	uniaxial ($2\ m^2$)			1800				
Eastern Suburb Rly. Line, Snowy Mt. Authority, Austrialia	Hawkesbury sandstone	plate load		600–3200					
Emosson Dam, Switherland	hornfels	plate			46000				
	smooth joint	shear test					1.0	22	
							1.4	26	
	rough joint	,, ,,					3.0	33	
							3.9	26	
Edmonston Pump Plant, USA	diorite-gneiss, granodiorite-gneiss	plate		3861					
		relaxation		3999					
Fallone Dam, Italy	quartz-feldspar-biotite granite gneiss	plate load (590 mm∅)		400	1000				
Fontana Dam, USA	quartzite with phyllite & shale			41369					
Frera Dam, Italy	schistose quartziferous phyllite								
	vert.					43000			
	horiz.					30000			
	left bank bottom								
	trans.					39000			
	long.					41500			
	right bank					25500			
Fedaia. Dam, Italy	limestone	seis., hydraulic chamber	25						
	right slope			7500	38500				
	left slope				38500				
Forte Buso, Italy	quartz-porphyry	hydraulic chamber	30		6000	32000			
Pfestiniog Pump Storage, U.K.	siltstone	hydraulic chamber	35		19000	46500–67500			parallel to strat.
Pfestiniog Pump Storage, U.K.	siltstone	pressure chamber			52000	60200			
					58500	63800			
					46500	52500			
Funil, Brazil	gneiss	jack							
	vert.		24	7300					
	horiz.		23	30300					
	vert.		20	30300					
	horiz.		20	23900					
Forte Buso Dam, Italy	quartz porphyry	pressure chamber	30		6000				
Finodal Pressure Shaft, Yugoslavia	limestone massive	plate	9		22300				
Flaming Gorge Dam, USA	quartzite					17440			

Field Properties (after grouting)						Laboratory Properties				References
E_d	E_c	E^*	C	ϕ	Remarks	E_d or E_c	E^*	σ_t	σ_c	
						95000		17	350	Gimm et al, 1966
						6000		12	160	Gimm et al, 1966
						600– 5900				Snowy Mountain Authority, 1969
						72000	60000	123.0		Schnitter & Schneider, 1970
						24338		99.16		Kruse et al, 1969
						24338		99.16		
										De Risso, 1974
										Blee & Meyer, 1955
										Scalabrini et al, 1964
	52100									
										Link, 1964
										Link, 1964
										Link, 1964
						26000	63500			Chapman, 1961
						26000	63500			
						26000	63500			
										Serafim & Da Costa Nunes, 1966
						34260				
						66530				
						34260				
						66530				
										Müller, 1960
										Müller, 1960
						16340				Wantland, 1963

(continued)

Location	Rock Type	Test Details: Type	P	Field Properties (before grouting): E_d	E_c	E^*	C	ϕ	Remarks
Girabolhos, Portugal	crystalline granite								
	top					1600–3300			
	middle					8500–20500			
	bottom					53000–57000			
Glagno Dam, Italy	stratified dolomite with evidence of fracture	hydraulic chamber	24						
		(gallery B)	30						
		hydraulic jack (gallery B)	100	2057–6217	7200–11000				
		hydraulic jack (gallery A)	100	480–3101	2400–4000				
Glagno Dam, Italy	dolomite	jack							
	vert.		100		7033				
	vert.		100		10760				
	dolomite	press chamber	25		5100				
		,, ,,	25		3700				
Grocio Dam, Italy	chlorite schist	jack	25	22600	36900	52400			
Grimsel Dam, Switzerland	granite	overall structure			15000				
Gerlos Press. Shaft, Austria	quartz-phyllite	overall structure			3500–5000				
Greoux Dam, France	limestone	jack		3240	5090				
Gage Dam, France	granite	jack			6100	30000			
Glen Canyon Dam, USA	Navajo sandstone (massive)	radial jack	40						
		vert.		1082		1175			2nd cycle
		horiz.		8148		752			
Glen Canyon Dam, USA	Navajo sandstone (left abutment)	hydraulic jack (61 cm ∅)							
		vert.		8960	9630	16890			
		vert.		8274		18620			
		horiz.		6620	7360	16890			
		horiz.		6900		18620			
		overall		3860–5309					
Greoux Dam, France	limestone	hydraulic chamber (2)							
		vert.			5700				
		horiz.			6200				
		hydraulic chamber (3)							
		vert.			7200				
		horiz.			12250				
Giovaretto, Italy	micaceous gneiss	hydraulic chamber	30		21500	85000			
Grado Dam, Spain	marls	jack	120				0.13	13–60	
	vert.				3100–17100				
	horiz.				24600				
	conglomerate						0.0	62	
	vert.				8500–19100				
	horiz.				6900–8400				

Field Properties (after grouting)						Laboratory Properties				References
E_d	E_c	E^*	C	ϕ	Remarks	E_d or E_c	E^*	σ_t	σ_c	
										SERAFIM and BOLLO, 1966
1990– 2028 2642– 3764	3800– 5100 3700 5400				with filling after consolidation					DEH, 1966
										DEH, 1966
	5400 3650									
										BARIOLI, 1966
										MÜLLER, 1960
										MÜLLER, 1960
										GTCNF, 1964a
						44250				GTCNF, 1964a
						5790				OLSEN, 1966
						4690				
										RICE, 1964
						12760 12760 12760 12760	17200 18620 17200 18620			
										GTCNF, 1964
	11500 11800									
	9050 15400									
										LINK, 1964
										JIMENEZ SALAS & URIEL, 1964

(continued)

Location	Rock Type	Test Details		Field Properties (before grouting)					
		Type	P	E_d	E_c	E^*	C	ϕ	Remarks
Grandval Dam, France	schist (micaceous)								
	site C_2					10000			
	site C_3					4800			
	buttress					10000			
						4800			
						5800			
	tunnel	jack			9000				
					12000				
Gordon Dam, Tasmania	quartzite and phyllite	jack	69						
	horiz.				18100				
	horiz.				11600				
	horiz.				28000				
	vert.				5600				
Hydroelectric Power Projects, USSR	upper cretaceous limestone (fissured with clay & calcite)	plate loading (80 cm ∅)		57520	81103				
	bituminous limestone (20–40° beds)			166340	172994				
	highly fissured (20–30° beds)			3334					
Harlan County Dam, USA	bentonite beam in chalk						0.04	8	
Hitotsuse Dam, Japan	sandstone + clay + shale	jack		3500	5200				
				3500	6500				
				2400	7500				
				2400	6000				
				3000	16300				
				3000	9300				
				4400	11800				
				4400	11900				
		overall structure		9300					
Inferno Dam, Italy	sandstone & conglomerate				14000				
		overall structure		13800					
IM Fount, Morocco		pressure chamber			2480				
	perpendic.	jack		965	5650				
	parallel	jack		1240	4270				
Iznajar, Spain	marl	jack	59						
	vert.			2270					
	horiz.			5033					
Ikari Dam, Japan	liparite			2300					
Inner Kirchen, Switzerland	granite, massive unweathered	pressure chamber			20000				
Iznajar Dam Spain	calcareous marl site-b	plate							
	vert.			9930					
	vert.			7475					
	vert.			7245					
	vert.			1020					
	horiz.			33820					
	horiz.			21620					
	vert.			11220					
	vert.			7225					
	horiz.			28050					

Field Properties (after grouting)						Laboratory Properties				References
E_d	E_c	E^*	C	ϕ	Remarks	E_d or E_c	E^*	σ_t	σ_c	
										GTCNF, 1964a
										Andric et al, 1976
16566										Sapegin et al, 1966
430806										
9670										
										Underwood, 1964
										Aoki and Tano, 1974
										DEH, 1966
										DEH, 1966
	9500									Mayer, 1963
						9790 17100				
										Jiminez Salas & Uriel, 1964
										Japanese National Committee on Large Dams, 1958
										Müller, 1960
										Bravo, 1967

(continued)

Location	Rock Type	Test Details	Field Properties (before grouting)						
		Type	P	E_d	E_c	E^*	C	ϕ	Remarks
Iznajar Dam, Spain (Cont.)	site b				40000				
					21000				
					19500				
					14500				
	site a				50000				
	site c			2920	17500–22500				
				5480	17500–22500				
				9780	17500–22500				
	site d				6000				
					12000				
	site e				6000				
					12000				
Irongate Dam, Romania	gneiss (altered)	jack (1.5–2.0 m ∅)							
	dry			1000–9500	2800–18500				
	wet			400–4500	600–9500				
		shear test (0.8 × 0.8 m)					0.30–0.75	44–73	
								42–60	residual
	gneiss	direct shear test (0.8 × 0.8 m)							
	rock-concrete dry						0.54–1.1	25–71	
								40–68	residual
	wet						0.1–0.78	30–57	
								32–41	residual
Jassa Dam, Italy	quartz-feldspar biotite-granite-gneiss with amphiboles	plate load (590 mm ∅)		600	1000				
Japanese Hydro-Projects	arkose sandstone	jack (0.18 m^2)			1240	6070			
					2200	8140			
					1310	6070			
					3170	5100			
					1240	5100			
	sandstone + slate	jack (0.18 m^2)			1450	5100			
					1520	9100			
					2000	10100			
					5310	25400			
	granodiorite	jack (0.06 m^2)			1520	8100			
					4140	23300			
		jack (0.46 m^2)			8620	13200			
	tuff	jack (0.18 m^2)			1030	2000			
					1930	2000			
					3240	20000			
					2830	25400			
	andesite	jack (0.18 m^2)			10100	32400			
					4900	30300			
					7200	35500			
					6100	23200			
					7200	32400			
Jablanica Tunnel, Yugoslavia	Werfener shale	pressure chamber				12500			
Kariba Dam, Rhodesia	gneiss biotite	overall structure		3447–6895		75840			
		jack			6205	75840			

Field Properties (after grouting)						Laboratory Properties				References
E_d	E_c	E^*	C	ϕ	Remarks	E_d or E_c	E^*	σ_t	σ_c	
										VASILIU, 1970
										DE RISSO, 1974
										ONODERA, 1962
										MÜLLER, 1960
						52050				CORDING et al, 1971
						52050				

(continued)

Location	Rock Type	Test Details		Field Properties (before grouting)					
		Type	P	E_d	E_c	E^*	C	ϕ	Remarks
Kariba Dam, Rhodesia (right bank)	jointed quartzite	jack	14						
	(fissured)					8200			
	pegmatite					14200			
	mica seam					4900			
Kawamata Dam, Japan	liparite (hard)	jack (0.8 m ∅)		6500–7200			3.5		
	fault			400–2000			0.8		
Kurubegawa Dam, Japan	shear						1.0–3.2	45–50	
	triaxial						0.5–3.0	40–50	
	chamber			1100–6000	3000–9500	3000–			
	jack			800–3900	1800–6800				
Karadj, Iran	diorite	jack			15300				
		" "			12500				
		" "			12900				
		" "			12500				
		" "			17100				
		" "			7446				
Kaunertal, Austria	calcareous schist	radial jack	56	5650	10100	18600			
	sericite	radial jack	9	1030	1650				
	calcareous schist	radial jack	19	3030	5100	18600			
	schistose phyllite	radial jack	37	4900	9500	18600			
	augen gneiss	radial jack	39	35200	36200	18600			
	calcareous schist	radial jack	56	6000	10700	18600			
Koshibu Dam, Japan	amphibolite granodiorite	pressure chamber	29	23100		39600			
Kamishiba Dam, Japan	sandstone				5000–10000				
Kuroba Dam, Japan	granite (weathered)	pressure chamber	10		1600–3200				
		plate		2300–3300					
	granite (jointed) vert.	plate		2000–3000					
	granite (unweathered) vert.	plate		3100–6400					
	granite (weathered, 3 mm joint filled, vert.)	plate		1500–1600					
	granite (unweathered jointed) vert.	plate		2900–3400					
	horiz.			1100–1500					

Field Properties (after grouting)						Laboratory Properties				References
E_d	E_c	E^*	C	ϕ	Remarks	E_d or E_c	E^*	σ_t	σ_c	
										LANE, 1964
										CMJG, 1964
										NOSE, 1964
						3700–9800		23–63		
9600	11000									LAUFFER & SEEBER, 1966
827	689									
1500	2620									
4960	11200									
37200	39300									
7900	16800									
							64800			Multipurpose Dam Testing Group, 1964
										Japanese National Committee on Large Dams, 1958
										MÜLLER, 1960

(continued)

Location	Rock Type	Test Details		Field Properties (before grouting)					
		Type	P	E_d	E_c	E^*	C	ϕ	Remarks
Khantaika Hydro-power plant, USSR	dolerite (triassic intrusion)	pressure tunnel test (2.2 m ∅)	15						
	vert.			10400					$\upsilon = 0.25$
	horiz.			20000					$\upsilon = 0.25$
	vert.			37000					$\upsilon = 0.25$
	horiz.			29000					$\upsilon = 0.34$
	cracked			5900					
Krasnoyarsk Dam, USSR	fine grained granite with syenite, porphyrite, vogesite fillings	direct shear test (96 m^2) shearing force inclined to 15°	34 50						
Kameyama Dam, Japan	mudstone	hydraulic jack (300 mm ∅)	80	2400	3200				
Kloof Gold Mining Co. No. 1. shaft, Transvaal, S.Africa	diabase, (weathered jointed)	jack loading (50 cm ∅)		2413	3447				
Kariba Dam, Rhodesia (left bank)	gneiss (massive weathered)					82000 31100			
Kariba Dam, Rhodesia (right bank)	quartzite (jointed)	jack	14	884	3100	7929			
				1655	2400	7929			
				5309	8756	22750			
				6067	9100	22750			
	granite (weathered jointed)	triaxial	820	1300	4400				
		,, ,,	1300	1100					
	granite (weathered highly jointed)	triaxial	1140	1200	3700				
		,, ,,	1530	1100					
	granite (unweathered, lightly jointed)	triaxial	600	3100	4500				
		,, ,,	1570	2000					
	granite (unweathered, jointed)	triaxial	290	3400	4700				
		,, ,,	1420	1600					
Kastenbell Press. shaft, Italy	limestone (jointed)	press. chamber	10		27000				
			4		160–200				
Kesanyama Dam, Japan	chert + sandy slate	plate			4826	58600			
		overall structure		27200					
Lake Delio Underground Station, Italy	gneiss, (fine grained)	plate loading (r = 0.25 m) hydraulic chamber (1.7 m ∅)		9500	14900	50000 ($\upsilon = 0.2$)			$RQD =$ 50 %
Oroville Dam Power plant, USA		hydraulic jack (266 cm^2)	1000	10342		70326			
		tunnel relaxation			17926				
Limberg, Austria	laminated limestone (slope)	jack	6–26		4000–15000	21000–53600			
	(gallery)					30200–58200			

Field Properties (after grouting)						Laboratory Properties				References
E_d	E_c	E^*	C	ϕ	Remarks	E_d or E_c	E^*	σ_t	σ_c	
										PANOV et al, 1970
(calculated assuming transversely isotropic medium)										
(calculated assuming elliptical zone of crack around a gallery)										
										PANOV et al, 1970
50	(1st cycle)				joint					
62	(2nd cycle)				stiffness					
									9.8	OKAMOTO, 1974
										WEBB, 1966
										LANE, 1964
						72400	60670			LANE, 1964
						72400	60670			
						47570	68950			
						47570	68950			
										MÜLLER, 1960
										CORDING et al, 1971
						21500		74	6.6	DOLCETTA, 1971
18200	24100									
						88900	110300			KRUSE, 1971
										LINK, 1964

(continued)

Location	Rock Type	Test Details		Field Properties (before grouting)					
		Type	P	E_d	E_e	E^*	C	ϕ	Remarks
La Bathie, France	gneiss	pressuremeter		34700					
Latiyan Dam, Iran	quartzite sandstone	jack	14	1800	4630	17700			
				1270	4570	5400			
				1330	3750	16400			
	sandstone	jack		5861	14680	24100			
				4400	9790	20000			
				1448	3930	1200			
				1724	1931	7239			
	sandstone & shale	jack		758	1930	11400			
Lucendro Press. Tunnel, Switzerland	Fibbia gneiss (massive)	press. chamber	10	6000–7500	6300–7900	defor-mation			
	shale, (thinly laminated)	press chamber		12000	39000	defor-mation			
Lyse Press. Shaft, Norway	granite gneiss	press chamber			8000–15000				
Lovero Tunnel, Italy	quartz seri-sitic shale (closed joints)	plate (0.9 × 1.7 m)	4–7	2500–3000	5300 4100				
Lenin Hydro-electric, Rumania	fissured sandstone	shear (0.4 × 0.4 m)	20				0.80	34	
							0.98	25	
	argillaceous schist ,, (fractured)						0.04	37	
							0.25	35	
							0.35	28	
Lanoux Dam, France	schist-	jack	40		10000				
	Ardasiers	jack	57		12000				
Lago Debo	gneiss (finegrained, foliated)	press. chamber			9650–26200	53800–59300			
		plate			6895–20680	53800–59300			
		overall structure			6550	53800–59300			
Mratje Dam, Yugoslavia	triassic massive limestone	jack loading (left bank)		4.5–15	8–30				6 tests
		(right bank)		2–16	2–28				8 tests
		in situ shear (1–3 mm joint) 80 × 80 cm surface					1.5	65	
		in situ shear (clean joint)					0.5	55	
		in situ shear test (5 mm open joint)					0	32	
Mae Dam, Italy	stratified limestone	hydraulic chamber	23						site A site B
	dolomite	hydraulic chamber			8300 6550	30300 25400			
Mis Dam, Italy	limestone dolomite	hydraulic chamber	18						
		hydraulic chamber							
		hydraulic	150	7393	11500				
		jack	50	4533	6800				
		hydraulic chamber	18		3900				

Field Properties (after grouting)						Laboratory Properties				References
E_d	E_c	E^*	C	ϕ	Remarks	E_d or E_c	E^*	σ_t	σ_c	
						40700				GTCNF, 1964
						35800				Lane, 1964
						45400				
						35800				
										Knill & Jones, 1965
										Müller, 1960
										Müller, 1960
										Müller, 1960
										Bancila et al, 1961
										GTCNF, 1964a
										Cording et al, 1971
										Bukovansky, 1970
8379	8500									DEH, 1966
6160	6700									
	9800									DEH, 1966
	56500									
2130	4000				with fillings					DEH, 1966
3808	5500				after consolidation					
	53780									DEH, 1966

(continued)

Location	Rock Type	Test Details		Field Properties (before grouting)					Remarks
		Type	P	E_d	E_c	E^*	C	ϕ	
Morasco Dam, Italy	mica schist	dilatometer	50		17500				
		dam displacement			210000				
Malga Bissona Dam, Italy	tonalite	hydraulic chamber (F.E. 7, 8, 9)	50		22000				
		(F.E. 7)			17000				
Malga Bissona Italy	quartz diorite	pressuremeter			21400				
		overall structure		16500					
Mount Holyoke Range, USA	triassic diabase					84.17	($\nu=0.24$)		depth 88.5 m
	triassic arkose					46.73	($\nu=0.26$)		142.2 m
Mica Project, Canada	biotite schist	plate loading ($645\ cm^2$)	272	965–6233	3640–26062 (av = 5102)				$\nu=0.35$
	granitic gneiss			7515–83978	15237–>85494 (av = 27648)				$\nu=0.2$
Morrow Point, USA	schist & gneiss (left abutment)	jack loading							
		vert.		11032	15538	21374			
		horiz.		8963	18293				
	schist & gneiss (right abutment)	jack loading							
		vert.		5930	7602	36542			
		horiz.							
Monar Dam, U.K.	granulite (highly metamorphosed) schist					33500	48.55		
Mae Dam, Italy	limestone	seismic, hydraulic chamber							
	valley bottom					87000			
	right slope		20		8500	31000			
	left slope		20		6500	26000			
Mulargia Dam Sardinia	slate, porphyritic	seismic, hydraulic chamber	12		4500	7500			
Mequinenza Dam, Spain	limestone	plate (0.5 × 1.0 m)	120		18200–27600				vert.
	lignite	plate & shear			200–260		0.07	34	vert.
	lignite & limestone				22400–26300				vert.
	,, ,,				19800–35800		0.7	47	horiz.
Mis Dam, Italy	dolomite + limestone	jack			6550				
Morrow Point, USA	gneiss pegmatite	jack (0.86 m ∅)	41						
				6820	8960	37230			vert.
				5930	9170	37230			horiz.
				10340	14340	47570			vert.
				7440	14410	47570			horiz.
				6205	9653	42060			vert.
				11720	15720	42060			horiz.

Field Properties (after grouting)						Laboratory Properties				References
E_d	E_c	E^*	C	ϕ	Remarks	E_d or E_c	E^*	σ_t	σ_c	
										DEH, 1966
										DEH, 1966
							79.6	($\upsilon = 0.24$)		MURPHY & HOLT, 1966
							41.7	($\upsilon = 0.26$)		
							6412	43.7		MEIDAL and DODDS 1971
							26889 ($\upsilon = 0.9$)	137.1		
										RICE, 1964
						28268				
						11721				
										HENKEL et al, 1964
						56500	69000	15.3		
						56000	74000	50.5		
										LINK, 1964
										LINK, 1964
										JIMENEZ SALAS & URIEL, 1964
										DEH, 1966
										USBR, 1965
						36540	55850			
						36540	55850			
						29600	55850			
						29600	55850			
						34470	55850			
						34470	55850			

APPENDIX III

(continued)

Location	Rock Type	Test Details		Field Properties (before grouting)					Remarks
		Type	P	E_d	E_e	E^*	C	ϕ	
Mossyrock Dam, USA	rhyolite	jack		5580					
				6200		33100			
				6826		31700			
				13800		51000			
				9653		37200			
				17926		30300			
				11000		35850			
Morasco, Italy	mica schist	pressure meter			17200				
		overall structure		18600 20700					
Mauvoisin Dam, Switzerland	calcareous shale	plate	90	4000	28000				
Marmorera Tunnel, Switzerland	sandy lime-stone, shaly	pressure chamber	11	21500	31000–34400				
Meadow Bank Dam, Australia	clay seams in sandstone	shear					0.007	14	drained
							0.021	20	undrained
Malpasset Dam, France	gneiss	jack		500	1000				
Morrow Point Dam, USA	quartzite (micaceous) + schist (micaceous)	overall structure		3447 14480 20684 2760					
		plate			8963				
Manapouri Power Project, New Zealand	gneiss	plate			4830				
		flat jack			22060				
	pegmatite	plate			6890				
		flat jack			23440				
Nagase Dam, Japan	conglomerate (with fissures)	jack loading 25 cm ∅			2300				
	conglomerate				9400				
	sandstone & shale				1700				
	sandy shale				3550				
		overall structure		4400					
Naruko Dam, Japan	shale	overall settlement		4000					
	shale	jack		4074					
Nevada Test Site, USA	vesicular dacite	pressure chamber	100	2620		19300			
	tuff, (massive bedded) (cavity I & II)	overall structure		689	3447	6895			
				3447	3447	6895			
				689	3447	6895			
	granite (cavity III)			27580	27580				
				20680	27580				
				13790	27580				
Neudaz Press. Shaft, Switzerland	shale	pressure chamber	120		5100				
Nagawado Dam, Japan	granite	in situ shear ($2\ m^2$)					2.5	50	
		jack			3000–10000				
	sand (loamy fault zone)	shear ($4\ m^2$)					0.5	20	

Field Properties (after grouting)						Laboratory Properties				References
E_d	E_c	E^*	C	ϕ	Remarks	E_d or E_c	E^*	σ_t	σ_c	
						17920				COON, 1968
6550						38600				
11000						13800				
13100						22100				
19300										
						17900				
										DEH, 1966
										MÜLLER, 1960
										MÜLLER, 1960
										MADDOX et al, 1967
										GTCNF, 1964a
						8963–27580				CORDING et al, 1971
						14480				ALEXANDER, 1969
						14480				
						22060				
						22060				
										Japanese National Committee on Large Dams, 1958
		20000								
										Japanese National Committee on Large Dams, 1958
										Japanese National Committee on Large Dams, 1958
							24800	36900		JUDD, 1965
						3447	10342	10.3		CORDING et al, 1971
						3447	10342	10.3		
						3447	10342	10.3		
						68948				
						68948				
						68948				
										MÜLLER, 1960
										FUJII, 1970
						30000–75000		90–300		

APPENDIX III

(continued)

Location	Rock Type	Test Details	Field Properties (before grouting)						
		Type	P	E_d	E_c	E^*	C	ϕ	Remarks
Niagara Tunnel, USA	bedded limestone + sandstone + shale	over all structure		17240					
Ogochi Dam, Japan	sandstone				60000–70000				
Ottenhein Power Plant, Austria	slate	pressure chamber		100					
Oroville Power Plant, USA	amphibolite	jack			60330	51700			
		overall structure			56540	51700			
Prag, Bridge Foundation, Czechoslovakia	ordovician shales	square plate loading, (1000–5000 cm^2) at 6–7 m depth	0–26	70–220	170–670				oblique to bedding
				290–340	890				⊥ bedding
				530–890	1200–1800				∥ bedding
		at 9 m depth		820–1860	900–2200				
		shear test						25–35	
Pieve di Cadore, Italy	highly fissured limestone	hydraulic chamber	16	3000	4034				
Pantano D'Avio Dam, Italy	granodiorite	hydraulic chamber	50		24000				
		” ”			12000				
		” ”			17000	57000			seismic
		dam displacement			17000				
		” ”			32000				
		hydraulic chamber	50						
		” ”			12000				
		seismic				59000			
Pantano D'Avio, Italy	granodiorite	pressuremeter			23400				
		overall structure		11700					
				11700					
				16500					
Prutz Press. Shaft, Austria	limestone	radial press	50	6000–6900	10300–12500				
Pongelapoort Dam, S.Africa	dacite	jack			43200				
Place Moulin Dam, Italy	kiesengite	pressure chamber							
	left bank		36		15000–25000	12500–78500			
	right bank		36		17000–24000	12500–78500			
Paltinul Dam, Romania	sandstone	jack		1000–2000	1800–5000	14000–18500			
Poiana Uzului Dam, Romania	sandstone	jack (1.5 m ∅)		4800	9000	17500–29000	1.2–1.4		
	sandstone + schist	jack (1.5 m ∅)		1000–3000	1700–7000	17500–29000	1.0–1.8		

Field Properties (after grouting)						Laboratory Properties				References
E_d	E_c	E^*	C	ϕ	Remarks	E_d or E_c	E^*	σ_t	σ_c	
						34470				CORDING et al, 1971
										Japanese National Committee on Large Dams, 1958
										FENZ et al, 1970
						89630	110320			CORDING et al, 1971
						89630	110320			
										DVORAK, 1966
									95–120	
5007	4200									DEH, 1966
										DEH, 1966
										DEH, 1966
8000–11000	13300 14000									MÜLLER, 1960
								112.5		PHELINES, 1967
										OBERTI & REBAUDI, 1967
						78300		106.3	10.1	
						71800		72.3	10.4	
						13500		43.0		PRISCU et al, 1970
										PRISCU et al, 1970

(continued)

Location	Rock Type	Test Details		Field Properties (before grouting)					
		Type	P	E_d	E_c	E^*	C	ϕ	Remarks
Poatina Power Plant, Australia	mudstone (thin to massive)	overall structure		11720					
		jack, ∥			16550				
		jack, ⊥			22060				
Prag, Czechoslovakia	ordovician shales	plate loading		1000–2000	up to 5000 800–1000				
Prag, Czechoslovakia	Letna formation weathered shales	plate loading		70–230	120–310				
Pieve di Cadore Dam, Italy	limestone	hydraulic chamber							
	right slope					46500			
	Pian delle Ere		12		3500	25000–51500			
	left slope					21000			
Pieve di Cadore, Italy	limestone	hydraulic chamber	12			2960			
Pietra del Pertusillo, Italy	sandstone & conglomerate	seismic, hydraulic chamber	8–15		2000	25000			
Port Morsby Power Station, Papua New Guinea		flat jack			23440				
		overall structure		17010					
Randall Dam, USA	niobrara chalk	uniaxial 9 × 0.76 × 0.9 m (H)		1369–2586 av = 1986			1.7	17	
Roujanel Dam, France	mica schist	hydraulic chamber (1)							
		vert.			2150				
		horiz.			6000–1800				
		hydraulic chamber (2)							
		vert.			2800				
		horiz.			7000–2200				
Rappbode Dam, Germany	banded slate	jack, seismic	3		700	19000			
Rothenbrunnen Press. Shaft, Switzerland	sandy calcareous shale, faulted	press. chamber	29		7900				
			32		11300				
Rossens Dam, Switzerland	molasse-sandstone f.g.	plate	30						
	∥ bedding			1000–2000	1800–3600				
	⊥ bedding			800–1000	1300–2600				
	mollasse sandstone, mg-fg.	plate	15						
	∥ bedding				7000–5800				
	⊥ bedding				4200–3800				
	molasse sandstone (tunnel),	hydraulic chamber	15						
	∥ bedding				7000–5800				
	⊥ bedding				4200–3800				

Field Properties (after grouting)						Laboratory Properties				References
E_d	E_c	E^*	C	ϕ	Remarks	E_d or E_c	E^*	σ_t	σ_c	
										Cording et al, 1971
						31030 43440				
										Záruba & Bukovansky, 1966
										Záruba & Bukovansky, 1966
										Link, 1964
	5200									
	4100									DEH, 1966
	1500									Link, 1964
						8960	30340			Alexander, 1969
						1241– 2753	3033– 3516	0.83– 1.66		Underwood, 1964
										GTCNF, 1964
	2750 8000– 3000									
	5750 9000– 2200									
										Link, 1964
										Müller, 1960
										Müller, 1960

(continued)

Location	Rock Type	Test Details		Field Properties (before grouting)					Remarks
		Type	P	E_d	E_c	E^*	C	ϕ	
Shimouke Arch Dam, Japan	pyroxene—hornblende andesite (hard, lightly closed cracks, joints = > 2/m)				9000–16000			50	
	(open joints with clay joints = 2–7/m)				3500–9000		15–25	50	
	(open joints = 7/m)				2000–3500		0.5–15	50	
	(disintegrated with clay)				< 2000		< 0.5	< 30	
Slavia, Czechoslovakia	cretraceous sandstone	plate loading (0.5 m^2)			83–100				
		radial jack (2 m)	100		68				
Susequeda Dam		hydraulic jack		7200	16000				
Sabbione Dam, Italy	calcareous schist	dilatometer	50		26000 35000				
Sabbione Dam, Italy	calcareous schist	overall structure		34500					
South Moravia, Czechoslovakia	biotite gneiss & amphibolites (fine grained)	plate loading (71 × 71 cm)	80						
	unweathered			2500–10000	6000–15000	6000–10000			
	partially weathered			1000–2500	2500–4000	2500–3500			
	highly decayed			250–1000	1000–2500	1000–2000			
Sao Simao Hydroelectric Plant, Brazil	sedimentory breccia	direct shear with inclined load (20°) (1 × 1 m)	30 (σ_n)				0.46	39.5	peak
								35	residual
	weathered	dilatometer		5000					
	sound	,, ,,		9500					
Srisailam Dam, India	quartzite	plate loading (60 × 60 m)		19900–34900			1.3 2.0	43	
	river bed shale	,, ,,		2500–3200			0.56	34	
	abutment shale	,, ,,		1100–1600			0.2–0.45	31 39	
	shear zone	shear test					0.0–0.15	14–17	
	shale quartzite surface	,, ,,					0.02	30	
	shale-concrete surface	,, ,,						35	
	quartzite with intercalated shale	,, ,,					0.56	33	
	quartzite-concrete surface	,, ,,					1.06	49	
Salamonde Dam, Portugal	granite + (biotite-feldspar), fg-cg., gallery 260	jack load (1 m^2)		2000					$\perp$
				2000					$\parallel$
	gallery 265			580					$\perp$
				1630					$\parallel$
	gallery 265	hydraulic chamber		800–6300					$\parallel$

Field Properties (after grouting)						Laboratory Properties				References
E_d	E_c	E^*	C	ϕ	Remarks	E_d or E_c	E^*	σ_t	σ_c	
										Soejima & Shidomoto, 1970
							240–370			Bukovansky, 1970
11000	17000									Argüelles et al, 1966
										DEH, 1966
										DEH, 1966
										Drozd & Louma, 1966
										Fujimura, 1974
										Brito et al, 1974
						64900			318	Mahendra, 1974
						1960			41.2	
						8000–12700				Fernandes & De Sousa, 1955
1100 2200						1700–5400				
										Rocha et al, 1955

(continued)

Location	Rock Type	Test Details		Field Properties (before grouting)					
		Type	P	E_d	E_c	E^*	C	ϕ	Remarks
Salamonde Dam, Portugal (Cont.)	gallery 250	jack load (1 m^2)		1280					⊥
				2900					‖
	gallery 225	jack load (1 m^2)		14600					⊥
				23000					‖
	gallery 245	jack load (1 m^2)		2800					⊥
				3400					‖
La Soledad Dam, Mexico	tuff	seismic							
	$\rho = 1.9$					1700			
	$\rho = 2.2$					27000			
Sufers Dam, Switherland	gneiss, (slightly laminated)	overall		10000	20000	17500–37500			
Sylvenstein Dam, Germany	dolomite								
	right slope	jack + sonic			7100	8500			
					14600				
	left slope					110000			
Speccheri, Italy	limestone slopes	sonic + jack			60000	55000			
St. Cassein, France	gneiss	jack		31990					vert.
				7171					horiz.
				4820					vert.
				7310					horiz.
	gneiss schistose	hydraulic chamber (2)	100		13000–52500				vert.
									,,
					4700–10000				horiz.
									,,
		horizontal (3)	100		5100–4800				vert
									,,
					5000–1000				horiz.
St. Jean du Gard, France	gneiss, altered	jack + seismic							
				965	2340				vert.
				2760	4140	28400			horiz.
				1793	2965	26890			vert.
Sasaogawa Dam, Japan	shalestone				2000				
Sudegai Dam, Japan	quartz porphyry				5000–10000				
Schwarzenbach Tunnel, Germany	granite (massive)	pressure chamber	6	26400	37500				
St. Antonio Press. Shaft, Italy	quartz porphyry	pressure chamber			40000				
Sampolo Dam, France				926	3660				
Silesian Beskids Power Plant, Poland	sandstone mudstone	plate (0.8 m ∅)	6	8160					roof
				549					walls
		plate (0.8 m ∅)	24	1490					roof
				941					walls
		plate (0.8 m ∅)	48	3890					roof
				2337					walls
San Bernardino Tunnel, USA						12760			
						10000			
						10690			

Field Properties (after grouting)						Laboratory Properties				References
E_d	E_c	E^*	C	ϕ	Remarks	E_d or E_c	E^*	σ_t	σ_c	
1460						3200				FERNANDES & DE SOUSA, 1955
3700										
14800						24600–				
30000						24500				
3700						6100–				
4450						10800				
										ULLOAO, 1964
						27300	16367	99.3		SCHNITTER, 1964
										LINK, 1964
										LINK, 1964
23925										GTCNF, 1964
10687										
5380										
9030										
36000–										
13000										
6300–										
15500										
5800–										
2200										
9200										
										MAYER, 1963
										Japanese National Committee on Large Dams, 1958
										Japanese National Commettee on Large Dams, 1958
										MÜLLER, 1960
										MÜLLER, 1960
										GTCNF, 1964a
										BILINSKI & SMOLKA, 1971
						20050				
						20050				
						20050				

(continued)

Location	Rock Type	Test Details		Field Properties (before grouting)					
		Type	P	E_d	E_e	E^*	C	ϕ	Remarks
Cedar City test site, USA	quartz-diorite				82700				
Tena Termini, Italy	limestone (massive)	jack + seismic (48 m ∅)		39300	46060	42060			
				41575	52400	49500			
				30680	46060	40800			
				59450	17380	4700			
				5700	10070	18700			
					42100	4300			
					48100	50500			
					39200	42000			
					16850	4800			
					8150	19000			
Tena Termini, Italy	limestone	hydraulic jack (C-374)	100	6224	15800				$\parallel$
				1523	6600				$\perp$
		hydraulic jack (C-374)	100	3760	4700				$\parallel$
				2880	5500				$\perp$
		seismic (C-353)				26500			
		hydraulic jack (C-353)	100	7500	45000				$\parallel$
				5771	20200				$\perp$
		seismic				52000			
		hydraulic jack	100	30030	48800	50000–59500			$\parallel$
		(C-310)		17337	36600	50000–59500			$\perp$
Tonoyama Dam, Japan	sandstone				5300–13000				
Tignes Dam, France	quartzite				35000	17000			
					10000	11000			
Telessio Dam, Italy	orthogneiss (very dense, low jointed)	plate	40		30000–40000				
Tihange Nuclear Power Plant, Belgium	silurian schist (fractured)	dilatometer (2–7 m depth)		250–750	750				
	(compact)	(3–13 m depth)			900–2000				
	(very compact)	(5–13 m depth)		600–3000	2000–5000				
Tawa Dam, India	Panchmari sandstones (Gondwana) with shale	jack test (315 mm ∅)	17	1.75	2.25				
		direct shear test					0.07	32	
		laboratory test					0.13	44	
Trona Dam, Iraly	sandstone + conglomerate	hydraulic chamber	50		13300				
		,, ,,			18000				
		pressuremeter			25500				
		overall structure		17900					
Tachian Dam, Taiwan	quartzite with interbedded shale	plate load (300 m ∅)		7000	15000				
	slate			4000	6000				
		radial jack (3.1 m ∅)		2000–10000	5000–25000				

Field Properties (after grouting)						Laboratory Properties				References
E_d	E_c	E^*	C	ϕ	Remarks	E_d or E_c	E^*	σ_t	σ_c	
										Cooper & Blouin, 1970
							38610 38610 38610 38610 57900			Ferratini, 1966
										DEH, 1966
										Japanese National Committee on Large Dams, 1958
							50000 34500			GTCNF, 1964a
										Müller, 1960
										Monjoie, 1970
									3.0– 70.0	Gole and Mokhashi, 1970
										DEH, 1966
										DEH, 1966
						30000– 60000				John & Gallico, 1974

(continued)

Location	Rock Type	Test Details	Field Properties (before grouting)						
		Type	P	E_d	E_e	E^*	C	ϕ	Remarks
Takane Dam, Japan	massive hard chert	plate loading			7700	50000			
	banded chert with argillaceous shale (4–10 m)				6700	30000–50000			
	Banded Chert	shear test (1.2 × 1.7 m)					0.5–1.2	40–45	
							5–6	50–60	
Tehachapi, USA	diorite-gneiss	jack	70	827	2620				
				3170	11860				
				1030	6200				
				689	3241				
				5380	12620				
				4690	13652				
				6963	10340				
				3790	11860				
				2890	7380				
				4890	9100				
				3034	5085				
				4200	8890				
				1034	2965				
				4137	9240				
				1655	5100				
				2070	5720				
				4130	6690				
				5520	11380				
				1720	4820				
				1790	5930				
				2550	7310				
				5380	8140				
Tsruga Dam, Japan	granite	jack			4760	33100			horiz.
					7170	31030			vert.
					2620	33100			vert.
Two Forks, USA		jack	70	8690	12070	29650			
				12960	15650	29650			
				12480	17100	29650			
				15930	19170	29650			
Tumut 1, Australia	granite + granite gneiss	jack			34470				
		pressure chamber			13800				
		overall structure			13790				
		overall structure			4826				near fault
Tumut 2, Australia		jack			41370				
		pressure chamber			20680				
		plate			6895				
		overall structure		34474					
		overall structure		13790					
		overall structure		5516					
		overall structure		48263					

Field Properties (after grouting)						Laboratory Properties				References
E_d	E_e	E^*	C	ϕ	Remarks	E_d or E_e	E^*	σ_t	σ_c	
										FUKUSHIMA & TAKAHASHI, 1970
										COON, 1968
										KITSUNEZAKI, 1965
						49640	71700			COON, 1968
						47570	76530			
						46200	77900			
						62740	84100			
						55160	55160			CORDING et al, 1971
						55160	55160			
						55160	55160			
						55160	55160			
						55160	55160			CORDING et al, 1971
						55160	55160			
						55160	55160			
						55160	55160			
						55160	55160			
						55160	55160			
						55160	55160			

(continued)

Location	Rock Type	Test Details		Field Properties (before grouting)					
		Type	P	E_d	E_e	E^*	C	ϕ	Remarks
Uzbekistan, Round Head Buttress Dam, USSR	paleo-meta-morphic schist	rigid plate (reinforced concrete, plate 0.9 × 1.1 × 0.8 m thick)							
	green schist Type I			2300	36300		1.0	34	
	green schist Type II			2600–4600	20400–36300		1.0	34	
	black schist			1200	2400				
USSR Hydro-projects	limestone + bituminous	jack	33	15170					
	limestone + fractured			5378					
	gneiss			483					
Tiefenkastel Press. Tunnel, Switzerland	slate	pressure chamber	7	9200–16800					
Teisnach Dam, Germany	paragneiss (unweathered)	plate	30	6900–8500	13000–13400				
			40	3100–4700	6100–7700				
Val Vestino, Italy	dolomite	jack	10		6550	38330			vert.
			10		3860	38330			horiz.
			100		3792	37370			vert.
			100		3723	37370			horiz.
Viandian Press. Shaft, Luxemburg	clay + shale, (compact)	plate + pressure chamber	30		15000–25000				$\parallel$
					9000				$\perp$
Villefort Dam, France	granite	jack		1920	6560				
Val Gallina, Italy	stratified limestone	hydraulic chamber	20	3892	4100				
	fractured		24	2684	2800				
Val Vestino, Italy	dolomite stratified with frac-tures and mylonite	hydraulic jack (site a)	100		16500				$\parallel$
					6500				$\perp$
		seismic (site b)				39000			
		hydraulic jack (site c)	10	4503	6700				$\parallel$
				3288	4100				$\perp$
		hydraulic jack (site d)	100		3900	38000			$\parallel$
					3800	38000			$\perp$
Vouglans Dam, France	limestone dolomite	hydraulic jack (280 mm ∅)	160	10000	160000				
				3200	6960				
		dilatometer (160 mm ∅) (1600 mm long)	160	19000	26000				
				6300	13000				
				33400	37500				
		shear					25		

Field Properties (after grouting)						Laboratory Properties				References
E_d	E_e	E^*	C	ϕ	Remarks	E_d or E_e	E^*	σ_t	σ_c	
										UKHOV & TSYTOVICH, 1966
25370										SAPEGIN et al, 1966
11400										
1580										
										MÜLLER, 1960
										MÜLLER, 1960
										DEH, 1966
										MÜLLER, 1960
										GTCNF, 1964a
										DEH, 1966
										DEH, 1966
										DUFFAUT and COMES, 1966
										GTNCF, 1964

(continued)

Location	Rock Type	Test Details	Field Properties (before grouting)						
		Type	P	E_d	E_e	E^*	C	ϕ	Remarks
Val Gallina Dam, Italy	limestone	hydraulic chamber							
	right slope		12			18500			
	left slope		15		3900	17500			
	limestone	pressure	12		4100				
		chamber	24		2800				
	stratified limestone,	hydraulic chamber	20	3892	4100				
	fractured		24	2684	2800				
Vaiont Dam, Italy	limestone	hydraulic chamber							
	upper slope		24		4000–5000	33000–46000			
	lower slope		40		12000	30700			
Valdecanas, Portugal	granite	jack		1450					
				1520					
				1590					
				1590					
				1560					
				1450					
				1790					
				2550					
				2620					
				5380					
				3240					
				4890					
				5100					
				5170					
				10000					
Waldeck II, Germany									
	shale	plate loading	200	4200					west
	sandy shale	direct-shear (1.9 m²)		9600			0.15	20	east, (∥ bedding)
							0.5	37	(⊥ bedding)
	shale with sandstone and graywacke	radial pressure-meter (86 mm)		4500–5000					
Germany	iron ore mines, (calcite iron ore)	uniaxial compression (40 cm cube)							$\sigma_c = 67$ (in situ)
Waldshut Press. Tunnel, Germany	sandstone (fine grained)	plate (0.5 × 0.5 m)	24		3000–19500				
Wallsee Dam, Austria	slate	pressure chamber			100				
Witbank Colliery, S.Africa	coal	uniaxial compression (H/D = 2)			3640				$\upsilon = 0.26$

Field Properties (after grouting)						Laboratory Properties				References
E_d	E_e	E^*	C	ϕ	Remarks	E_d or E_e	E^*	σ_t	σ_c	
										LINK, 1964
										DEH, 1966
										LINK, 1964
						1450 1790 1590 1560 1380 2480 2480 2340 2480 1790 5780 5100 5860 7310 15650				ROCHA, 1964
									500–800	PAHL, 1974
										PAHL & ALBRECHT, 1970
										HABETHA, 1974
										PAHL & ALBRECHT, 1970
						37000–67000				JAHNS, 1966
										MÜLLER, 1960
								1.9–3.4		FENZ et al, 1970
										BIENIAWSKI & VOGLER, 1970

(continued)

Location	Rock Type	Test Details		Field Properties (before grouting)					
		Type	P	E_d	E_e	E^*	C	ϕ	Remarks
Yahagi Dam, Japan	weathered granite	shear test							
	fresh joint, interval > 50 cm, joint closed						9.25	45	
	joint slightly weathered, closed						7.75	45	
	joint considerably weathered,						5.63	48	
	along joint weathered						1.75	46.5	
Yellow Tail	Madison limestone left abutment	hydraulic jack	40	11000	16500	26200			
				21400	31700	26200			
				33100	40700	42400			
				26200	38529				vert.
	right abutment			13790	22241				vert.
		overall structure		2800–9700					
Sylvenstein, Germany	dolomite compact	plate	40–160		6700–13500 (11500–32000)				(after creep)
Yugoslavian Hydroplat, Yugoslavia	limestone			2890	12750				
				4410	17100				
				4890	24500				
				6830	25900				
				12270	29370				
				11700	30300				
				6830	38200				
				7860	41100				
				14700	41100				
				22500	4700				
				11800	56700				
				19600	63570				
				37200	56700				
				41600	56700				
				9790	23580				
				7300	23000				
				8825	33300				
				14200	35700				
				8825	37200				
				11200	44000				
				17100	50400				
				20500	63100				
Zillierback Dam, Germany	diabase	overall structure			1000				
Zinnoun Tunnel, Morocco	quartzite (compact)	plate	100		40000				vert.
					45000				
					38000–40000				horiz.
Not known	not known	uniaxial (70 cm ∅)	0–47	2700					
			12–72	5500					
			12–177	1800					
		triaxial (all-side uniform pressure)	25–100	2400–6900			70		
		plate loading	143	9900–10800 ($\upsilon = 0.15$) 9200–10100 ($\upsilon = 0.3$)					

Field Properties (after grouting)						Laboratory Properties				References
E_d	E_c	E^*	C	ϕ	Remarks	E_d or E_c	E^*	σ_t	σ_c	
										IIDA, OKAMOTO & YASUE, 1970
						65500	79290			RICE, 1964
						57200	77200			
						50300	71700			
						63432				
						71016				
										MÜLLER, 1960
										KUJUNDZIC & GRUJIC, 1966
										MÜLLER, 1960
										MÜLLER, 1960
										GILG, 1966

(continued)

Location	Rock Type	Test Details		Field Properties (before grouting)					Remarks
		Type	P	E_d	E_e	E^*	C	ϕ	
Not known	black schist (schistosity > 200/m, fractured)	inclined load shear test (1 × 1 m)					0.7	48	
Not known	weathered gneiss	plate loading		7.0					
		seismic (distance = 60 m)				39–43			
		settlement (foundation width = 30 m)		36					
Not known	mica schist	seismic (distance = 60 m)				16–22			
	jointed quartzite	(distance = 30 m)	68 (av.)			69			
	clay shale					22			
	clay shale (slightly weathered)	plate loading	0.32–0.47			2.9–6.3			
		settlements (width of foundation = 6 m)	0.32–0.56			2.9–6.3			
Not known	not known	plate loading A	0.43		5100				
		B	1.05		570				
			1.05–1.57		850				
		C	0–20		730				
			14–20		540				
		D	0–0.48		110				
Not known	limestone (very compact)	flat jack (1 m^2)	100	52500 56500					
Brazil	limestone shale	pressiometer	100						anisotropy
		depth 7.75 m		17650					1.38
		14.1 m		15555					1.69
		15.5 m		18125					1.04
		28.00 m		20150					1.15
		29.00 m		23275					1.20
		36.6 m		25825					1.74
Not known	limestone	pressiometer	100		11100				$\upsilon = 0.2$
Not known	conglomerate	hydraulic chamber		3062 2812					
Mine near Antioch, California, USA	sandstone	jack			3720	5170			
Not known	limestone	plate			6200–10000				
		dilatometer			30700–38300				
Not known	shale + quartzite-shale	jack		965 7520	2400 21400				
	weathered shale			1030	5800				

Field Properties (after grouting)						Laboratory Properties				References
E_d	E_c	E^*	C	ϕ	Remarks	E_d or E_c	E^*	σ_t	σ_c	
										Takano & Furujo, 1966
										Dvorak, 1966
										Dvorak, 1966
						48200				Kimishima, 1970
						5700				
						5700				
						6300				
						6300				
						1800				
										Rocha, 1969
										Rocha, 1969
										Rocha, 1969
4625 4162										Lotti and Beamonte, 1964
						5380	3720			Goodman & Ewoldsen, 1965
										Mary et al, 1967
										Dvorak, 1957

APPENDIX III

(continued)

Location	Rock Type	Test Details			Field Properties (before grouting)				
		Type	P	E_d	E_e	E^*	C	ϕ	Remarks
Not known	schist	jack							
					7240				⊥
					18600				∥
	quartzite				20500				⊥
					37200				∥
	schist	pressure chamber							
					19600				⊥
					35300				∥
					44300				⊥
					65800				∥
	quartzite								
					47900				⊥
					31900				∥
					10800				⊥
					29400				∥
	gypsum	pressure chamber							
					4400				⊥
					4400				∥
					4900				⊥
					23500				∥
Not known	granite	plate dilatometer			4000 9300– 27000				
Nuclear Power Plant, Japan	conglomerate	jack (repeated)			3670– 4390	32030			
	shale	jack (repeated)			420– 580	6670			
	sandstone				500– 1650	13000– 16000			
	granite	jack (repeated)			300– 1000	10900– 20400			
	mudstone	jack (repeated)			830– 855	4920			
	mudstone	jack (repeated)			1300– 1750	6130			
	mudstone	jack (repeated)			1240– 1990	7140			
	gabbro	jack (repeated)			1100				
	diabase	jack (repeated)			1990	56000			
	diorite	jack (repeated)			1800	43400			
	tuff	jack (repeated)			10000– 16000	43400			
	tuff	jack (repeated)			10000	44500			
	tuff	jack (repeated)			3000– 3500	41500			
	tuff	jack (repeated)			18000– 24000	30500			
	slate	jack (repeated)			15000– 25000	38500			
	slate	jack (repeated)			9000	44500			
Laboratory	marble blocks	slot closure flat jack			34470 2689				
Laboratory	trachyte blocks	slot closure flat jack			35850 37920				
Laboratory	marble blocks	slot closure flat jack			42750 40680				
Laboratory	trachyte blocks	slot closure flat jack			42060 41370				

Field Properties (after grouting)						Laboratory Properties				References
E_d	E_e	E^*	C	ϕ	Remarks	E_d or E_e	E^*	σ_t	σ_c	
										BERNARD, 1953
										MARY et al, 1967
										HAYASHI et al, 1974
						61360				HOSKINS, 1966
						61360				
						44130				
						44130				
						59980				
						59980				
						41370				
						41370				

References to Appendix III

1. Alexander, L.G.: Some observations on mechanical properties and deformation of rock in situ, Proc. Symp. Rock. Mech., Univ. Sydney, 1969, pp. 98–103.
2. Andric, M., Roberts, G.I. and Tarvydas, R.K.: Engineering geology of the Gordon dam, South-West Tasmania, Q. J. Eng. Geol., Vol. 9, 1976, pp. 1–24.
3. Aoki, K. and Tano, J.: Long term measurement of bed rock deformation of arch dam, Rock Mech. in Japan, Vol. 2, 1974, pp. 190–192.
4. Argüelles, H. Marinier, P., Navalon, N. and Sanz Saracho, J.M.: Amélioration des massifs rocheux cristallins par injections, Proc. 1st Cong. Int. Soc. Rock Mech., Lisbon, 1966, Vol. II, pp. 675–679.
5. Băncilă, I., Diacon, A., Georgescu, M. and Radulescu, D.: Les conditions de fondation d'un barrage dans le flysch des Carpethes Orientales, Trans. 7th cong. Large Dams, Rome, 1961, Vol. II, Q.25, R.25, pp. 331–350.
6. Barioli, E.: Physical properties of rock mass and research of a correlation between static and dynamic modulus of elasticity, Proc. 1st Cong. Int. Soc. Rock Mech., Lisbon, 1966, Vol. I, pp. 237–242.
7. Bernard, P.: Mesure des modules élastiques et application en calcul des galéries en charge, Proc. 3rd Int. Conf. Soil Mech. Found. Eng., 1953, Vol. 2, pp. 145–156.
8. Bhatnagar, P.S. and Shah, S.R.: Experiments on Bhakra dam foundations, Trans. 8th cong. Large Dams, Edinburgh, 1964, Vol. I, R.58, Q.28, pp. 1081-1108.
9. Bieniawski, Z.T. and Vogler, U.W.: Load deformation behaviour of coal after failure, Proc. 2nd Cong. Int. Soc. Rock Mech., Belgrade, 1970, Vol. 1, pp. 345–351.
10. Bilinski, A. and Smolka, J.: Proprietés anisotropes de déformation des roches finement stratifiées, Proc. Symp. Rock Fracture, Nancy, 1971, Paper II-28.
11. Blee, C.E. and Meyer, A.A.: Measurements of settlement at certain dams of the TVA system and assumptions for earth loadings for dams in the TVA area, Trans. 5th Cong. Large Dams, Paris, 1955, Vol. III, Q.18, R.5.
12. Bravo, G.: La fondation du barrage de Iznajar, Trans. 9th Cong. Large Dams, Istanbul, 1967, Vol. I, Q.32, R.25, pp. 573–582.
13. Brito, S., Moller, W. and Goncalves, E.: Geological investigations of Sao Simao dam, Proc. 2nd Int. Cong. Int. Assoc. Eng. Geol., Sao Paulo, 1974, Vol. II, Paper VI-23.
14. Bukovansky, M.: Determination of elastic properties of rocks using various on-site and laboratory methods, Proc. 2nd Cong. Int. Soc. Rock Mech., Belgrade, 1970, Vol. I, pp. 329–332.
15. Chapman, E.J.: Pressure tests on rock galleries for the Ffestiniog pumped storage plant. Trans. 7th Cong. Large Dams, Rome, 1961, Vol. 2, R.22, Q.25, pp. 237–260.
16. CMJG (Construction Ministry, Japan Government): Foundation treatment of Kawamata dam. Trans. 8th Cong. Large Dams, Edinburgh, 1964, Vol. I, R.10, Q.28, pp. 187–207.
17. Cogan, J.: The creep of weak rocks in the Burgin mine, Stanford Univ., Ph. D. Thesis, 1975, 315 pp.
18. Coon, R.: Correlation of engineering behaviour with the classification of in situ rock. Univ. Illinois, Ph. D. Thesis, 1968, 236 p.
19. Cooper, H.F. and Blouin, S.E.: Dynamic in situ rock properties from buried high explosive arrays. Proc. 12th Symp. Rock Mech., Rolla, Missouri, 1970, pp. 45–70.

20. Cording, E.J., Hendron, A.J. and Deere, D.U.: Rock engineering for underground caverns, Proc. Symp. Underground Rock Chambers, Phoenix, Arizona, 1971, pp. 567–600.

21. De Risso, R.: Hydraulic and geomechanical characteristics of the foundation rock of some dams in Calabria (Southern Italy). Proc. 2nd Int. Cong. Int. Assoc. Eng. Geol., Sao Paulo, 1974, Vol. II, paper VI-19.

22. DEH (Direction Equipments Hydrauliques), (Rome): Déformabilité de la roche de fondation dans le cas de quelques barrages Italiens, Proc. 1st Cong. Int. Soc. Rock. Mech., Lisbon, 1966, Vol. 2, pp. 603–615.

23. Dodds, R.K.: Measurement and analysis of rock physical properties on the Dez project, Iran, Proc. Symp. Testing Techniques for Rock Mech., Seattle, Wash., 1965, pp. 52–72.

24. Dolcetta, M.: Problems with large underground stations in Italy, Proc. Symp. Underground Rock Chambers, Phoenix, Ariz., 1971, pp. 234–286.

25. Drozd, K. and Louma, B.: The correlation of moduli of elasticity determined by microseismic measurement and static loading test, Proc. 1st Cong. Int. Soc. Rock Mech., Lisbon, 1966, Vol. I, pp. 291–293.

26. Duffaut, P. and Comes, G.: Comparaison de la déformabilité statique d'un matériau de fondation mesurée en place sur le parement d'une excavation et à la paroi d'un sondage, Proc. 1st Cong. Int. Soc. Rock Mech., Lisbon, 1966, Vol. I, pp. 399–403.

27. Dvǒrák, A.: Field tests of rocks on dam sites, Proc. 4th Int. Conf. Soil Mech. Found. Eng., London, 1957, Vol. I, pp. 221–224.

28. Dvǒrák, A.: Tests of anisotropic shales for foundation of large bridge, Proc. 1st Cong. Int. Soc. Rock Mech., Lisbon, 1966, Vol. II, pp. 537–541.

29. Fenz, R., Kobilka, G. and Makovec, F.: Problems encountered in the slate foundations of Wallsee and Ottenheim power plants on the Danube in Austria. Trans. 10th Cong. Large Dams, Montreal, 1970, Vol. II, Q.37, R.31, pp. 551–569.

30. Fernandes, L.H.G. and De Sousa, A.A.C.: Le comportement du barrage de Salamonde dans les premiers 15 mois du remplissage de sa retenue, Trans. 5th Cong. Large Dams, Paris, 1955, Vol. III, Q.16, R.18, pp. 261–272.

31. Ferratini, G.: Seismic tests and hydraulic jack tests to determine the modulus of elasticity of the foundation rock of Tana: Termini dam, Proc. 1st Cong. Int. Soc. Rock Mech., Lisbon, 1966, Vol. I, pp. 549–557.

32. Fujii, T.: Fault treatment at Nagawado dam, Trans. 10th Cong. Large Dams, Montreal, 1970, Vol. II, Q.37, R.9, pp. 147–169.

33. Fujimura, F.: Direct in situ shear test on São Simão sedimentary breccia, Proc. 2nd Int. Cong. Int. Assoc. Eng. Geol., São Pãulo, 1974, Vol. I, paper IV-21.

34. Fukushima, H. and Takahashi, H.: Foundation treatment of Takane dam, Rock Mech. in Japan, Vol. 1, 1970, pp. 145–147.

35. Gilg, B.: Verformung und Bruch von Gesteinsproben unter dreiaxialer Belastung, Proc. 1st Cong. Int. Soc. Rock Mech., Lisbon, 1966, Vol. I, pp. 601–606.

36. Gimm, W.A.R., Richter, E. and Rosetz, G.P.: Über das Verformungs- und Festigkeitsverhalten des Gebirges bei Grossversuchen im Eisenerzbergbau, Proc. 1st Cong. Int. Soc. Rock Mech., Lisbon, 1966, Vol. I, pp. 457–463.

37. Gole, C.V. and Mokhashi, S.L.: Some studies on the foundation rock of Tawa dam (Madhya Pradesh) India, Proc. 2nd Cong. Int. Soc. Rock Mech., Belgrade, 1970, Vol. 2, pp. 351–356.

38. GOODMAN, R. and EWOLDSEN, H.: The relationship between field and laboratory measurements of the mechanical properties of sandstone, Univ. California, 1965.

39. GTCNF (Groupe de Travail du Comité National Français, France): Measure des qualités mécaniques des massifs rocheux avant et après la consolidation par injection, Trans. 8th Cong. Large Dams, Edinburgh, 1964, Vol. I, R.18, Q.28, pp. 351–376.

40. GTCNF (Groupe de Travail du Comité National Français, France): La deformabilité des massifs rocheux analyse et comparaison des resultats. Trans. 8th Cong. Large Dams. Edinburgh, 1964a, Vol. II, R.15, Q.28, pp. 287–312.

41. HABETHA, E.: Large scale shear tests for the Waldeck-II pump fed storage station construction project, Proc. 2nd Int. Cong. Int. Assoc. Eng. Geol., São Pãulo, 1974, Vol. I, Paper IV-35.

42. HAYASHI, M., FUZIWARA, Y. and KOMODA, H.: Dynamic deformability and viscosity of rock mass and rock fill material under repeated loadings in situ and triaxial test, Rock Mech. in Japan, Vol. II, 1974, pp. 84–86.

43. HENKEL, D.J., KNILL, J.L., LLOYD, D.G., and SKEMPTON, A.W.: Stability of the foundations of Monar dam. Trans. 8th Cong. Large Dams, Edingburgh, 1964, Vol. I, R.22, Q.28, pp. 425–441.

44. HOSKINS, E.R.: An investigation of the flat jack method of measuring rock stress. Int. J. Rock Mech. Min. Sci., Vol. 3, 1966, pp. 249–264.

45. IIDA, R., OKAMOTO, R. and YASUE, T.: Geological rock classification of dam foundation, Rock Mech. in Japan, Vol. I, 1970, pp. 161–163.

46. ISHII, F., IIDA, R., YUGETA, H., KISHIMOTO, S. and TENDOU, H.: On the strength characteristics of a bedded rock, Proc. 1st Cong. Int. Soc. Rock Mech., Lisbon, 1966, Vol. I, pp. 525–529.

47. JAHNS, H.: Messung der Gebirgsfestigkeit in situ bei wachsendem Massstabsverhältnis, Proc. 1st Cong. Int. Soc. Rock Mech., Lisbon, 1966, Vol. I, pp. 477–482.

48. Japanese National Committee on Large Dams: Present status of measurement of structural behaviour of dams in Japan, Trans 6th Cong. Large Dams, New York, 1958, Vol. 2, R.25, Q.21, pp. 301–357.

49. JIMENEZ SALAS, J.A. and URIEL, S.: Some recent rock mechanics testing in Spain, Trans. 8th Cong. Large Dams, Edinburgh, 1964, Vol. I, R.53, Q.28, pp. 995–1022.

50. JOHN, K.W. and GALLICO, A.: Engineering geology of the site of the Upper-Tachen project, Proc. 2nd Int. Cong. Int. Assoc. Eng. Geol., São Pãulo, 1974, Vol. II, Paper VI-9.

51. JUDD, W.R.: Some rock mechanics problems in correlating laboratory results with prototype reactions, Int. J. Rock Mech. Min. Sci., Vol. 2, 1965, pp. 197–218.

52. KAWABUCHI, K.: A study of strain characteristics of a rock foundation. Trans. 8th Cong. Large Dams, Edinburgh, 1964, Vol. I, R.11, Q.28, pp. 209–218.

53. KIMISHIMA, H.: A study on failure characteristics of foundation rock through a series of tests in situ, Rock Mech. in Japan, Vol. I, 1970, pp. 91–93.

54. KITSUNEZAKI, C.: In situ determination of variation of POISSON's ratio in granite accompanied by weathering effect and its significance in engineering projects, Bull. Disaster Prevention Res. Inst., Kyoto Univ., Vol. 15, 1965, Part 2, No. 92.

55. KNILL, J.L. and JONES, K.S.: The recording and interpretation of geological conditions in the foundation of the Roseires, Kariba and Latiyan dams, Geotechnique, Vol. 15, 1965, pp. 94–124.

56. Kruse, G. H.: Power plant chamber under Oroville dam, Proc. Symp. Underground Rock Chambers, Phoenix, Arizona, 1971, pp. 333–379.

57. Kruse, G.H., Zerneke, K.L., Scott, J.B., Johnson, W.S. and Nelson, J.S.: Approach to classifying rock for tunnel liner design, Proc. 11th Symp. Rock Mech., Berkeley, California, 1969, pp. 169–192.

58. Kujundzic, B. and Grujic, N.: Correlation between static and dynamic investigations of rock mass in situ, Proc. 1st Cong. Int. Soc. Rock Mech., Lisbon, 1966, Vol. I., pp. 565–570.

59. Kujundzic, B. and Stojakovic, M.: A contribution to the experimental investigation of changes of mechanical characteristics of rock massives as a function of depth, Trans. 8th Cong. Large Dams, Edinburgh, 1964, Vol. I, R.26, Q.28, pp. 1051–1067.

60. Lane, R.G.T.: Rock foundations: Diagnosis of mechanical properties and treatment, Trans, 8th Cong. Large Dams, Edinburgh, 1964, Vol. I, R.8, Q.28, pp. 141–165.

61. Lauffer, H. and Seeber, G.: Die Messung der Felsnachgiebigkeit mit der TIWAG – Radialpresse und ihre Kontrolle durch Dehnungsmessungen an der Druckschachtpanzerung des Kaunertalkraftwerkes, Proc. 1st Cong. Int. Soc. Rock Mech., Lisbon, 1966, Vol. II, pp. 347–356.

62. Link, H.: Evaluation of elasticity moduli of dam foundation rock determined seismically in comparison to those arrived at statically, Trans. 8th Cong. Large Dams, Edinburgh, 1964, Vol. I, R.45, Q.28, pp. 833–858.

63. Locher, H.G. and Rieder, U.G.: Shear tests on layered jurassic limestone, Proc. 2nd Cong. Int. Soc. Rock Mech., Belgrade, 1970, Vol. 2, pp. 1–5.

64. Lotti, C. and Beamonte, M.: Execution and controls of consolidation works carried out in the foundation rock of an arch gravity dam, Trans. 8th Cong. Large Dams, Edinburgh, 1964, Vol. I, R.37, Q.28, pp. 671–698.

65. Maddox, J.M., Kinstler, F.L. and Mather, R.: Fondation studies for Meadowbank buttress dam, Trans. 9th Cong. Large Dams, Istanbul, 1967, Vol. I, Q.32, R.9, pp. 123–141.

66. Mahendra, A.R.: The low shear strength foundations and design of Srisailam dam, Andhra Pradesh, India, Proc. 2nd Int. Cong. Int. Assoc. Eng. Geol., São Pãulo, 1974, Vol. II, Paper VI-25.

67. Mary, M., Duffaut, P. and Comes, G.: Connaissance mécanique du rocher par sondages et saignées, Trans. 9th Cong. Large Dams, Istanbul, 1967, Vol. I, Q.32, R.44, pp. 727–746.

68. Mayer, A.: Recent work in rock mechanics, Geotechnique, Vol. 13, 1963, pp. 99–118.

69. Meidal, P. and Dodds, R.K.: Rock mechanics studies, Mica Project, B.C., Proc. Symp. Underground Rock Chambers, Phoenix, Arizona, 1971, pp. 203–242.

70. Monjoie, A.: Propriétés mécaniques des schistes siluriens a Tihange (Belgique), Proc. 2nd Cong. Int. Soc. Rock Mech., Belgrade, 1970, Vol. I, pp. 213–220.

71. Müller, L.: Der Felsbau, Band I, Stuttgart, Ferdinand Enke Verlag, 1960, 624 p.

72. Multipurpose Dam Testing Group, Japan: Contribution to the 8th Congress on Large Dams, Edinburgh, 1964.

73. Murphy, V.J., and Holt, R.J.: Seismic velocities and elastic moduli measurements, Mount Holyoke Range, Massachusetts, U.S.A., Proc. 1st Cong. Int. Soc. Rock Mech., Lisbon, 1966, Vol. I, pp. 711–712.

74. NOSE, M.: Rock test in situ, conventional tests on rock properties and design of Kurobegawa No. 4 dam based thereon, Trans. 8th Cong. Large Dams, Edinburgh, 1964, Vol. I, R.12, Q.28, pp. 219–252.

75. OBERTI, G. and FUMAGALLI, E.: Propriétés physico-mecaniques des roches d'appui aux grands barrages et leur influence statique documentée par les modèles., Trans. 8th Cong. Large Dams, Edinburgh, 1964, Vol. I, R.35, Q.28, pp. 637–657.

76. OBERTI, G. and REBAUDI, A.: Bedrock stability behaviour with time at the Place Moulin arch-gravity dam., Trans. 9th Cong. Large Dams, Istanbul, 1967, Vol. I, Q.32, R.52, pp. 849–872.

77. OKAMOTO, R.: Investigation of neogene-tertiary dam foundation in Japan. Proc. 2nd Int. Cong. Int. Assoc. Eng. Geol., São Pãulo, 1974, Vol. II, Paper VI-7.

78. OLSEN, O.: Contribution to Theme 3. Proc. 1st Cong. Int. Soc. Rock Mech., Lisbon, 1966, Vol. III, pp. 298–299.

79. ONODERA, T.F.: Dynamic investigation of foundation rocks in situ, Proc. 5th Symp. Rock Mech., Minneapolis, Minn., 1962, pp. 517–533.

80. PAHL, A.: The cavern of the Waldeck II pump storage station, geomechanical investigations and critical analysis of control measurements, Proc. 2nd Int. Cong. Int. Assoc. Eng. Geol., São Pãulo, 1974, Vol. II, Paper VII-16.

81. PAHL, A. and ALBRECHT, H.: Felsmechanische Untersuchungen zur Beurteilung der Standfestigkeit grosser Felskavernen, Proc. 2nd Cong. Int. Soc. Rock Mech., Belgrade, 1970, Vol. II, pp. 767–772.

82. PANOV, S.I., SAPEGIN, D.D. and KHRAPKOV, A.A.: Some specific features of deformability of rock masses adjoining a gallery, Proc. 2nd Cong. Int. Soc. Rock Mech., Belgrade, 1970, Vol. I, pp. 485–490.

83. PHELINES, R.F.: Measures and procedures adopted to ensure the stability and safety of foundations for the Pongolapoort dam, Trans. 9th Cong. Large Dams, Istanbul, 1967, Vol. I, Q.32, R.39, pp. 619–646.

84. PIGOT, C.H. and MACKENZIE, I.D.: A method for an in situ bedrock shear test, Trans. 8th Cong. Large Dams, Edinburgh, 1964, Vol. I, R.25, Q.28, pp. 495–512.

85. PRISCU, R., BANCILA, I., TEODORESCU, A. and FLEGONT, G.: La construction de deux grands barrages dans le flysch Carpathique de Roumanie, Trans. 10th Cong. Large Dams, Montreal, 1970, Vol. II, Q.37, R.39, pp. 729–754.

86. RADOSAVLJEVIĆ, Z., ČOLIĆ, B. and LOKIN, P.: Stabilité des talus des excavations à ciel ouvert des mines de Cuivre a Bor, Proc. 2nd Cong. Int. Soc. Rock Mech., Belgrade, 1970, Vol. III, pp. 403–410.

87. RICE, O.L.: In situ testing of foundation and abutment rock for dams, Trans. 8th Cong. Large Dams, Edinburgh, 1964, Vol. I, R.5, Q.28, pp. 87–100.

88. ROCHA, M.: Mechanical behaviour of rock foundations in concrete dams, Trans. 8th Cong. Large Dams, Edinburgh, 1964, Vol. I, pp. 785–831.

89. ROCHA, M.: New techniques in deformability testing of in situ rock masses, Proc. Symp. Determination of the In Situ Modulus of Deformation of Rock, Denver, Colo., 1969. A.S.T.M. Spec. Tech. Pub. 477, 1970, pp. 39–57.

90. ROCHA, M.: SERAFIM, J.L. and DA SILVEIRA, A.F.: Deformability of foundation rocks, Trans. 5th Cong. Large Dams, Paris, 1955, Vol. III, Q.18, R.75.

91. SAPEGIN, D.D. and SHIRYAEV, R.A.: Deformability characteristics of rock foundations before and after grouting. Proc. 1st Cong. Int. Soc. Rock Mech., Lisbon, 1966, Vol. I, pp. 755–760.

92. SARAMENTO, G. and VAZ, L.: Cambambe dam; problems posed by the foundation ground and their solution, Trans. 8th Cong. Large Dams, Edinburgh, 1964, Vol. I, R.23, Q.28, pp. 443–464.

93. SCALABRINI, M., CARUGO, G. and CARATI, L.: Determination in situ of the state of the Frera dam foundation rock by the sonic method, its improvement by consolidation grouting and verification of the result by again using the sonic method, Trans. 8th Cong. Large Dams, Edinburgh, 1964, Vol. I, R.31, Q.28, pp. 585–600.

94. SCHNITTER, N.J.: Properties and behaviour of the foundation rock at Sufers arch dam, Trans. 8th Cong. Large Dams, Edinburgh, 1964, Vol. I, R.39, Q.28, pp. 717–729.

95. SCHNITTER, N.J. and SCHNEIDER, T.R.: Abutment stability investigations for Emosson arch dam, Trans. 10th Cong. Large Dams, Montreal, 1970, Vol. II, R.37, Q.4, pp. 69–87.

96. SERAFIM, J.L and BOLLO, M.F.: Importance of studies with elastic waves in the geotechnical exploration of the foundations of concrete dams, Proc. 1st Cong. Int. Soc. Rock Mech., Lisbon, 1966, Vol. II, pp. 655–660.

97. SERAFIM, J.L. and DA COSTA NUNES, A.: Studies of dam foundations under a residual cover, Proc. 1st Cong. Int. Soc. Rock Mech., Lisbon, 1966, Vol. II, pp. 639–644.

98. SERAFIM, J. L., ULATE, C. A. and UMANA, J. E.: Geological exploration for a dam site, Proc. 1st Cong. Int. Soc. Rock Mech., Lisbon, 1966, Vol. II, pp. 667–674.

99. SHANNON and WILSON, Inc.: Report on in situ rock tests, Dworshak Dam site for U.S. Army Engineers District, Walla Walla, Corps of Engineers, Seattle, Wash., 1964.

100. Snowy Mountain Authority: Eastern Suburb Railway Line, (Unpublished report), 1969.

101. SOEJIMA, T. and SHIDOMOTO, Y.: Foundation improvement of an arch dam by special consolidation grouting. Proc. 2nd Cong. Int. Soc. Rock Mech., Belgrade, 1970, Vol. III, pp. 167–174.

102. TAKANO, M. and FURUJO, I.: Deformation and resistance in in situ block shear test on a black schist and a characteristic loading pattern, Proc. 1st Cong. Int. Soc. Rock Mech., Lisbon, 1966, Vol. I, pp. 765–768.

103. TIMUR, A.E.: A study on correlation of dynamic and static elasticity modulus determined by in situ and laboratory tests, Proc. 2nd Cong. Int. Soc. Rock Mech., Belgrade, 1970, Vol. I, pp. 543–549.

104. UKHOV, S.B. and TSYTOVICH, N.A.: Some principles of mechanical properties of chloritic schists, Proc. 1st Cong. Int. Soc. Rock Mech., Lisbon, 1966, Vol. I, pp. 781–786.

105. ULLOAO, A.: Field observations at la Soledad dam. Trans., 8th Cong. Large Dams Edinburgh, 1964, Vol. II, Q.29, R.45, pp. 821–839.

106. UNDERWOOD, L.B.: Chalk foundations at four major dams in the Missouri river basin, Trans. 8th Cong. Large Dams, Edinburgh, 1964, Vol. I, R.2, Q.28, pp. 23–47.

107. U.S. Bureau of Reclamation: Foundation bearing tests at Davis dam. St. Res. Lab. Rep. SP-18, and SP-18A, Denver, Colo. 1948 and 1951.

108. U.S. Bureau of Reclamation: Morrow Point dam and power plant foundation investigations, A Water Resources Tech. Publication, Denver, 1965.

109. VASILIU, M.F.: Fondation du barrage des Portes de Fer, Trans. 10th Cong. Large Dams, Montreal, 1970, Vol. II, Q.37, R.38, pp. 709–728.

110. WANTLAND, D.: Geophysical measurements of rock properties in situ, Proc. Int. Conf. State of Stress in the Earth's Crust, Santa Monica, California, 1963, pp. 408–443.

111. WEBB, D. L.: Stress-strain characteristics of a diabase mass, Proc. 1st Cong. Int. Soc. Rock Mech., Lisbon, 1966, Vol. I, pp. 673–678.

112. WEYERMANN, W.: Resultats de l'observation des sous-pressiors du barrage de l'Alto Rabagao (Portugal), Trans. 10th Cong. Large Dams, Montreal, 1970, Vol. II, Q.37, R.22, pp. 369–390.

113. ZARUBA, Q. and BUKOVANSKY, M.: Mechanical properties of Ordovician shales of Central Bohemia, Proc. 1st Cong. Int. Soc. Rock Mech., Lisbon, 1966, Vol. I, pp. 421–424.

APPENDIX IV

Crack Propagation Velocity in Rock

It is becoming more and more clear that the fracture of rocks in compression or tension occurs with crack initiation, stable growth and unstable propagation leading to its collapse. The time required for the collapse of a specimen is dependent upon the rate of crack growth. The knowledge of the velocity with which cracks propagate in rock is important to predict both its creep and time dependent strength.

Very little work on crack propagation velocities in rocks has been done so far. A large body of literature in this field exists, however, on metals, ceramic materials and plastics.

Mott (1948), while considering the energy balance for a moving crack in a body in the absence of any external forces concluded that the kinetic energy of propagating crack can be given by

$$W = \frac{k\rho C^2 V^2 \sigma^2}{2E^2} \qquad \text{(IV.1)}$$

where W = kinetic energy per unit thickness of plate subjected to uniform tension
E = modulus of elasticity of the plate
ρ = density of the material
C = half crack length
V = velocity of propagating crack
σ = applied stress
k = constant.

DULANEY and BRACE (1960) have incorporated the energy balance concept of MOTT and have shown that

$$V = \sqrt{\frac{2\pi E}{k\rho}}\,\{1 - C_o/C\} \qquad \text{(IV.2)}$$

where C_o = original half crack length

As crack grows, $C_o/C \rightarrow 0$, the terminal velocity V_T will be $= \sqrt{\frac{2\pi E}{k\rho}}$.

Eq. IV.2 can be written as

$$V = V_T(1 - C_o/C) \qquad \text{(IV.3)}$$

The plot of V with $1/C$ gives a straight line (Fig. 1) and the velocity intercept of the line gives V_T and the reciprocal half-crack length axis intercept is C_o^{-1}. This type of plot is convenient for comparing velocity measurements with predicted behaviour. Fig. 2 gives V versus C/C_o showing a hyperbolic asymptotic approach to terminal velocity.

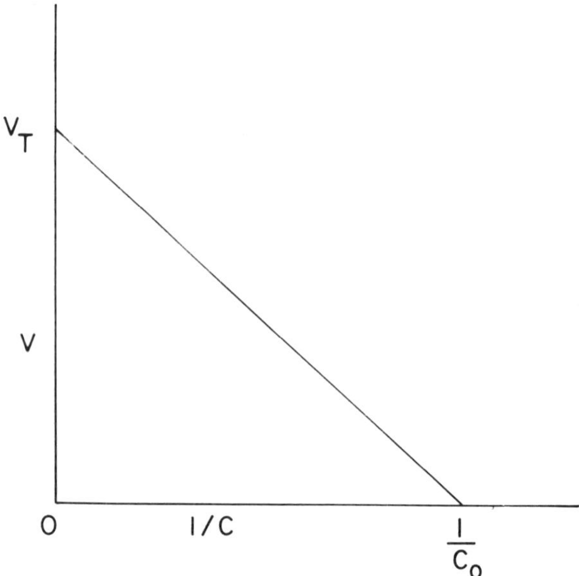

Fig. 1. Predicted velocity of a growing crack as a function of inverse crack-half length. V_T is terminal velocity and C_o is the initial crack-half length.

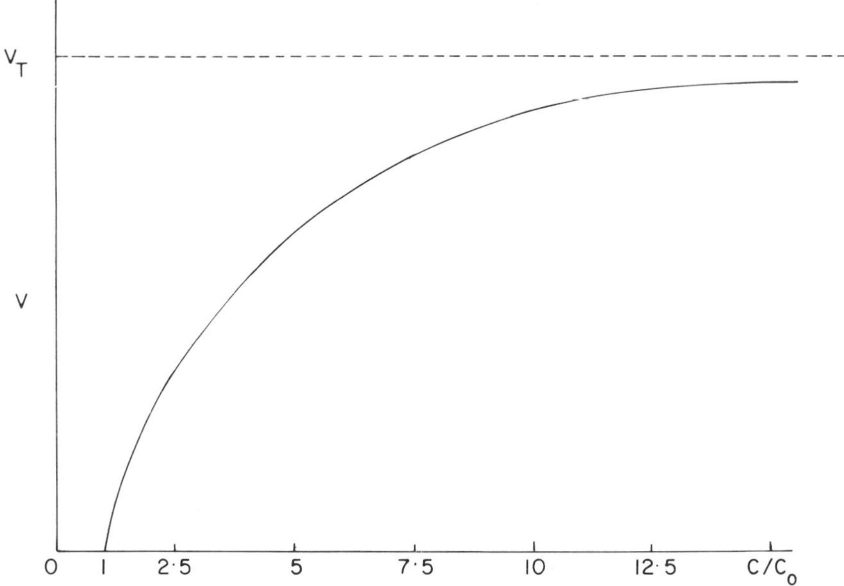

Fig. 2. Predicted velocity of a growing crack as a function of crack-half length.

BROBERG (1960) has shown that the terminal velocity for a propagating crack is a fraction of the RALEIGH surface wave velocity and is dependent upon the POISSON's ratio of the material. The RALEIGH wave velocity, however, must be taken as the velocity in the strained medium. The RALEIGH wave velocity in a strained medium decreases with increase in strain and can be related by

$$V_{Rs} = \frac{V_{Ro}}{1+f(\varepsilon)} \tag{IV.4}$$

$$\simeq 0.7\, V_{Ro} \qquad \text{(for POISSON's ratio} = 0.25)$$

where V_{Rs} = velocity of RALEIGH waves in strained medium
V_{Ro} = velocity of RALEIGH waves in unstrained medium
ε = strain

For values of POISSON's ratio lower than 0.25, V_{Rs}/V_{Ro} will be slightly lower. A 50% difference in POISSON's ratio results in a difference of about 5% in the value of 0.7.

The Eq. IV.3 does not take into account the condition of the crack tip and therefore cannot predict crack velocities that start from notches having varying crack tip root radii. As cracks from blunt notches reach their terminal velocity at crack lengths much shorter than cracks from sharp notches, some factor accounting for this has to be introduced. For example, empirically, it can be given by (BRADLEY & KOBAYASHI, 1971)

$$\frac{V}{V_T} = \left[\left(1 - \frac{C_o}{C}\right)\left(\frac{1}{He}\right)\right] \tag{IV.4}$$

when He = empirical factor for the crack tip condition.

Tests on brittle materials such as glass have shown that terminal velocity is a characteristic of brittle materials (SCHARDIN, 1959). As this velocity is approached, the propagating cracks bifurcate (phenomenon of branching or forking) and coalescence of bifurcating cracks leads to rupture of material. This usually occurs at 0.4–0.5 the RALEIGH wave velocity. Terminal velocity in norite was found to be 1875 m/s (BIENIAWSKI, 1968).

The above discussion is valid only when no energy is applied to the system from outside. Different conditions occur under a constant stress or constant strain rate tests. Studies on ceramic materials (WIEDERHORN 1967; WIEDERHORN & BOLZ, 1970) have shown that the crack propagation velocity is dependent upon stress intensity factor (K) for a given microstructure and environmental conditions. A typical K, V curve is schematically given in Fig. 3.

There are 3 principal regions: In region I in which the crack motion is controlled by the rate of reaction of the crack tip, $V \alpha \exp(BK)$, where B is a constant. This region commences at the minimum K value for the system K_o, the slow crack growth limit (WIEDERHORN & BOLZ, 1970). In region II, crack velocity is essentially constant, crack motion is essentially constant and is controlled by diffusion of corrosive species. In region III, crack velocity increases very rapidly with increase in K. At the onset of region III, K is close to critical stress intensity factor (K_{ic}) for crack propagation in the absence of slow crack growth.

Crack propagation is associated with acoustic wave emission. Crack propagation increment is related to stress wave emission amplitude. A number of techniques have been developed for the determination of velocity of propagating cracks in materials. BIENIAWSKI (1966) used BARR and STROUD roating mirror camera capable of taking 1.59×10^6 frames per second. Wave fractographic technique where running crack is modulated by passing a high intensity ultrasonic shear wave, and a slight ripple is produced on the fracture surface which can be seen using reflected light has been used in glasses. The speed can be calculated by multiplying the ripple spacing by the frequency of the ultrasonic wave (KERKHOF, 1972).

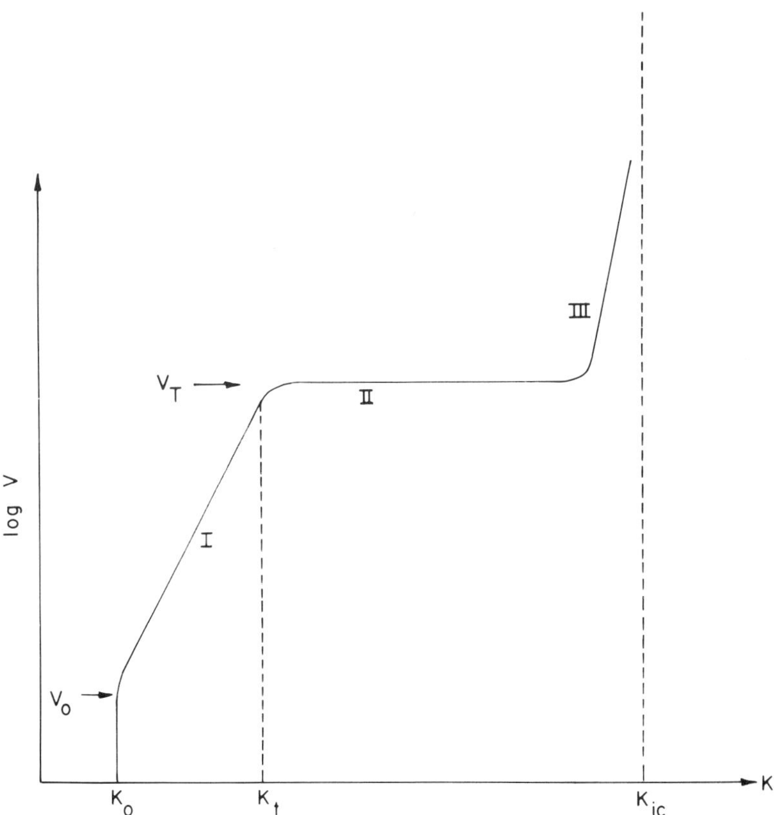

Fig. 3. Schematic representation of a typical K, V curve. K_o is the stress corrosion limit for the system and K_t is the stress intensity factor at the onset of Region II. V_o is the velocity of crack when K reaches K_o. V_T is terminal velocity in the terminal velocity Region II.

Double cantilever methods (WIEDERHORN, 1967; WIEDERHORN and BOLZ, 1970) are very useful for determination of slow crack velocities in material. SWAIN and LAWN (1973) have used microprobe technique where a flat machined sphere is pressed against a specimen to maintain a constant contact circle. The method is simple and useful for conducting studies under saturated or other artificial conditions.

References to Appendix IV

1. BIENIAWSKI, Z. T.: Fracture velocity of rock. Rep. S. African C.S.I.R., No. MEG517, 1966.
2. BIENIAWSKI, Z. T.: The phnomenon of terminal fracture velocity in rock. Felsmech. – Ingenieurgeologie, Vol. 6, 1968, pp. 113–125.
3. BRADLEY, W. B. and KOBAYASHI, A. S.: Fracture dynamics – a photoelastic investigation. Eng. Fract. Mech., Vol. 3, 1971, pp. 317–332.
4. BROBERG, K. B.: The propagation of a brittle crack. Arkiv for Fysik, Vol. 18, 1960, pp. 159–192.
5. DULANEY, E. N. and BRACE, W. F.: Velocity behaviour of a growing crack. J. Appl. Phys., Vol. 31, 1960, pp. 2223–2236.
6. KERKHOF, F.: Wave fractographic investigations of brittle fracture dynamics. Proc. Int. Conf. Dynamic Crack Progagation, Lehigh Univ., Bethlehem, Pa, July, 1972.
7. MOTT, N. F.: Fracture of metals: Theoretical considerations. Engineering, Vol. 165, 1948, pp. 15–18.
8. SCHARDIN, H.: Velocity effects in fracture. In Fracture, Ed. B. L. AVERBACH et al, John Wiley & Sons, New York, 1959, pp. 297–329.
9. SWAIN, M. V. and LAWN, B. R.: A microprobe technique for measuring slow crack velocities in brittle solids. Int. J. Fracture, Vol. 9, 1973, pp. 481–483.
10. WIEDERHORN, S. M.: Influence of water vapor on crack propagation in soda-lime glass. J. Am. Cer. Soc., Vol. 50, 1967, pp. 407–414.
11. WIEDERHORN, S. M. and BOLZ, L. H.: Stress corrosion and static fatigue of glass. J. Am. Cer. Soc., Vol. 53, 1970, pp. 543–548.

Author Index

Subject Index

HANDBOOK ON MECHANICAL PROPERTIES OF ROCKS

CONTENTS

Volume I

HANDBOOK ON MECHANICAL PROPERTIES OF ROCKS

CONTENTS

Volume II

HANDBOOK ON MECHANICAL PROPERTIES OF ROCKS

CONTENTS

Volume IV